Smart Grids

This reference text presents a threat to Internet of Things (IoT) security and their mitigation strategies in smart grids. It further discusses aspects of adaptive demand response management, multiport converter topology, and voltage stability in smart grids. An IoT-based prioritized load management technique for photovoltaic (PV) battery-powered buildings is illustrated. The importance of zero energy buildings, sustainable farming, and electric vehicles is discussed. Case studies on optimal distributed generation (DG) planning are presented.

This book:

- Discusses threats to IoT security and their mitigation strategies in the smart grid.
- Adaptive demand response management, multiport converters, and voltage stability in smart grids.
- Zero energy buildings, sustainable farming, and electric vehicle deployment in smart grids.
- Case studies on optimal DG planning.
- Covers a multi-objective optimal power flow algorithm for optimal performance in the distribution system and a techno-economical optimal solution for the distribution system.

It is primarily written for senior undergraduates, graduate students, and academic researchers in the fields of electrical engineering, electronics and communications engineering, computer science and engineering, and environmental engineering.

Smart Grids

Sustainable Energy Systems

Edited by
O.V. Gnana Swathika and K. Karthikeyan

CRC Press
Taylor & Francis Group
Boca Raton London New York

CRC Press is an imprint of the
Taylor & Francis Group, an **informa** business

Designed cover image: shutterstock

First edition published 2026
by CRC Press
2385 NW Executive Center Drive, Suite 320, Boca Raton FL 33431

and by CRC Press
4 Park Square, Milton Park, Abingdon, Oxon, OX14 4RN

CRC Press is an imprint of Taylor & Francis Group, LLC

ISBN: 9781032774947 (hbk)
ISBN: 9781032884561 (pbk)
ISBN: 9781003537878 (ebk)

DOI: 10.1201/9781003537878

Typeset in Sabon
by Newgen Publishing UK

Contents

About the Editors

O.V. Gnana Swathika (Member'11–Senior Member'20, IEEE) received a B.E. degree in Electrical and Electronics Engineering from Madras University, Chennai, Tamil Nadu, India, in 2000, MS in Electrical Engineering from Wayne State University, Detroit, MI, USA, in 2004, and PhD degree in Electrical Engineering from VIT University, Chennai, Tamil Nadu, India, in 2017. She also completed her post-doc at the University of Moratuwa, Sri Lanka. Her current research interests include microgrid protection, power system optimization, embedded systems, and photovoltaic systems. She currently works as a Professor at the Centre for Smart Grid Technologies, Vellore Institute of Technology, Chennai, India.

K. Karthikeyan is an Electrical & Electronics engineering graduate with a Master's in Personnel Management from the University of Madras. With his two decades of rich experience in electrical design, he has immensely contributed toward the building services sector comprising airports, information technology office space (ITOS), tall statues, railway station/depots, hospitals, educational institutional buildings, residential, hotels, steel plants, and automobile plants in India and abroad (Sri Lanka, Dubai, and the UK). He is currently working as a Chief Engineering Manager – Electrical designs for Larsen & Toubro construction, commonly known as L&T, which is an Indian multinational engineering, procurement, and construction (EPC) contracting company. Additionally, he has worked at other major companies, like Voltas, ABB, and Apex Knowledge Technology Private Limited. His primary role involved the preparation/review of complete electrical system designs up to 110 KV voltage levels, including the detailed engineering stage covering various electrical design calculations, design basis reports, suitable-for-construction drawings, and mechanical, electrical, and plumbing (MEP) design coordination. He is the point of contact for both client and internal project teams; leads and manages a team of design and divisional personnel; engages in day-to-day interaction with clients, peer review, and managing project deadlines, and project time estimation; and assists in staff appraisals, training, and recruitment.

Contributors

Aadi Ashutosh Chauhan
Vellore Institute of Technology, Chennai, India

Aayush Karthikeyan
Schulich School of Engineering
Univeristy of Calgary, Canada

Akhtar Kalam
Victoria University, Australia

Ameesh Singh
Vellore Institute of Technology, Chennai, India

Anshul Yadav
Vellore Institute of Technology, Chennai, India

Apurv Malhotra
Vellore Institute of Technology, Chennai, India

Arun Mozhi S.
Department of EEE,
Thiagarajar College of Engineering, Madurai, India

Asha Anu Kurian
Vellore Institute of Technology, Chennai, India

Balasundar C.
Department of EEE,
Thiagarajar College of Engineering, Madurai, India

Charles Raja S.
Department of EEE,
Thiagarajar College of Engineering, Madurai, India

Deepak K. M.
Vellore Institute of Technology, Chennai, India

Faazila Fathima S.
Vellore Institute of Technology, Chennai, India

Gayan Hemal Colamba Hewage
Department of Electrical Engineering
Univeristy of Moratuwa, Sri Lanka

Gnana Swathika O.V.
Centre for Smart Grid Technologies
Vellore Institute of Technology, Chennai, India

Hannah Grace G.
Vellore Institute of Technology, Chennai, India

Hari Krishnan M. B.
Vellore Institute of Technology, Chennai, India

Harish Raaghav S. S.
Vellore Institute of Technology, Chennai, India

Harsh Gupta
Vellore Institute of Technology, Chennai, India

Hemapala K. T. M. U.
Department of Electrical Engineering
Univeristy of Moratuwa, Sri Lanka

Jeslin Drusila Nesamalar J.
Tamilnadu Government Polytechnic College, Madurai, Tamil Nadu, India

Jeyaraman R.
CET Power Solutions India Pvt Ltd, Chennai, India

Kanimozhi G.
Centre for Smart Grid Technologies
Vellore Institute of Technology, Chennai, India

Karthick T.
Quantanics Techserv Pvt Ltd., Madurai, India

Karthikeyan K.
L&T Chennai, India

Kshitiz Srivastava
Vellore Institute of Technology, Chennai, India

Manivasagam R.
Department of Electrical and Electronics Engineering
K. Ramakrishnan College of Engineering, Trichy, India

Mohit Sharan
Vellore Institute of Technology, Chennai, India

Obeng-Denteh W.
Kwame Nkrumah University of Science and Technology

Prantika Das
Vellore Institute of Technology, Chennai, India

Premalatha L.
Vellore Institute of Technology, Chennai, India

Puppala Anusha
Vellore Institute of Technology, Chennai, India

Rajapaksha Mudiyanselage Tharangani Lakmali
Department of Electrical Engineering
Univeristy of Moratuwa, Sri Lanka

Ramani Kannan
Universiti Teknologi Petronas, Malaysia

Ravi Samikannu
Botswana International University of Science and Technology
Palapye, Botswana

Revathi G. K.
Vellore Institute of Technology, Chennai India

Rohan Bhojwani
Vellore Institute of Technology, Chennai, India

Savani M. Kulkarni
Vellore Institute of Technology, Chennai, India

Sivakumar M.
Department of EEE
Thiagarajar College of Engineering, Madurai, India

Teenu Raj M.
Vellore Institute of Technology, Chennai, India

Tushar Singh
Vellore Institute of Technology, Chennai, India

Varun Mukesh Lal
Vellore Institute of Technology, Chennai, India

Veeram Priyanka
Vellore Institute of Technology, Chennai India

Vishnu Dharssini A. C.
Department of EEE, Thiagarajar College of Engineering, Madurai, India

Chapter 1

Emerging Threats to IoT Security and their Mitigation Strategies in Smart Grids

Aayush Karthikeyan, K. Karthikeyan, O.V. Gnana Swathika

1.1 INTRODUCTION

According to IBM, the percentage of World Wide Web devices was due to surpass the human population by 2020, with connectivity continuing to evolve, with about 50 billion devices being connected by 2020. The Web of Things, also known as the Internet of Things (IoT), was born from the spread of smart devices in a linear actuator network. A platform that seamlessly integrates sensors and actuators with the surroundings throughout in order to exchange data and create a unified operating picture is a necessity. An IoT system starts with a single item that is identified by a globally addressable unique global identifier. The quantity of data procured by unlocking the object may be low in this case. Radio frequency identification (RFID) tags store static data. Thus, the IoT is characterized as particles with a unique code that have internet access and are (adaptively) obtainable through other particles, which are alluded to as "stuff" in this context. Technology emerges out of its infant stage to become the subsequent groundbreaking technology in the transformation of the internet into a completely integrated future IoT. The recent uptick in wireless internet technology use and integration, such as wireless sensor networks (WSN), RFID, and actuating nodes, is driving this trend. The IoT connects a wide variety of gadgets, services, and applications. This category includes intelligent computers; cellular phones; office equipment; Bluetooth cars; lighting; heating, ventilation, and air conditioning (HVAC); and appliances. Long-term evolution, 3G, area networks, Bluetooth, ultra-wideband (UWB), Z-wave, and Sigfox, as well as other communications infrastructure technologies (infrastructures), provide internet connectivity for IoT applications across a wide range of services in social media channels. As the IoT advances, cloud computing (CC) will continue to be the backbone of globally connected diffusion, database management (or information processing), and storage. Cloud solutions like Microsoft Azure, Amazon Web Services (AWS), Google Docs, and many more are expected to provide normal entry points for connecting solid

DOI: 10.1201/9781003537878-1

goods to computation and communication abilities across a broad range of applications, systems, and technologies.

Smart grids (SGs) use digital technologies to wisely distribute fuel, for example, through two different correspondences, while also meeting sustainability targets through eco-friendly information and communications technology (ICT) usage. The inherent weakness of communication technology has exposed the framework to a slew of security risks. Numerous published papers have addressed and proposed solutions to these issues. However, the vast bulk of these papers categorized the basis of symptoms on privacy and security, integrity, and allocation rather than accountability. Furthermore, existing preventative measures are targeted at specific attacks or constituents, but there is no comprehensive solution to safeguarding the entire system.

In this chapter, we discuss security specifications and significance, and suggest a blueprint for the future to discover and confront airstrikes [1]. Safety initiatives are essential for enhancing cyberattack solutions in applications related to the SG. We present a thorough survey aided by the rigorous evaluation of prior work. In addition, we discuss recent developments and prevention methods for SG computer hackers. We also investigate the dangers and prospective solutions of an IoT-based SG. We concentrate on information security types and conduct an in-depth examination of the intelligent base load cybersecurity status. We specifically discuss and examine network vulnerabilities, attack defense systems, and security requirements. We hope to provide a holistic knowledge of security threats and quick fixes, along with a blueprint for future attack research in applications involving SGs [2].

SGs are undoubtedly the future power infrastructure of the energy world. An SG can connect millions of consumers and devices in a network, which necessitates that it be highly robust, reliable, and secure. Because of long-distance communication on open networks, security is regarded as one of a few of the great barriers in today's technology. Cybercriminals, hackers, and terrorists attempt to attack this national infrastructure for malicious purposes or to gain control of automated energy monitoring and remote controlling for personal gain. This chapter provides a comprehensive survey of SGs, architecture, the methodologies used, and communication protocols, but the focus is primarily on the cyberattacks carried out and the SG solutions recommended. Finally, we discuss the various challenges in cybersecurity, the issues that remain unresolved in the literature, and the current solution space, as well as research gaps [3].

Rising load and utilization needs involve the power supply, with concerns like power shortages, surges, and overvoltage, alongside current energy connections increasing the crucial carbon output and, most relevantly, attempting to deal with computer hackers. However, power system software is full of shortcomings and consequences, notably in regard to information security, which is its most pressing concern. These are the same dangers and possible ideas related to an intelligent grid system that uses the IoT, as examined in this chapter. We focus on different types of cyberwarfare and provide extensive data [4]. To continue improving the program's resiliency, a great deal of work has been performed in systems engineering, designing and building implementations, and implementing new machine learning. This work, on the flip side, has principally introduced various ranking techniques to determine links (transmission and distribution) or access points (stations) and then effectively choose the best detection nodes or widely accepted links as the critical ones. These techniques do not solve the problem of component connectedness or the position of consumers, which are among the essential variables affecting SG risk assessments [5].

The IoT age's continual new technologies deliver stronger alternatives to more effectively organize certain hurdles and enact demand response measures. Smart meter facilities and smart metering (SM) are devices that can revolutionize the classic electricity system by trying to expose the background treasures of electrical energy through multiple communication systems introduced during power payment transactions with both consumers and electricity providers. The literature not only on advanced metering infrastructure (AMI) but also SM systems is extensive. However, few researchers have investigated the significance of SM in voltage stability and reliability surveillance in IoT-enabled SGs. As a result, the objective of this chapter is to comprehensively investigate the feasibility of using SM to monitor power quality and reliability[6]. The most essential IoT system is the SG. Grid computing is a personal information network technology that is assimilated with both the power network to acquire data from pipelines, transfer switchgears, and consumers. In this chapter, we start by debating the IoT and SG, as well as their relationship. We address a few small IoT layouts in Singapore, including specifications that use the IoT in Singapore; CC applications, as well as services in Singapore; challenges; and future projects [7]. We examine intelligent grid power that uses the IoT and edge computing (EC) in thorough detail. Moreover, we emphasize emerging technology advancements, the blueprint for an IoT-SG, and the demands for utilizing the EC-IoT-based SG framework. We investigate the EC-IoT-based SG framework and describe basic features for instituting the EC-IoT-based SG structure. Ultimately, we define some crucial problems and obstacles discovered during the application of the EC-IoT-based SG framework expressed in the declaration. Additionally, we provide a few of the best open issues that have been recognized recently [8].

Digitalization has made good inroads for population development and a higher standard of living. Technology fashions have increased competition for efficient device-to-device correspondence that doesn't demand human interference, creating an enormous business for the IoT network. Given the amount of growth potential, each civilized country, and even a few emerging nations, has been creating its own version of somebody else's adequately funded IoT projects. This chapter encapsulates the function of IoT in the creation of smart cities, as well as urban planning structures for efficient road consumption, the hurdles of IoT economic development, and solutions. Additionally, it outlines IoT services centered on one domain [9]. Each connected device in the IoT-SG certainly has the potential to be a private data hub to information systems. Data privacy and security worries are unpreventable, but due to further data richness, flaws in safety, and security flaws in case scenarios such as integration and the impact on individuals, IoT threats have expanded to unprecedented levels. This chapter focuses primarily on IoT-SG research, advantages, and upcoming challenges, as well as viable methods to overcome these difficulties. Furthermore, broad-sense recommendations are made for confronting and handling some of the biggest hurdles, including protection and big data [10].

1.2 THE IMPORTANCE OF CYBERSECURITY IN TODAY'S DIGITAL AGE: WHY EVERY BUSINESS NEEDS TO PAY ATTENTION

To accomplish the above important advantages of digitization, home automation is renowned for its ability to merge coding, interaction, and powerful stimulus into present electricity supply networks. It will improve local information sharing, geographically dispersed power, two-input transmitters, and supply-side efficiency. Despite this, because of the essential nature of online connectivity, this arising platform introduces an additional security threat, namely cybersecurity. This chapter provides a summary of smart energy virtual security in general. It begins with an assessment of a typical energy system safety framework before moving on to evolving security problems in the SG. Additionally, a few models are proposed to mitigate computer hacking [11].

The IoT has also come to be an essential facilitator of such intelligent transmission networks. Because these machines are on the internet, creating a sensor network defenseless to considerable assault, one of the main challenges as the number of cellular devices expands at a fast speed is protection. A technology power system has the world's biggest security vulnerabilities of any IoT-focused framework because of the potential danger of vast numbers of vertices. Computer hackers on a sensor network would also have a disastrous impact on the accuracy of prevalent connectivity because of the possible canyon side effects of locking down the electricity system. Most of the machines in our homes, departments, clinics, and train services depend on electrical power. When a single system is compromised, the entire infrastructure becomes vulnerable to cyberattacks [12]. In pursuit of growing artificial intelligence (AI) algorithms, one very detailed state estimation is obligatory as a requirement for numerous essential processes to enable effective control and oversight in a smart energy grid system. As electricity systems and websites become even more interrelated, normal political predictors will become sensitive to large data integrity attack vectors such as incorrect data infusion, also known as false data injection (FDI), which overrides current poor information tracking (behavior-driven development (BDD)) systems. Scholars have recommended FDI detection equipment, but such countermeasures ignore changes in the power state caused by possible scenarios. Because this dramatic physiological change does indeed have a critical effect on currently operational direct investment detection equipment, new instances will be classified incorrectly. In the hope of resolving the issue, we performed analysis to acquire an essential understanding of the distinctions between corporeal electricity alteration and numerical computation that keeps changing [13].

The globe is quickly transitioning from the conventional power network to electric cities. A sensor network, like any type of entrepreneurial endeavor, has weak points. Among the most significant problems is going to be protecting the network architecture from the possibility of cybercrime. The most difficult problem is having to manage a vast network, which has huge numbers of sensor systems continuously transferring and sending datagrams across it. Any ransomware attack has the ability to jeopardize the SG's critical components of privacy and security, authenticity, and ease of access. The general national grid network consists of a customer base who utilize the connectivity web of sensors and smart gadgets, and the individuals managing the network (influencers); all tiers seem to be at risk from cyberattacks [14]. Trying to assure demand response cybercrime is vital to the long-term viability of this new type of electrical infrastructure. Highly standardized procedures and methods, according to experts, should have been adopted first and foremost. Multiple new microgrid standards have been produced in recent years, making it difficult to locate a useful paper among the cascade of data. This chapter outlines the results of projects that attempted to solve this problem by listing all guidelines that outline power system cybersecurity criteria. A thorough examination of the available literature yielded 17 standards, which are detailed in this chapter, with an emphasis on needs and classified according to types of assessment criteria [15]. This chapter provides a quick assessment of observable factors affecting power system safety improvements and strategy acceptability. Essentially, it is vital to evaluate the extent to which electric grid computer vulnerability can also be addressed utilizing outlay robustness augmentation strategies. A number of projects are proposed to strengthen the robustness of the electrical grid. This chapter examines recently proposed cybersecurity strategies for detecting and recognizing fake information injection assaults and assesses respective accuracies, the high computational requirement, and robustness against external influences. To date, there isn't such a thing as a solution to all electric grid concerns. To address this, new ways to identify and recognize computer security are systematically examined and debated [16].

Virtual refers to the complete activity of keeping threats from compromising the security, authenticity, and accessibility of digital properties by exploiting a vulnerability in those materials. As a corollary, the critical components of computer hackers, privacy, authenticity, and access (confidentiality, integrity, and availability (CIA)), must be adequately provided. One of the greatest difficulties associated with creating yesterday's critical systems was information security dangers and computer security. Microgrids are distinguished by their performance in this regard in terms of the creation, storage, and distribution of electricity, which is an important factor for infrastructure. The wide use of automated infrastructures involves threat recognition and categorization, as well as prevention execution [17]. The cybersecurity of network technologies in distributed generation is a key concern for manufacturers and controllers. Numerous new power system specifications were suggested in past years, rendering it hard to discover a detailed link in this multitude of data. The conclusions of a comprehensive review aiming to solve that problem by setting standards that provide robust cybersecurity assessment recommendations are offered in this study. This should help healthcare professionals to choose requirements that are applicable to our field. Additionally, the specifications' data can also be utilized to inform the audit of power system devices [18]. More connectedness is necessary to adopt newer collaboration and oversight mechanisms. Several reasons, such as device design capacity, universal connection, and powerful methods, have led to the widespread adoption of transceivers. Unfortunately, for efficient functioning, the core network needs safety safeguards. Demand response cyberattacks are a growing industry, as opposed to bank transfers, which have had equivalent safety precautions in place for centuries. To secure information in all layers, functional software assurance is indeed essential [19]. The electrical energy supply is at risk of operations and assaults. This chapter highlights key power system issues that may be utilized to launch both offline and online attacks. A thorough examination of a trimming study to boost physical crime and cybercrime in a national grid setting is carried out. An incursion concept also serves to highlight the practical and digital connection. The emphasis is still on electric grid digital safety and cybercrime. The conclusions of a research report in a simulated universe are provided [20].

The inquiry, research, and creation of strategies to reduce or fix these issues relies significantly on precise models, methodologies, and an assessment of the interacting systems involved with renewable energy computer hackers. The purpose of this study is to comprehensively outline numerous methodologies and strategies of computer hackers that target power sources in a network. These aspects are presented, accompanied by a thorough examination and overview of the conceptualization and procedures to counteract such attacks. The technological aspects and relevance of several internet strategies are also analyzed and examined [21].

The digital networks' globally connected network is a danger to the electric grid, especially in the realm of information technology (IT). A mere request can force the existing network to shut it down inside of an automated service because control points can also be created using a simple step, even from a considerable distance. This is because when the power source stops running, all vital services are suspended. As a reason, in the environment of SGs, considerable computer hacking is evident, in contrast to the previous power infrastructure, whereby practically all control activities were handled manually or via the limited usage of local nodes [22]. Depending on safety objectives and weaknesses, we divide various products into four types: authenticate options, anonymity ideas, key distribution technologies, and systems for detecting intrusions. We define the fundamentals of each procedure genre and categorize them according to protection criteria. As an outcome, we categorize security intrusion detection system solutions into nine classes based on the algorithms for machine learning employed, which include depth learning, adaptive intelligence networks-based, relevance vector hardware, decision shrub, rule-based, Bloom filter-based, irregular forest-based, and spontaneous hidden Markov model learning-based [23].

Gameplay theory refers to a mathematical paradigm that assesses multiple individuals' plans and choices in order to cope with exchanges within them. Gamification is increasingly widely employed in subjects such as business, history, and cultural studies due to its adaptable character and uses in a large variety of disputes and problems. Because of its adaptability and durability, this strategy has been adapted to data science. We examine behavioral economics implementations in the disciplines of cyberdefense and virtualization in this chapter, alongside an organization's competitive game models employed in these fields.

We also discuss mixture theory and its potential to be applied to complex challenges [24]. The Web of Things is fast transforming our culture into one where each and every device is linked to the network, making computers more prevalent than ever before. Virtualization has definitely altered how intellectual assets and offerings are utilized and accessible, adopting the per use principle, the merits of which are evident for any enterprise. Regarding security measures, the IoT and cloud digitalization could be terrible, as any gadget may have become intrinsically globally vulnerable to hackers and, as a result, exploitable by bad actors. We show where security risks emerge and the possible repercussions. In a nutshell, we believe that safeguarding sensor nodes is untenable given the computer security starts in the Net [25]. Cloud technology is a developing virtualization technology that places compute and data storage closer to end users in the hope of improving reaction times and freeing up the transmission rate. Mobile CC uses parallel processing to provide services to cellular phones. CC and cloud technology, on the other hand, have security concerns, including client exposure and affiliation confirmation, slowing the development of grid computing. The science of automated systems that improve spontaneously as a consequence of repeated use is called machine learning.

In this literature review, we provide an examination of CC cyberattacks, challenges, and fixes that have been used in data mining methods [26]. CC is a computing grid that enables the use of a communal pool of computer resources by cloud customers. CC offers customers and corporations various advantages in the form of both capital and maintenance spending savings. This protection concept's lack has a detrimental influence on the computer persona, generating human, ethical, and financial loss. We study and emphasize the cybersecurity threats that online organizations confront. The cloud service provider, the client side, and remote server are examples of all these organizations. Concentrating on bitcoin, which consists of several transmission, arithmetic, and contractual agreements regarding service levels, is essential. After examining the origins and impact of different cyberthreats, it will deliver the appropriate improvements [27]. CC allows organizations to offshore their whole ICT operations, permitting them to concentrate a great deal on their business model, and boost efficiency and creativity in offering consumer services. Yet, many older and developing businesses have hesitated to utilize CC due to a variety of safety and privacy issues. As a part of this research, several discussions with service providers and specialists were conducted, and [27] was studied. This analysis makes it possible to comprehend present as well as potential safety and privacy issues concerning CC by providing effective responses to potential barriers. The results of this analysis identified an overall of 18 current as well as potential known vulnerabilities involving various parts of cloud applications [28].

1.3 SECURING THE INTERNET OF THINGS: CHALLENGES AND SOLUTIONS

The present power gauge method has several flaws, among them the shortage of full aural transmission. A smart electricity gauge powered by the Internet of Information is presented as a solution to this problem (loT). The suggested smart electricity meter uses a Wi-Fi module to manage and analyze energy before sending the result to the Net, where consumers or production can examine the readings. As a corollary, user analysis is becoming much quicker and far more manageable. This technology also helps to identify standby power. As a result, by applying the internet and permitting connectivity, this advanced metering aids in remote monitoring, making it a big step toward Startup India [29]. SGs use a wide variety of equipment located at energy plants and distributors for grid observation, assessment, and administration. Hence, SG must be interconnected, mechanized, and findable. These are done with the help of connected things. The IoT supports SG technologies in enabling multiple distributed tasks across the creation, transportation, administration, and usage of energy by embedding devices and providing interconnection, monitoring, and control for such equipment. We provide a detailed assessment of sensor SG systems in this chapter, spanning old designs, techniques, and demonstrators. In addition, this evaluation outlines unresolved concerns, limitations, and upcoming study objectives for technology SG systems [30].

The addition of the IoT in distributed generation offers massive opportunities in today's political home energy world due to its multifarious benefit in numerous domains. The national grid structure, on the contrary, has grabbed the imagination of the world's academic researchers, and the concept of merging IoT and electric buildings has huge potential for development. In an analysis paper, researchers concentrate on a number of significant research studies on the IoT in regard to SGs. This paper also discusses a number of unique methodologies utilized in the IoT and intelligent infrastructures, along with their applicability to a variety of disciplines. This study aims to aid scientists and beginners in the domain of the IoT and intelligent utilities by making people aware of innovative research activities [31].

The Web of Things is a unique notion born of great breakthroughs in the field of IT and electronics (IoT). When sensor innovations are used, many devices and pieces can be joined to one another via online and perhaps other new media platforms. Operating devices are computer units that are created for specialized applications, like those that exist in technological products, intelligent homes, aircraft, and cars, and incorporate powerful processors, detectors, and monitors that can exchange data over IT infrastructures. In this respect, the name "link" refers to any sort of site or mentor network as opposed to the global web. In a study, researchers look at the use of Wi-Fi in distributed generation [32]. The network is deemed personal because the evidence obtained through SM customers is secure and anonymous. Rigorous security study utilizing its arbitrary operator approach, step encryption techniques, and operating system comprehensive safety assurance all reveal that IoT is immune to different assaults. The test setup of several encryption algorithms was used for a similar evaluation and the generally used definitiveness numeric and arbitrary mathematics cryptography module (MIRACL) were used. A detailed comparison demonstrates that IoT-SG provides higher feature aspects and provides greater protection than the proposed related schemes, in spite of less communication as well as computation costs [33]. The new system, which incorporates technology and detectors in electrical infrastructure, transforms the system of electricity produced, delivery, metering, and administration. The smarter system's security is a tough and time-sensitive undertaking. Using algorithms, in a study, researchers propose a safe request managing engine for such an IoT network. According to the objectives, the suggested machine is on duty to retain energy efficiency. Modern energy managing and interaction managing mechanisms are presented to handle data with the aim to increase energy utilization. In accordance with the results obtained, the new engine is less resistant to incursion and successful enough to lower power system electricity consumption [34].

We finish our data integrity attack (DIA) classes in this chapter by inventing the minimal DIA, which lets attackers execute covert data transformation techniques while revealing branching factors. Only topology knowledge about cutterhead is necessary for constructing such an assault. We show that only an assailant can randomly adjust the background model of just one decker, which again is related to the external world via just a lone cutterhead, in addition to the coaches in a yet another super-bus, which is a collection of vehicles related to the outside world only through a top edge, using an identical absurd skew. Lastly, we undertake comprehensive computations using IEEE standards transmission lines to corroborate the simulated predictions [35]. The article presents a blockchain-based architecture plus a clever contracts-based architecture to allow dependable and outlay failover gas, with functions provided as decentralized and distributed software. Fuel exchanges are uploaded to the blockchain, with established norms level, which provides greater generally pro assurances. Energy bids follow clear regulations that are applied as a consensus mechanism and thus available to all affected parties. Clever metered attack vectors are handled by temporarily banning affected buyers and rebooting their units as early as secure fixes become released [36]. Fuel cells (FCs) improve CC by putting distributed hardware and software processing capacity toward the channel's border, enabling expedition SG systems to have a destination, fast data, and ping statistics. They utilize online info and transmit it all to SG clients via rapid personal support by building specialized computer resources at clients' premises. We look at the present situation of virtualized SG systems in this study after outlining the motivation for deploying FC as a technical facilitator for genuine SG analyses. We also describe a multiple FC-based SG design, underlining its suitability for incorporating a substantial percentage of connected internet devices into the SG [37].

Inside this work, we first explore the information needs of the electric city prior to covering various IoT interfaces and their standards. We undertake a literature review that utilizes a smart network requirement to examine IoT network features and efficiency in the electric city. The faults in these approaches that prohibit them from collecting overall recommendations for implementing effective IoT rules that can match the power system context contact criteria are highlighted in this methodology. We use local assets, including the open Web, to organize higher usage criteria for intelligent transportation architecture that includes the multi-agent system (MAS) and tri grid [38]. This study proposes a peripheral computer network for sensor microgrids to overcome the deficiencies of the present CC model in energy systems, where several difficulties remain unknown, including completely meeting the needs for large speed with latencies. The unique solution integrates cloud services into the existing virtualized electrical network while still building system and software components. As a corollary, now at the channel's interface, a considerable quantity of data created in the electric system will indeed be reviewed, digested, and kept. Cellular distributed generation will also be able to link and operate a significant number of ports, deliver authentic analysis and handling of large datasets, and encourage digitalization with the use of the computation architecture [39].

Given the rise of technology processes, humanity is currently upon that threshold of additional computer changes. The primary cornerstones of this revolution comprise a vast range of intelligent items coupled by the internet, which are constructed on the foundation of Wi-Fi (Internet of Everything). The number of device data series is massive, and the uncooked trend analysis is all very probable regarding having incomplete data as a result of faulty networking transfers. It is presently challenging to store the vast volume of a raw dataset while still delivering strong evidence and specific time series statistics. We provide a structure for durable dataset analytics for the IoT in this study [40].

The IoT is becoming extremely prevalent in smart electricity tracking, advanced robotics, and a wide range of other purposes. Sensor nodes are used at varying phases of green grid (SG) deployment to track and regulate grid data so as to provide dependable and sustainable electricity distribution. Because IoT inclusion in the SG arena has various advantages, the hurdles in IoT must be addressed for the matrix to work efficiently. The first part of this document provides a summary of SG and cellular SG systems. This project outlines an IoT-based powered tracking system capable of measuring and analyzing electrical behavior, such as output power, flow, load demand, and load energy costs [41].

The Web of Things and electricity cities are trying to cut informational and industrial solutions. The marriage of the Web of Things and electric buildings will boost national grid data transfer and assistance dramatically. To enable the nation's digital utility's top echelons, important connected device solutions will be deployed to the DC microgrid, and an electricity management and data management awareness repair facility will be constructed [42].

The consumption, assessment, and administration of different green energy supplies constitutes one of the most important elements of a hybrid energy storage system for crop monitoring. The installation of a sensor wirelessly for energy administration and climate parameter monitoring is a key aspect described in this study. A graphical user interface (GUI) and dedicated page are constructed for the genuine operation of the suggested model. This article examined a field network persistent connectivity prototype. Several sensor units were deployed for data gathering in order to enhance precision and eliminate estimation errors in the deployment of local area network scanning. The designed system makes use of a wireless instrument and has already been verified for applications within the IoT [43].

The Web of Things (IoT) is now being rapidly used, enabling smart urban ideas and efforts all around the planet. The World Wide Web of Things is a generic way to incorporate sensing devices into all ICT applications. Around a billion things are to be integrated and used in the green infrastructure. The IoT is important for enhancing amenities for city corporations and inhabitants by harnessing cutting-edge technologies. This article delves into the idea of the IoT and intelligent societies, and discusses their motives and uses. This study also examines the key problems and disadvantages of deploying intelligent systems that use intelligent appliances [44].

The Web of Things for the green infrastructure includes numerous disciplines and is supported by a number of relevant aspects. We present a thorough assessment of the connected things in pervasive computing in this article. The key parts of the Technology City – wide scene are explored immediately, along with the developments that allow the regions to flourish in respect of designs, internet, and cognitive programs utilized in Technology Innovative Technology devices. Following that, there will be a review of the most frequent approaches as implementation in several smart city fields. Lastly, the problems of deploying IoT technologies for green infrastructure, plus mitigating methods, are discussed [45].

People's increased online use will boost the number of subscribers, and the IoT has an essential role in this. To imitate intelligence and result in a precise and reliable conclusion sans user intercession, "smart computers " must mix IoT with intelligence. Integrating AI with IoT computer networks is now a prerequisite for computer network sustainability. Identifying the variables that impact an intranet is essential to its accomplishment. The purpose of this study is to pinpoint the features that govern the effective installation of an informational system that facilitates IoT and AI in Taiwan's envisaged Pervasive Computing [46]. The advent of the modern city predominantly involves IoT-aided power grids.

Nevertheless, no extensive definitions of IoT's work together with projects on sustainable urban control, expansion, and advancement are presently accessible. As a response, the recent paper analyzes the importance of IoTs on the technical landscape of a smart world (technical reference model (TRM)). Research by over 200 specialists was also conducted to study the impact and relevance of particular features of IoTs for a knowledge economy. In light of the poll responses, the important points and critical factors for the construction of a modern city are outlined [47]. The illuminating system has been interconnected with sophisticated technology and communication lines to establish intelligent street lights in a smart world. A smart lighting system (SLS) is designed using new lighting methodologies. In this work, we provide a summary of both the SLS and examine IoT network technologies that can be utilized to implement an SLS in the environment of something like a modern city. Additionally, we looked at different usage situations for IoT for both indoor and outdoor applications. IoT smart street lighting solutions, according to our research, can lower power usage by up to 33.33% in both interior and exterior settings. Finally, we discussed prospective research areas for SLS in pervasive computing [48].

1.4 CONCLUSION

In conclusion, cybersecurity and IoT have become intertwined, and it is essential to address the challenges posed by their integration. The expansion of IoT has revolutionized how we live, work, and play, but it has also created new vulnerabilities that hackers can exploit. The number of connected devices is increasing, and as a result, cyberattacks are becoming more frequent and sophisticated. The key to successful IoT implementation is to develop a comprehensive security strategy that can effectively mitigate the risks. This strategy should encompass measures such as device authentication, secure data transmission, and encryption. Security must be considered from the inception of IoT projects to their implementation and maintenance, and it should be a priority for all stakeholders, including device manufacturers, service providers, and end-users. It is also essential to keep abreast of the latest cyberthreats and to continuously update security measures to stay ahead of the game. Moreover, the integration of IoT and cybersecurity has brought about significant benefits, but it also has significant risks. As such, it is crucial to develop comprehensive security measures that can protect the growing number of connected devices from cyberthreats. By working together, all stakeholders can create a safer and more secure environment for IoT devices and the people who use them.

REFERENCES

[1] El Mrabet, Z., Kaabouch, N., El Ghazi, H. and El Ghazi, H., 2018. Cybersecurity in smart grid: Survey and challenges. *Computers & Electrical Engineering*, 67, pp.469–482.

[2] Gunduz, M.Z. and Das, R., 2020. Cyber-security on smart grid: Threats and potential solutions. *Computer Networks*, 169, p.107094.

[3] Kawoosa, A.I. and Prashar, D., 2021, January. A review of cyber securities in smart grid technology. In *2021 2nd International Conference on Computation, Automation and Knowledge Management (ICCAKM)* (pp. 151–156). IEEE.

[4] Faquir, D., Chouliaras, N., Sofia, V., Olga, K. and Maglaras, L., 2021. Cybersecurity in smart grids, challenges and solutions. *AIMS Electronics and Electrical Engineering*, 5(1), pp.24–37.

[5] Nguyen, T.N., Liu, B.H., Nguyen, N.P. and Chou, J.T., 2020, June. Cyber security of smart grid: Attacks and defenses. In *ICC 2020-2020 IEEE International Conference on Communications (ICC)* (pp. 1–6). IEEE.

[6] Al-Turjman, F. and Abujubbeh, M., 2019. IoT-enabled smart grid via SM: An overview. *Future Generation Computer Systems*, 96, pp.579–590.
[7] Ghasempour, A., 2019. Internet of things in smart grid: Architecture, applications, services, key technologies, and challenges. *Inventions*, 4(1), p.22.
[8] Mehmood, M.Y., Oad, A., Abrar, M., Munir, H.M., Hasan, S.F., Muqeet, H.A.U. and Golilarz, N.A., 2021. Edge computing for IoT-enabled smart grid. *Security and Communication Networks*, 2021, pp.1–16.
[9] Tanwar, S., Tyagi, S. and Kumar, S., 2018. The role of internet of things and smart grid for the development of a smart city. In *Intelligent Communication and Computational Technologies: Proceedings of Internet of Things for Technological Development, IoT4TD 2017* (pp. 23–33). Springer Singapore.
[10] Davoody-Beni, Z., Sheini-Shahvand, N., Shahinzadeh, H., Moazzami, M., Shaneh, M. and Gharehpetian, G.B., 2019, December. Application of IoT in smart grid: Challenges and solutions. In *2019 5th Iranian Conference on Signal Processing and Intelligent Systems (ICSPIS)* (pp. 1–8). IEEE.
[11] Zhao, Z. and Chen, G., 2018, June. An overview of cyber security for smart grid. In *2018 IEEE 27th International Symposium on Industrial Electronics (ISIE)* (pp. 1127–1131). IEEE.
[12] Kimani, K., Oduol, V. and Langat, K., 2019. Cyber security challenges for IoT-based smart grid networks. *International Journal of Critical Infrastructure Protection*, 25, pp.36–49.
[13] Mohammadpourfard, M., Weng, Y., Pechenizkiy, M., Tajdinian, M. and Mohammadi-Ivatloo, B., 2020. Ensuring cybersecurity of smart grid against data integrity attacks under concept drift. *International Journal of Electrical Power & Energy Systems*, 119, p.105947.
[14] Tufail, S., Parvez, I., Batool, S. and Sarwat, A., 2021. A survey on cybersecurity challenges, detection, and mitigation techniques for the smart grid. *Energies*, 14(18), p.5894.
[15] Leszczyna, R., 2018. A review of standards with cybersecurity requirements for smart grid. *Computers & Security*, 77, pp.262–276.
[16] Mohammadi, F., 2021. Emerging challenges in smart grid cybersecurity enhancement: A review. *Energies*, 14(5), p.1380.
[17] Gunduz, M.Z. and Das, R., 2018, September. Analysis of cyber-attacks on smart grid applications. In *2018 International Conference on Artificial Intelligence and Data Processing (IDAP)* (pp. 1–5). IEEE.
[18] Leszczyna, R., 2018. Standards on cyber security assessment of smart grid. *International Journal of Critical Infrastructure Protection*, 22, pp.70–89.
[19] Ustun, T.S., Hussain, S.S., Ulutas, A., Onen, A., Roomi, M.M. and Mashima, D., 2021. Machine learning-based intrusion detection for achieving cybersecurity in smart grids using IEC 61850 GOOSE messages. *Symmetry*, 13(5), p.826.
[20] Xie, J., Stefanov, A. and Liu, C.C., 2019. Physical and cybersecurity in a smart grid environment. *Advances in Energy Systems: The Large-scale Renewable Energy Integration Challenge*, pp.85–109.
[21] Ghiasi, M., Niknam, T., Wang, Z., Mehrandezh, M., Dehghani, M. and Ghadimi, N., 2023. A comprehensive review of cyber-attacks and defense

mechanisms for improving security in smart grid energy systems: Past, present and future. *Electric Power Systems Research*, 215, p.108975.

[22] Saini, S., Beniwal, R.K., Kumar, R., Paul, R. and Saini, S., 2018. Modelling for improved cyber security in Smart distribution system. *International Journal on Future Revolution in Computer Science & Communication Engineering*, 4(2), pp.56–59.

[23] Ferrag, M.A., Babaghayou, M. and Yazici, M.A., 2020. Cyber security for fog-based smart grid SCADA systems: Solutions and challenges. *Journal of Information Security and Applications*, 52, p.102500.

[24] Kakkad, V., Shah, H., Patel, R. and Doshi, N., 2019. A comparative study of applications of Game Theory in Cyber Security and Cloud Computing. *Procedia Computer Science*, 155, pp.680–685.

[25] De Donno, M., Giaretta, A., Dragoni, N., Bucchiarone, A. and Mazzara, M., 2019. Cyber-storms come from clouds: Security of cloud computing in the IoT era. *Future Internet*, 11(6), p.127.

[26] Butt, U.A., Mehmood, M., Shah, S.B.H., Amin, R., Shaukat, M.W., Raza, S.M., Suh, D.Y. and Piran, M.J., 2020. A review of machine learning algorithms for cloud computing security. *Electronics*, 9(9), p.1379.

[27] Subramanian, N. and Jeyaraj, A., 2018. Recent security challenges in cloud computing. *Computers & Electrical Engineering*, 71, pp.28–42.

[28] Khan, N. and Al-Yasiri, A., 2018. Cloud security threats and techniques to strengthen cloud computing adoption framework. In *Cyber Security and Threats: Concepts, Methodologies, Tools, and Applications* (pp. 268–285). IGI Global.

[29] Mugunthan, S. and Vijayakumar, T., 2019. Review on IoT based smart grid architecture implementations. *Journal of Electrical Engineering and Automation*, 1(1), pp.12–20.

[30] Saleem, Y., Crespi, N., Rehmani, M.H. and Copeland, R., 2019. Internet of things-aided smart grid: Technologies, architectures, applications, prototypes, and future research directions. *IEEE Access*, 7, pp.62962–63003.

[31] Reka, S.S. and Dragicevic, T., 2018. Future effectual role of energy delivery: A comprehensive review of Internet of Things and smart grid. *Renewable and Sustainable Energy Reviews*, 91, pp.90–108.

[32] Shahinzadeh, H., Moradi, J., Gharehpetian, G.B., Nafisi, H. and Abedi, M., 2019, January. IoT architecture for smart grids. In *2019 International Conference on Protection and Automation of Power System (IPAPS)* (pp. 22–30). IEEE.

[33] Bera, B., Saha, S., Das, A.K. and Vasilakos, A.V., 2020. Designing blockchain-based access control protocol in IoT-enabled smart-grid system. *IEEE Internet of Things Journal*, 8(7), pp.5744–5761.

[34] Babar, M., Tariq, M.U. and Jan, M.A., 2020. Secure and resilient demand side management engine using machine learning for IoT-enabled smart grid. *Sustainable Cities and Society*, 62, p.102370.

[35] Zhang, Z., Deng, R., Yau, D.K. and Chen, P., 2021. Zero-parameter-information data integrity attacks and countermeasures in IoT-based smart grid. *IEEE Internet of Things Journal*, 8(8), pp.6608–6623.

[36] Lombardi, F., Aniello, L., De Angelis, S., Margheri, A. and Sassone, V., 2018. A blockchain-based infrastructure for reliable and cost-effective IoT-aided

smart grids. In *Living in the Internet of Things: Cybersecurity of the IoT-2018* (pp. 1-6). IET.

[37] Hussain, M.M. and Beg, M.S., 2019. Fog computing for internet of things (IoT)-aided smart grid architectures. *Big Data and Cognitive Computing*, 3(1), p.8.

[38] Tightiz, L. and Yang, H., 2020. A comprehensive review on IoT protocols' features in smart grid communication. *Energies*, 13(11), p.2762.

[39] Chen, S., Wen, H., Wu, J., Lei, W., Hou, W., Liu, W., Xu, A. and Jiang, Y., 2019. Internet of things based smart grids supported by intelligent edge computing. *IEEE Access*, 7, pp.74089–74102.

[40] Wang, C., Zhu, Y., Shi, W., Chang, V., Vijayakumar, P., Liu, B., Mao, Y., Wang, J. and Fan, Y., 2018. A dependable time series analytic framework for cyber-physical systems of IoT-based smart grid. *ACM Transactions on Cyber-Physical Systems*, 3(1), pp.1–18.

[41] Khan, F., Siddiqui, M.A.B., Rehman, A.U., Khan, J., Asad, M.T.S.A. and Asad, A., 2020, February. IoT based power monitoring system for smart grid applications. In *2020 International Conference on Engineering and Emerging Technologies (ICEET)* (pp. 1–5). IEEE.

[42] Li, B., Lv, S. and Pan, Q., 2018, February. The internet of things and smart grid. In *IOP Conference Series: Earth and Environmental Science* (Vol. 113, No. 1, p. 012038). IOP Publishing.

[43] Chhaya, L., Sharma, P., Kumar, A. and Bhagwatikar, G., 2018. IoT-based implementation of field area network using smart grid communication infrastructure. *Smart Cities*, 1(1), pp.176–189.

[44] Rajab, H. and Cinkelr, T., 2018, June. IoT based smart cities. In *2018 International Symposium on Networks, Computers and Communications (ISNCC)* (pp. 1–4). IEEE.

[45] Syed, A.S., Sierra-Sosa, D., Kumar, A. and Elmaghraby, A., 2021. IoT in smart cities: A survey of technologies, practices and challenges. *Smart Cities*, 4(2), pp.429–475.

[46] Chatterjee, S., Kar, A.K. and Gupta, M.P., 2018. Success of IoT in smart cities of India: An empirical analysis. *Government Information Quarterly*, 35(3), pp.349–361.

[47] Park, E., Del Pobil, A.P. and Kwon, S.J., 2018. The role of Internet of Things (IoT) in smart cities: Technology roadmap-oriented approaches. *Sustainability*, 10(5), p.1388.

[48] Sikder, A.K., Acar, A., Aksu, H., Uluagac, A.S., Akkaya, K. and Conti, M., 2018, January. IoT-enabled smart lighting systems for smart cities. In *2018 IEEE 8th Annual Computing and Communication Workshop and Conference (CCWC)* (pp. 639–645). IEEE.

Chapter 2

IoT-Interfaced Protection Strategy using a Z-Source Breaker for DC Microgrid Applications

Faazila Fathima S., Premalatha L., Anshul Yadav, Tushar Singh, Kshitiz Srivastava

2.1 INTRODUCTION

In the past few decades, research and development toward renewable energy sources (RESs) have been highly concentrated due to carbon traces in the atmosphere. Additionally, the energy demand is another concern regarding integrating micro sources such as photovoltaic (PV) and wind turbines into distribution networks that leads to microgrid and smart grid initiatives. The DC microgrid system is a small-scale network that consists of distributed energy sources with energy storage. It supplies the required load, which operates in the islanded mode of operation at low-voltage and medium-voltage levels. Based on the configuration, it can be a radial or ring type with an AC-bus and DC-bus system. AC bus-based microgrids can be directly fed into the existing AC distribution, but the DC network is more feasible, with high efficiency, less transmission loss, and compatibility to connect sources to the DC bus. However, addressing DC network protection issues is an arduous task because no zero current crossings exist in DC. Hence, the determination of the scope of development for protecting the DC microgrid is still in progress [1–5].

The main advantages of the DC microgrid are no synchronization issues with renewables and no concentration is required for frequency regulation. Power conversion for DC can be simplified, but protecting a DC microgrid is a tedious task due to no zero-crossing current in the DC system [6,7]. Additionally, the DC circuit breaker (DCCB) for a medium voltage direct current (MVDC) is in the experimental stage. A conventional solid-state DCCB (SSDCCB) has an auxiliary silicon controlled rectifier (SCR) in series with passive components to reverse bias the main switch. During a fault, the gate activated SCR detects the fault and the main switch is reverse biased before the fault current runs over the main switch's turn-off capability. To reset the capacitor for the next turn-off cycle, an SCR is added to the circuit. Therefore, using the traditional circuit breaker, fault detection and operating time are tedious [8]. Hence, solid-state Z-source breakers were proposed in [9] and further modifications to the circuit were performed in [10,11].

DOI: 10.1201/9781003537878-2

Protecting a DC microgrid using a unidirectional breaker is inefficient because of the presence of bidirectional power flow in the microgrid. Hence, a bidirectional Z-source breaker is preferred in the detection of forward and reverse faults in the DC microgrid. A breaker should satisfy a few conditions during the reclosing operation in order to operate in the DC microgrid. Monitoring and neglecting the negative current circulation in the load and high starting current flowing through the main SCR during reclosing operation are essential. A hybrid breaker with a main mechanical switch integrated by a solid-state switch is one solution for protecting the grid from a fault. Although the solid-state breaker has higher losses, it can provide a faster response. A solid-state circuit breaker with bidirectional protection can be achieved using a Z-source breaker that can auto restore the system after the fault to overcome the hindrance in conventional breakers [12,13]. Hence, interfacing it near the source and load is mandatory to provide efficient protection.

A DC microgrid is used in a ring configuration [14]. With the local measurement of voltage and current data, fault path parameters are determined by the least square method (LSQ). The fault direction is determined from the result inferred from the LSQ. But if this method is incorporated with some applications, such as the Internet of Things (IoT) and other communication tools, it can provide higher efficiency. Additionally, no attention has been paid to reverse fault current issues due to the presence of distributed generation (DG) units. In a radial microgrid configuration, a fault can perturb consumers linked to one bus system. This results in other configurations, such as the mesh type or ring type. A ring-type low voltage DC microgrid is proposed with a probe power unit (PPU), which locates the faulted zone where the other zones were active. In general, the ring-type configuration works in multizone protection with two or more intelligent electronic devices (IEDs) and PPUs. IEDs detect the fault current and isolate the faulted zone, which avoids the entire network shutting down. Simultaneously, PPUs are also installed to locate a fault at each node and thereby monitor the bus near the fault for automatic reclosing. In the case of a permanent fault, the reclosing of breakers leads to unforeseen damage to the system [15]. This can be rectified if each zone is monitored by the IoT, and the IoT detects and communicates to the platform based on our requirements. This issue is keenly addressed in our proposed work, where the protection circuit of the entire grid is enabled with the IoT using a bidirectional Z-source breaker.

In this chapter, a solution is provided for a microgrid network facing a pole-to-pole fault in an IoT-enabled islanded DC ring bus microgrid, which intends to achieve the automatic location of forward and reverse fault identification and auto-restoration using a Z-source breaker, with the continuous monitoring of grid parameters. Section 2 elaborates on the modeling of the Z-source breaker. Section 3 explains the proposed ring DC microgrid configuration using the Z-source breaker in IoT applications.

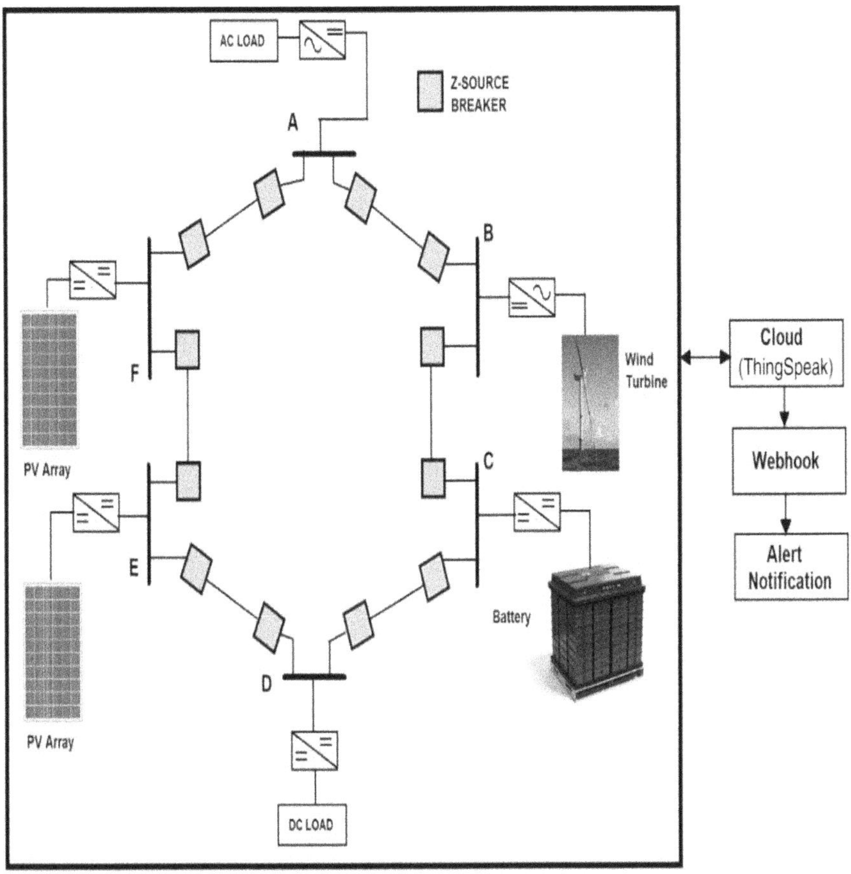

Figure 2.1 Block diagram of the IoT-interfaced DC microgrid

2.2 IOT-INTERFACED DC MICROGRID

A DC microgrid incorporates several DG units. It is possible to operate the DC microgrid autonomously. Multiple energy sources using wind, solar, and other types of energy storage help a great deal to run the DC microgrid autonomously and meet the required demand. In Figure 2.1, the microgrid system is modeled with RESs and some storage system. A 1 kW wind turbine, two 1 kW solar panels, and 0.5 kWh battery as energy storage supply the AC and DC load. The parameters of each source and the storage unit are provided in Table 2.1. Using those data, a simulation is described later in the chapter. Figure 2.1 explains the IoT enabled DC microgrid system, which has energy sources such as a PV array and permanent magnet synchronous generator (PMSG) fixed-speed wind turbine with a battery as a backup device. The power produced is 4 kW, which supplies the AC and DC load after the required conversion. It is further connected to an IoT application to monitor the grid in islanded mode.

2.3 MATHEMATICAL MODELING OF SOURCES AND ENERGY STORAGE

2.3.1 Modeling of the PV

The most abundant and sustainable source of renewable energy is solar energy. Because the PV requires less maintenance and is easy to install, it is widely used by consumers. PV output purely depends on the presence of the insolation level in the solar cell and the temperature of the cell. The maximum power point tracking (MPPT) design for PV system applications is mandatory because of the PV's nonlinear characteristics. Generally, the PV module can be modeled using the Shockley diode equation. The basic single diode model of the PV cell is shown in Figure 2.2.

Considering $V_{pv} = V_{oc}, N_p = 1,$and $N_s = 36$, the PV module is designed based on the given equations:

$$I_{ph} = \left[I_{scr} + K_i\left(T - 298\right)\right] * \frac{\lambda}{1000} \tag{2.1}$$

$$I_{rs} = I_{scr} / \left[exp\left(qV_{OC} / 36kAT\right) - 1\right] \tag{2.2}$$

$$I_0 = I_{rs}\left[\frac{T}{T_r}\right]^3 exp\left[\frac{q^* E_{g0}}{Bk}\left\{\frac{1}{T_r} - \frac{1}{T}\right\}\right] \tag{2.3}$$

Because N_p =1, the equation for I_{PV} can be written as

$$I_{PV} = I_{ph} - I_0\left[expexp\left\{\frac{q^*\left(V_{PV} + I_{PV}R_s\right)}{N_S AkT}\right\} - 1\right], \tag{2.4}$$

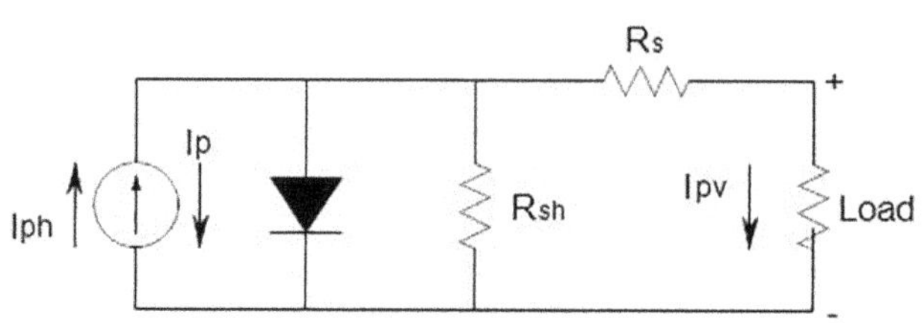

Figure 2.2 Single diode model of a PV cell [16]

where I_{ph} is the photocurrent of a module, I_{rs} is the reverse saturation current of a module, I_{PV} is the output current of a PV module, I_o is the saturation current of a module, and N_s and N_p are the numbers of cells connected in series and parallel, respectively. I_{ph} denotes the photocurrent of a cell, and R_{sh} and R_s are the shunt and series resistances, respectively. Because $R_{sh} > R_s$, they may be neglected for simplification. Additionally, I_o varies with the temperature of a cell [16].

PV cells are clustered in a single system called a PV module. PV modules are further interconnected in series and parallel combinations that result in a PV panel. Based on the requirements of the voltage and current, PV panels are arranged to form a PV array. In this chapter, the proposed PV array consists of 20 PV modules [16] that are integrated in series to obtain an output voltage of 380 V. Solar irradiance is approximately 1000 W/m^2 and a temperature of 25°C is provided to the PV array. The ThingSpeak block (IoT application) monitors the PV from its voltage and current measurement data, and sends the data to the cloud for monitoring purposes. During abnormal operating conditions, overcurrent is detected from the data monitored in the cloud and the Z-source breaker acts accordingly.

2.3.2 Modeling of a PMSG Fixed-Speed Wind Turbine

In a PMSG, the permanent magnets in the rotor provide the excitation field. To achieve maximum efficiency, the air gap between the stator and rotor is reduced [17]. Considering a fixed-speed wind turbine, the wind speed is taken as 12 m/s and the pitch angle is 0° (because it is a fixed-speed wind turbine). The aerodynamic model of a wind turbine is given by

$$P = 1/2\rho ACp(\lambda)v^3 \tag{2.5}$$

$$\lambda = \frac{\omega_m}{v}R \tag{2.6}$$

$$C_p = C_1\left(\frac{C_2}{\gamma} - C_3\beta - C_4\right)e^{\frac{C_5}{\gamma}}C_6 \tag{2.7}$$

$$T_m = \frac{P_\omega}{\omega_m} = \frac{0.5\rho A_r C_p(\alpha,\beta)v^3}{\omega_m}, \tag{2.8}$$

where P is the wind power generated, ρ is air density, A is the area covered by the blades, v is the wind speed, C_p is the power coefficient, and λ is the tip speed ratio. The tip speed ratio of a wind turbine is given by

$$\lambda = Rn\pi/30v, \tag{2.9}$$

where n is wind turbine rotor speed in revolutions per minute (r/min).

The fixed-speed PMSG wind turbine is designed to produce an output of 380 V. The torque generated by the wind turbine is then directed to the PMSM generator, which converts the mechanical energy to three-phase electrical energy. Now a rectifier is used to convert the incoming AC input to DC output. Finally, the DC output from the rectifier has some pulsating effect that is rectified by passing it through an LC filter. The values of L and C are taken to be 47 uH and 220 uF, respectively [17].

2.3.3 Bidirectional Battery Energy Storage System (BBESS)

In a conventional system, power transmission is unidirectional in nature. However, in the case of a microgrid, conventional energy storage devices cannot be used because of the existence of bidirectional power flow. Thus, a bidirectional battery is modeled with a synchronous buck-boost converter based on the requirements and interfaced in the DC bus [18, 19].

The output current of a grid can be expressed as

$$I = I_{\mathrm{g}} - I_{\mathrm{d}}, \tag{2.10}$$

where I_g is the grid current and I_d is the output current of the battery as per the Shockley equation:

$$I_d = I_s \{\exp (q/AkT_c) - 1\}, \tag{2.11}$$

where I_s is the diode saturation current, q is the charge of an electron (1.60×10^{-19} C), k is the Boltzmann constant (1.38×10^{-23} J/K), and A is the ideality factor in buck mode. The state of charge (SoC) gradually increases from 50%, depending on the run time, where the battery voltage never exceeds 54 V. Similarly in the boost operation, the battery voltage reduces from a fully charged state of 48 V or 54 V to 36 V, where it cuts off. The SoC also reduces its initial value gradually, where the battery acts as a source for the microgrid.

2.4 Z-SOURCE BREAKER DESIGN

2.4.1 Classical Z-Source Breaker

The classical Z-source breaker operates unidirectionally. As shown in Figure 2.3, when a fault occurs near the load or Z-source output terminal, a fault current flows in two directions: from the downstream system

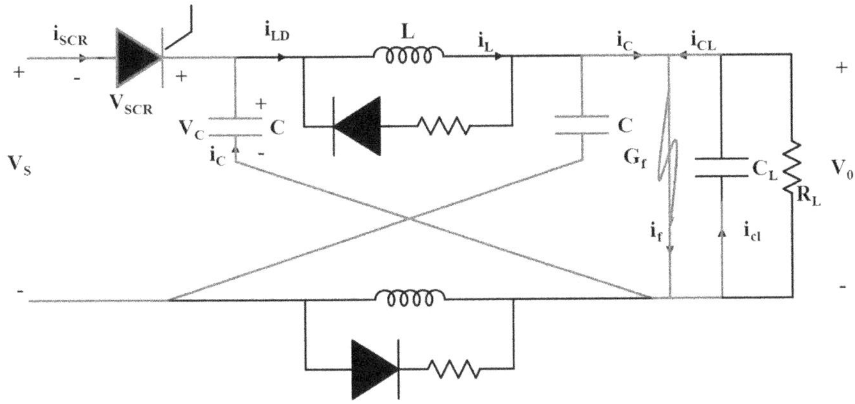

Figure 2.3 Schematic diagram of a unidirectional Z-source breaker [9]

capacitance C_L and the Z-source capacitances C as the fault conduction path. The Z-source LC branch commutates automatically through an SCR and there is no way for direct short capacitor voltages due to the presence of inductors L in the Z-source circuit. The breaker components mitigate faults in a DC system quickly [9].

The steady-state SCR current is

$$I_{SCR} = \frac{V_S}{R_L}. \tag{2.12}$$

When a fault occurs at load side, assume i_L remains constant. Then, the fault path will be through the capacitances C. The fault current can be expressed as

$$i_f = G_f v_0 = \frac{1}{dt.R_{f0}} v_0 t \tag{2.13}$$

By Kirchhoff's current law (KCL),

$$i_{SCR} = I_{SCR} - v_s C_s K t + \frac{V_s C_R K}{12 L C_1}\left(6 L C_s K + 1\right) t^3, \tag{2.14}$$

where v_0 is the output voltage, v_s is the source voltage, R_L and C_L are the load resistance and capacitance, respectively, I_{SCR} is the steady-state SCR

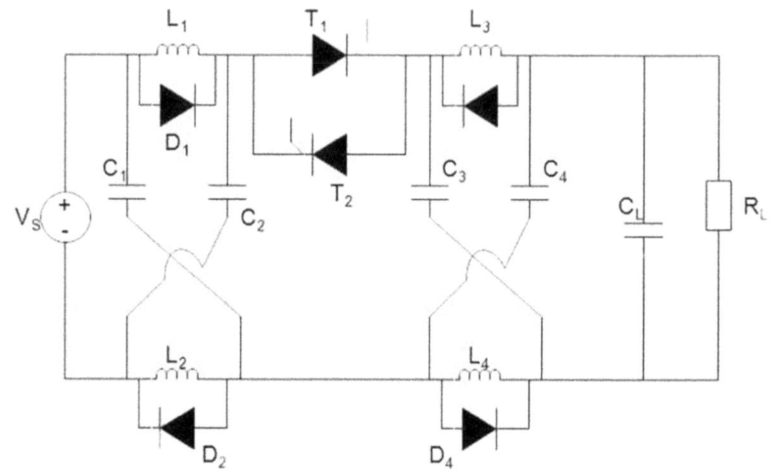

Figure 2.4 Schematic diagram of a bidirectional Z-source breaker [13]

current, i_f is the fault current dissipated by conductance G_f, and K is the derivative of conductance [9].

2.4.2 Bidirectional Z-Source Breaker

In a microgrid, the presence of bidirectional power flow may lead to a reverse fault current due to the presence of DGs in the system. This issue can be resolved if a bidirectional breaker comes into existence [8,9,11]. The classical Z-source breaker cannot be used in a microgrid. For single-source and load applications, the power flow direction may be unidirectional in a microgrid. However, in a complex network, the power flow direction may be complicated to notify and hence continuous variation may be present due to the load distribution. For such applications, the breaker must be capable of working bidirectionally [13]. Therefore, an improved bidirectional Z-source DC breaker creates a natural zero current crossing using a thyristor, which provides a method of interrupting load-carrying currents [13]. This improved Z-source circuit has many advantages, for example, it is cost effective and reliable, handles a large current, and enables voltage capability, faster turn-off, and simple control. The circuit consists of eight passive components and two controlled switches, and produces zero reflected fault current magnitude, as shown in Figure 2.4. During the overcurrent condition, the breaker isolates the DC bus autonomously, thus protecting the grid from severe damage. It does not have a common neutral or perform input filtering for power converters.

2.5 IMPLEMENTATION OF AN IOT-INTERFACED RING DC MICROGRID USING A Z-SOURCE BREAKER

A 4 kW DC microgrid [16–18] works at 380 V, 7.25 A during normal operating conditions, which supplies household appliances with a resistive load, as mentioned in Figure 2.5. The MATLAB Simulink model has been connected to ThingSpeak (IoT cloud) to obtain real-time data for overcurrent detection and grid protection. The ThingSpeak channel has been programmed such that when a line-to-line fault occurs, the rapid rise in current is detected via an alert notification in the form of an email or Short Message Service (SMS) [20–23]. The design specification of a DC microgrid in the ring configuration is shown in Table 2.1. The grid voltage and grid current during normal operating conditions is shown in Figure 2.5. The simulation results are shown in Figure 2.6.

A line-to-line fault F1 occurs between the transmission line connected to the PV array, as shown in Figure 2.7. The forward fault current that occurs due to F1 is denoted by $i1$ and the reverse fault that occurs due to an excess of generation with less load requirement is denoted by $i2$. The line-to-line fault F1 results in overcurrent and a voltage drop in the network. By using the IoT-interfaced overcurrent protection strategy, if a sudden spike of current occurs or the current value crosses a threshold value (set as 15 A) at any source, the Z-source breaker isolates the faulty section and suppresses the fault, thereby restoring the system.

The details and time instances related to the fault are described as follows: A line-to-line (L-L) fault F1 occurs at t = 3.02 s in DC bus 5. The fault clears at t = 3.12 s by the action of the unidirectional Z-source breaker; this value is under industrial set standards [18]. It restores the system to 380 V after 3.12 s, as shown in Figure 2.8. The Z-source breaker isolates the faulty source from the microgrid, and the microgrid keeps itself stabilized at approximately 380 V using the energy provided from other connected renewable sources and the battery. Thus, the microgrid voltage is not affected by the fault. Only the voltage of the concerned (faulty) DC bus is involved, which drops to zero, as stated above. The unidirectional Z-source breaker restores the system back to its normal state. As shown in Figure 2.8, during fault F1, the ThingSpeak (IoT) application senses the fault and the result clearly depicts the action of the unidirectional Z-source breaker.

The presence of the DG units results in the production of the reverse fault current when the generation is higher than the load demand. Because the protective device used here is unidirectional in nature, it cannot detect the reverse current. Therefore, if the reverse current, it will not get suppressed which leads to the greater damage to the DC microgrid. As a result, we can conclude that the unidirectional Z-source breaker is inefficient in detecting the reverse fault. Hence, a bidirectional Z-source breaker is implemented

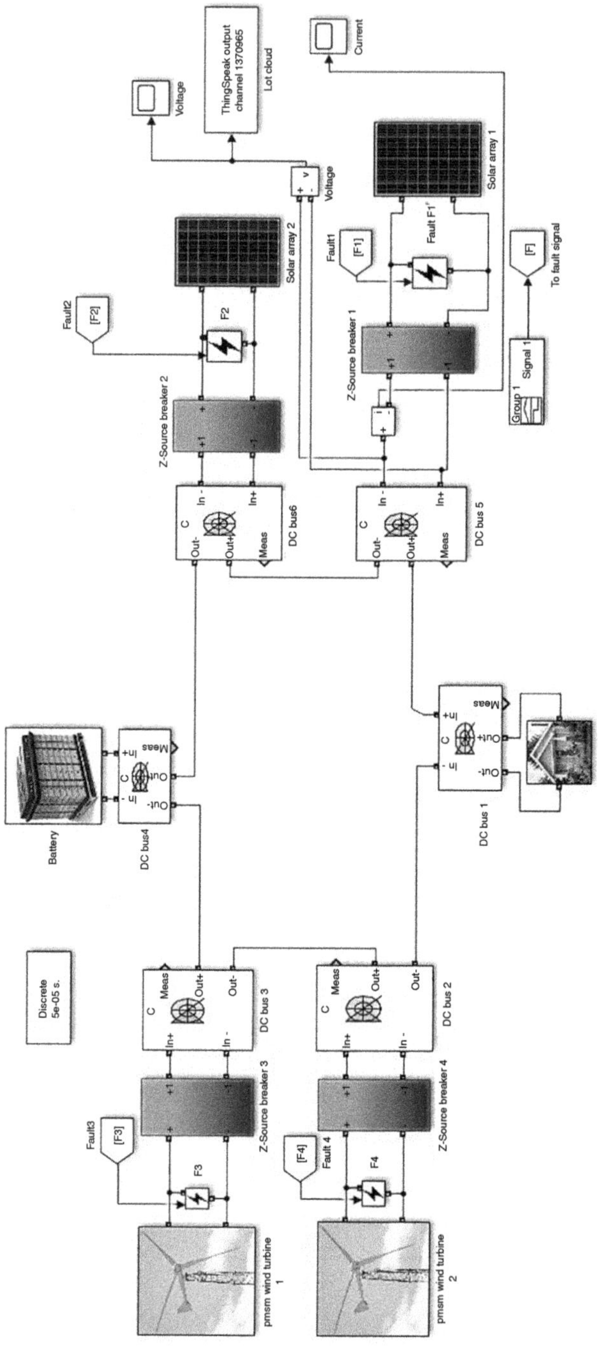

Figure 2.5 Proposed ring DC microgrid with an IoT application

Table 2.1 Design specification of the ring DC microgrid system

Parameter	*Value*
PV output power	900 kW
PV voltage (V_{pv})	380 V
PV current (I_{pv})	2.25 A
Number of series cells (N_s)	22
Number of parallel cells (N_p)	1
PMSG fixed-speed wind turbine (Power)	1 kW
Voltage of the wind turbine (V_{wt})	380 V
Current of the wind turbine (I_{wt})	2.5 A
Rated wind speed	12 m/s
Pitch angle (fixed speed)	0°
Inductance and capacitance value for filter circuit	47 μH and 220 μF
Battery charging duty cycle	12.63%
Initial state of charge (SoC) taken	50%
Battery capacity	10 Ah

in the same ring DC microgrid, which results in the detection of the reverse fault, which is proved by the results in Figure 2.9. The results of the IoT using ThingSpeak and the alert notification through SMS due to fault occurrence are shown in Figures 2.10 and 2.11.

2.6 CONCLUSION

This chapter analyzed a line-to-line fault in a 4 kW DC ring bus microgrid system consisting of solar and wind as energy sources and a battery as energy storage by providing an IoT-interfaced overcurrent protection strategy using unidirectional and bidirectional Z-source breakers. The server monitored the grid during all events and sensed the fault. It cleared and restored the system using a Z-source breaker. The benefit of the IoT is that it does not rely only on the operator, it also protects the entire grid, thereby providing a safer environment for consumers. The bidirectional breaker justified its purpose by detecting the reverse fault due to the presence of DG units. It demonstrated the difference between preferring a bidirectional breaker over a unidirectional breaker. Faults F2, F3, and F4 were analyzed at different locations and demonstrated the same efficiency as fault F1. Thus, implementing the proposed IoT-interfaced protection strategy in a small-scale microgrid network proved its reliability and efficiency. In the future, analysis can be conducted for a large-scale DC microgrid system with adaptive protection operating using communication protocols such as IEC 61850 and other applications.

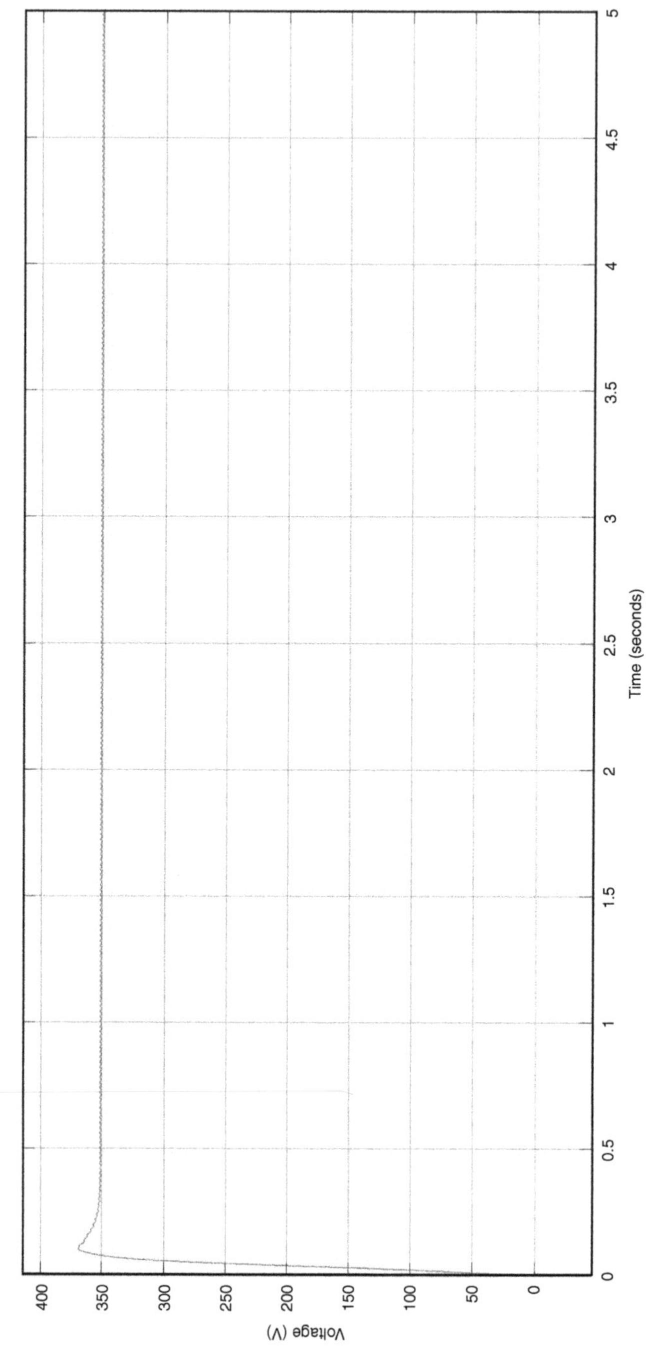

Figure 2.6 Simulation results of the ring DC microgrid during normal operating conditions

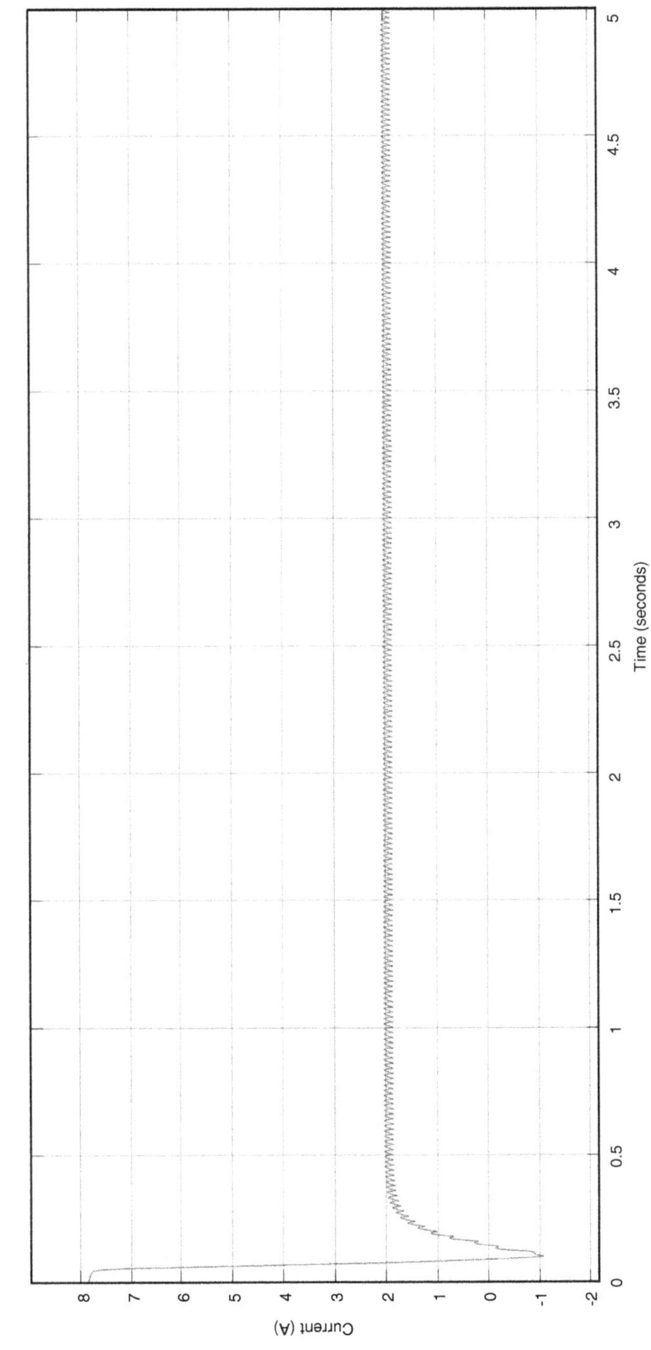

Figure 2.6 (Continued)

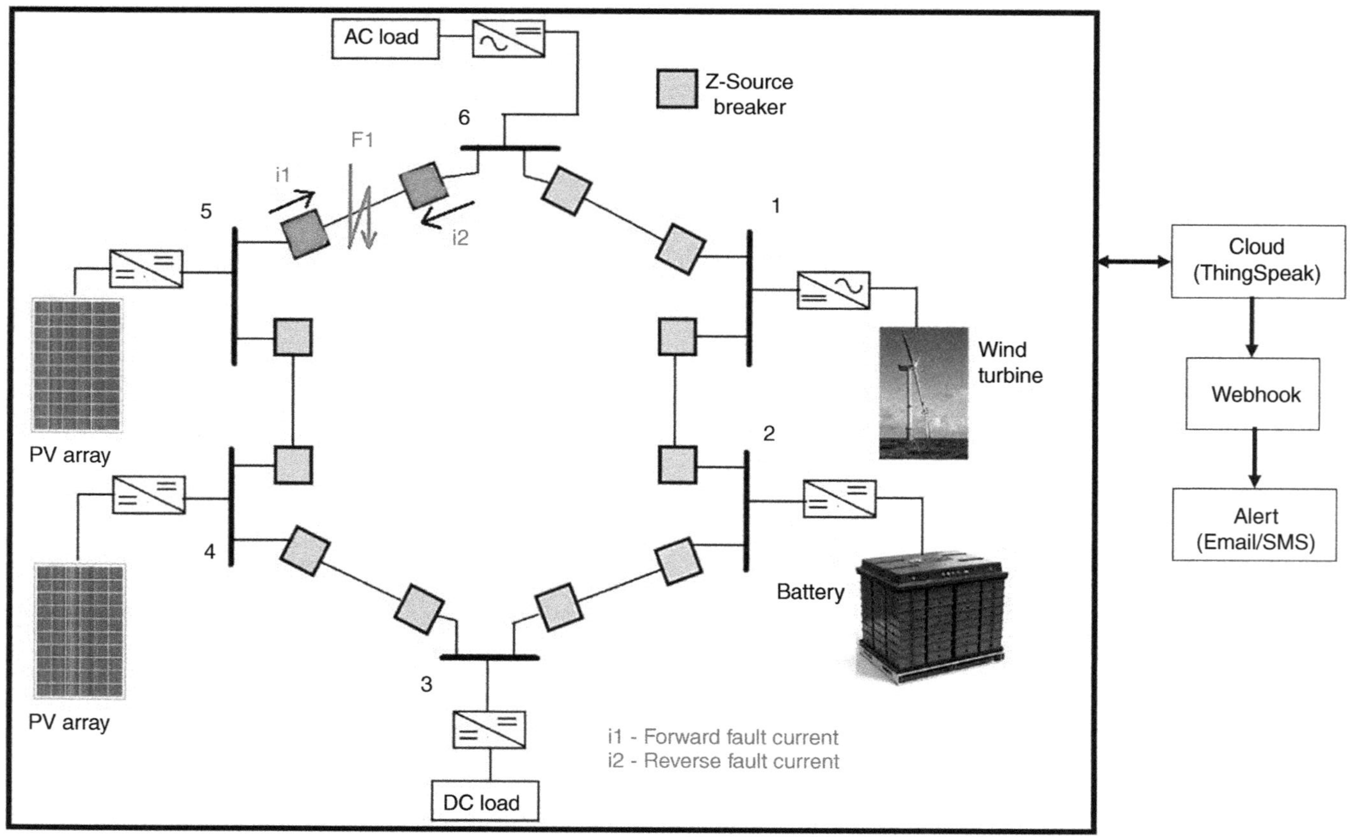

Figure 2.7 Ring DC microgrid during line-to-line fault F1 at DC bus 5

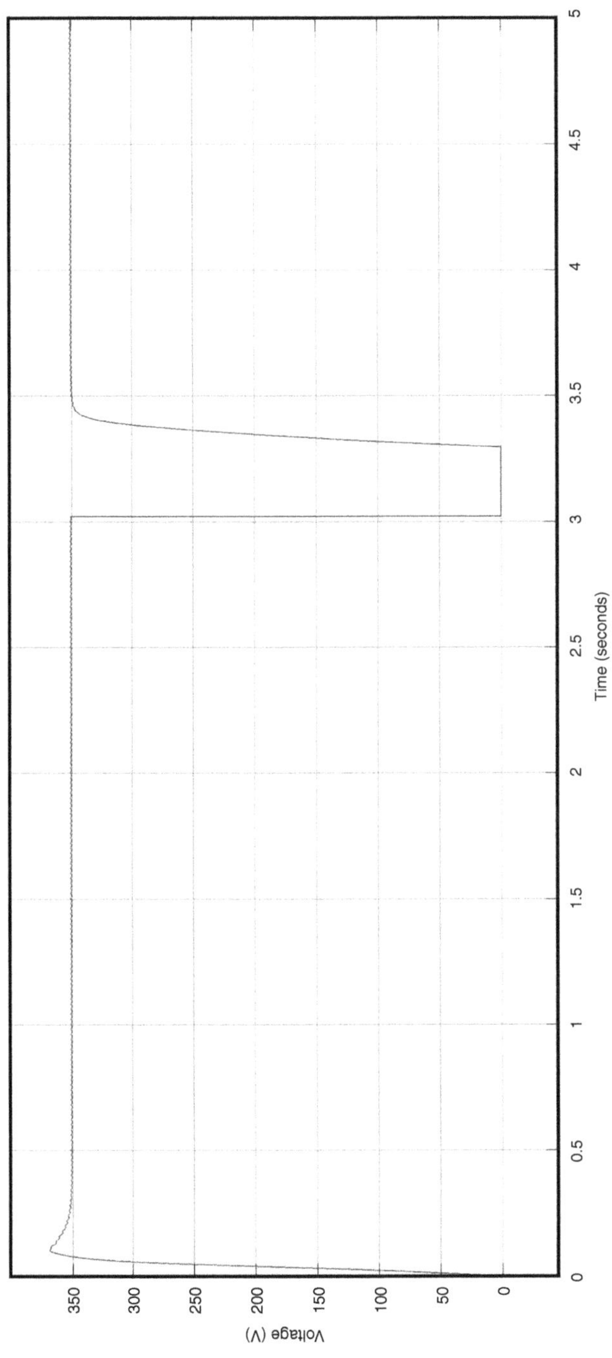

Figure 2.8 Simulation results of the ring DC microgrid during line-to-line fault F1 at DC bus 5 using a unidirectional Z-source breaker

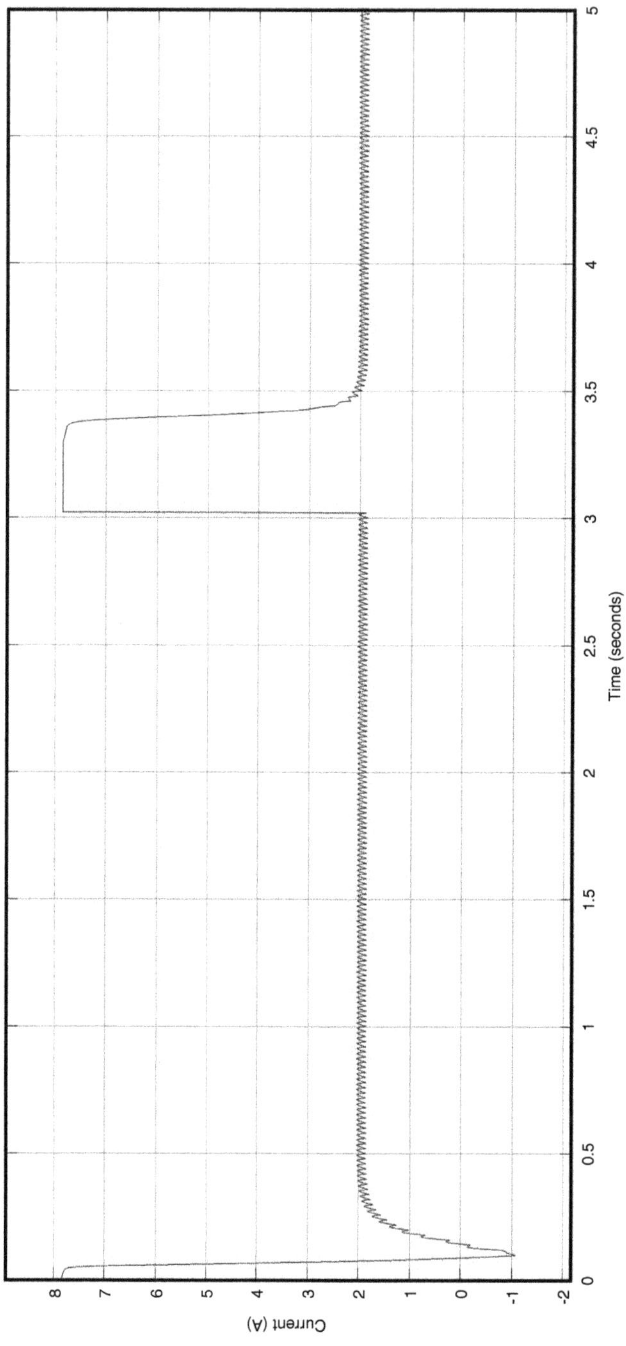

Figure 2.8 (Continued)

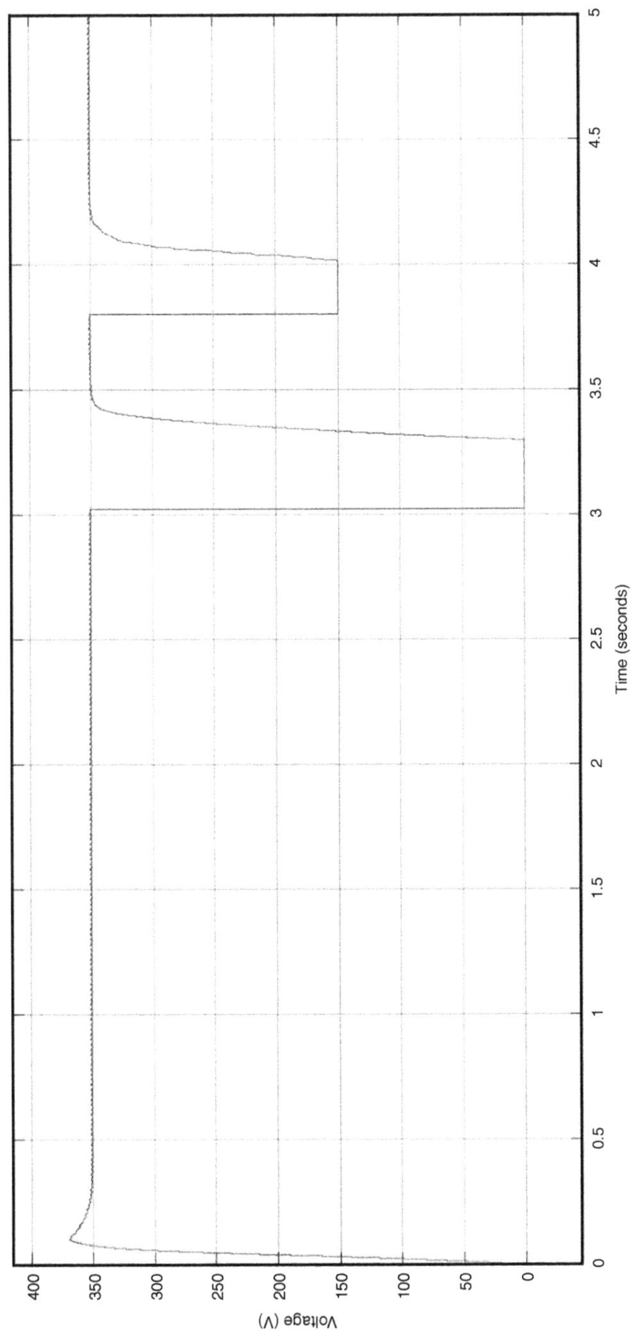

Figure 2.9 Output voltage and output current of reverse current fault rectification caused by DGs using the bidirectional Z-source breaker

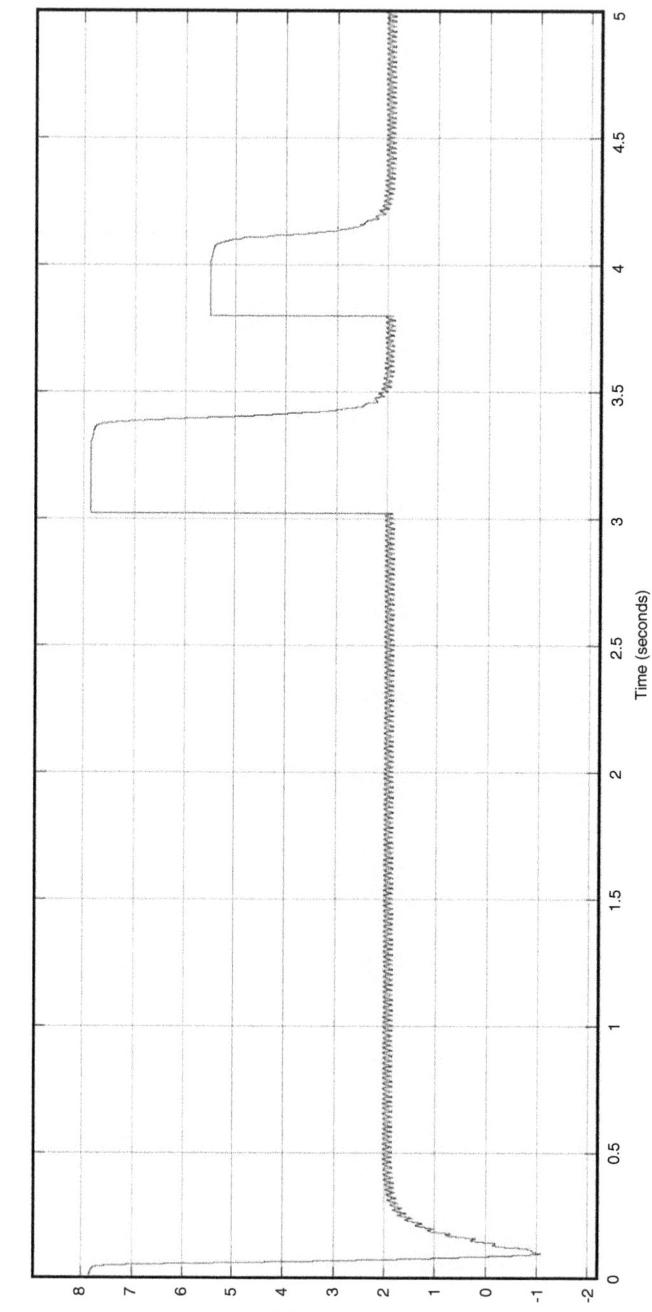

Figure 2.9 (Continued)

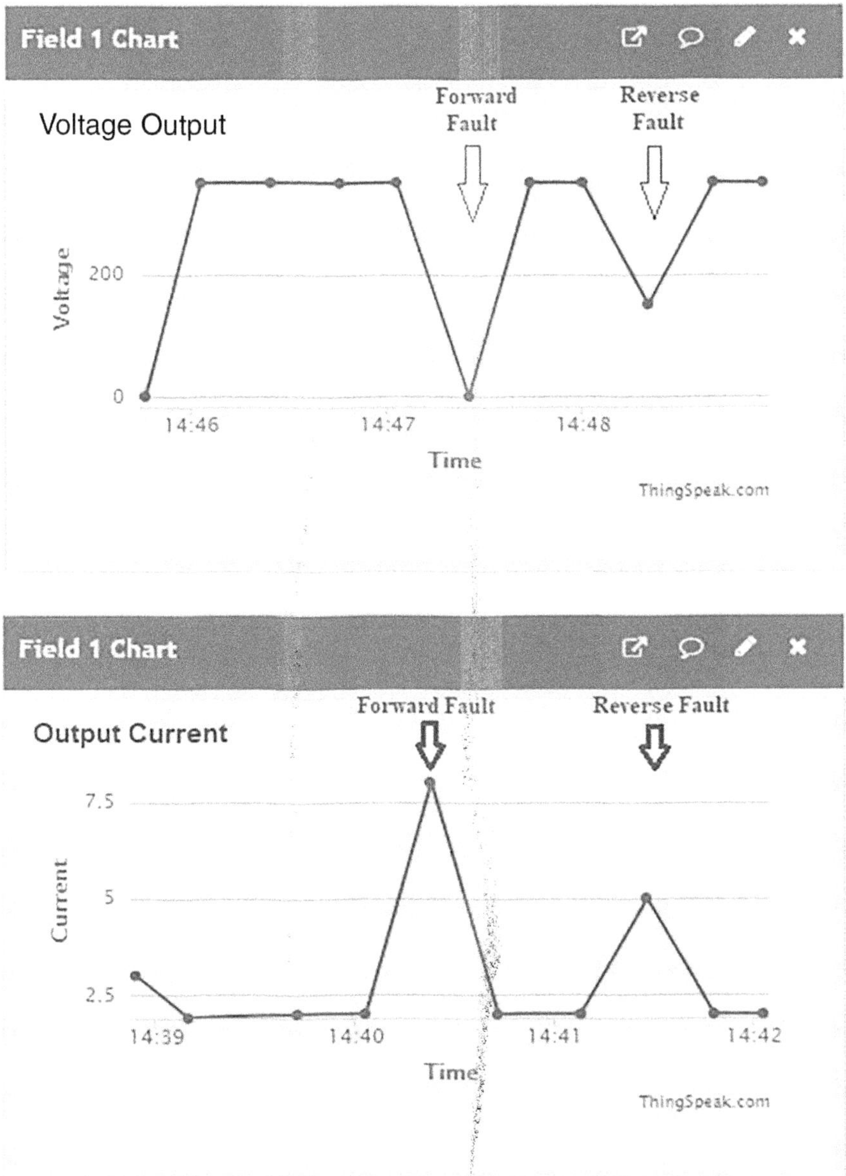

Figure 2.10 Results of the ring DC microgrid during forward and reverse faults at DC bus 5 in the ThingSpeak server

Figure 2.11 Alert notification of fault F1 in the ring DC microgrid

REFERENCES

[1] A. Qazi Atika Qazi, F. Hussain, N. ABD. Rahim, G. Hardaker, D. Alghazzawi, K. Shaban, "Towards sustainable energy: A systematic review of renewable energy sources, technologies, and public opinions," *IEEE Access*, vol. 7, pp. 63837–63851, 2019, doi: 10.1109/ACCESS.2019.2906402.

[2] D. Kumar, and A. Ghosh, "DC Microgrid technology: System architectures, communication networks, applications and standardizations aspects," *IEEE Conference Proceedings*, 2017.

[3] L. Che, M. Khodayar, and M. Shahidehpour, "Only connect: Microgrid for distribution system restoration," *IEEE Power and Energy Magazine*, vol. 12, pp. 70–81, Jan 2014.

[4] H. Nikkhajoei, and R.H. Lasseter, "Distributed generation interface to the certs Microgrid," *IEEE Transactions on Power Delivery*, vol. 24, pp. 1598–1608, July 2009.

[5] S. Faazila Fathima, and L. Premalatha, "Protection strategies for AC and DC Microgrid – A review of protection methods adopted in recent decade," *IETE Journal of Research*, vol. 69, no. 9, pp. 6573–6589, 2023, doi: 10.1080/03772063.2021.1990140.

[6] T. Hakala, T. Lahdeaho, and P. Jarventausta, "LV DC distribution – Utilization potential in a large distribution network company," *IEEE Transactions on Power Delivery*, vol. 30, pp. 1694–1701, Aug. 2015.

[7] S.A. Khaparde, "Infrastructure for sustainable development using renewable energy technologies in India," *2007 IEEE Power Engineering Society General Meeting*, Tampa, FL, 2007, pp. 1–6, doi: 10.1109/PES.2007.385738.

[8] S. Yamaguchi, and H.H. Kobe, Zero-current arc-suppression dc circuit breaker, U.S. Patent 04 740 858, Apr. 1988.

[9] K.A. Corzine, and R.W. Ashton, "A new Z-Source DC circuit breaker," *IEEE Transactions on Power Electronics*, vol. 27, no. 6, pp. 2796–2804, June 2012, doi: 10.1109/TPEL.2011.2178125.

[10] K.A. Corzine, and R.W. Ashton, "Structure and analysis of the Z-source MVDC breaker," *2011 IEEE Electric Ship Technologies Symposium*, 2011, pp. 334–338, doi: 10.1109/ESTS.2011.5770893.

[11] K. Corzine, "Dc micro grid protection with the z-source breaker," *IECON 2013 – 39th Annual Conference of the IEEE Industrial Electronics Society*, 2013, pp. 2197–2204, doi: 10.1109/IECON.2013.6699472.

[12] X. Pei, O. Cwikowski, D.S. Vilchis-Rodriguez, M. Barnes, A.C. Smith, and R. Shuttleworth, "A review of technologies for MVDC circuit breakers," *IECON 2016 – 42nd Annual Conference of the IEEE Industrial Electronics Society*, 2016, pp. 3799–3805, doi: 10.1109/IECON.2016.7793492.

[13] A. Maqsood, and K. Corzine, "The Z-source breaker for fault protection in ship power systems," *2014 International Symposium on Power Electronics, Electrical Drives, Automation and Motion*, 2014, pp. 307–312, doi: 10.1109/SPEEDAM.2014.6872131.

[14] R. Mohanty, and A.K. Pradhan, "Protection of smart DC microgrid with ring configuration using parameter estimation approach," *IEEE Transactions on Smart Grid*, vol. 9, no. 6, pp. 6328–6337, Nov. 2018, doi: 10.1109/TSG.2017.2708743.

[15] J. Do Park, J. Candelaria, L. Ma, and K. Dunn, "DC ring-bus microgrid fault protection and identification of fault location," *IEEE Transactions on Power Delivery*, vol. 28, pp. 2574–2584, 2013, doi: 10.1109/TPWRD.2013.2267750.

[16] N. Pandiarajan, and R. Muthu, "Mathematical modeling of photovoltaic module with Simulink," *2011 1st International Conference on Electrical Energy Systems*, 2011, pp. 258–263, doi: 10.1109/ICEES.2011.5725339.

[17] L. Barote, and C. Marinescu, "PMSG wind turbine system for residential applications," *SPEEDAM*, vol. 2010, pp. 772–777, 2010, doi: 10.1109/SPEEDAM.2010.5542166.

[18] K. JM, S. Mohammed, T.P Ahmed, and M. Shafeeque, "Design and simulation of stand-alone DC microgrid and energy storage system," *IEEE Conference on Optimal Signal Process*, 2019.

[19] S.H. Zacchaeus, W P.Q. Tong, and R.T. Naayagi, "Modelling and simulation of DC microgrid," *2018 IEEE 4th Southern Power Electronics Conference (SPEC)*, 2018, pp. 1–5, doi: 10.1109/SPEC.2018.8636004.

[20] B. Pattanaik, A. Taye, A.A. Umer, and H. G, "IoT based over current protection control system using for Micro Grids," *2021 6th International Conference on Inventive Computation Technologies (ICICT)*, 2021, pp. 235–239, doi: 10.1109/ICICT50816.2021.9358511.

[21] A. Majee, "IoT based automation of safety and monitoring system operations of mines," *SSRG International Journal of Electrical and Electronics Engineering (SSRG – IJEEE)*, no. 9, pp. 17–21, September 2016.

[22] S. Sujeeth, and O.V.G. Swathika, "IoT based automated protection and control of DC microgrids," *2018 2nd International Conference on Inventive Systems and Control (ICISC)*, 2018, pp. 1422–1426, doi: 10.1109/ICISC.2018.8399042.

[23] A. Hirsch, Y. Parag, and J. Guerrero, "Microgrids: A review of technologies, key drivers, and outstanding issues," *Renewable and Sustainable Energy Reviews*, vol. 90, pp. 402–411, 2018, ISSN 1364-0321, doi: 10.1016/j.rser.2018.03.04

Chapter 3

Voltage Unsteadiness Investigation of the IEEE 30 Bus System with Voltage Margin

R. Manivasagam

3.1 INTRODUCTION

Voltage insecurity is a significant issue. To lessen this voltage flimsiness, it is necessary to learn about voltage security. Voltage flimsiness is referred to as static and dynamic. Exploration procedures for voltage dependability can, by and large, be assigned into two types: a static methodology subject to control stream assessment and a unique technique reliant on differential conditions [1]. Static developed examination techniques incorporate a sensitivity exploration strategy, continuation power stream technique, and curious price decay technique. Dynamic analysis strategies for voltage stability incorporate a little flag examination technique, bifurcation investigation strategy (bifurcation hypothesis, e.g., saddle hub bifurcation and Hopf bifurcation) [2], and time-domain simulation method.

In recent years, power generation has primarily relied on coal, nuclear, and natural gas. However, the increasing integration of solar PV has introduced dynamic voltage instability issues [3]. Power systems have been widely reported in the literature as being highly susceptible to voltage collapse [4]. Moreover, voltage stability remains a critical challenge in modern power networks. An alternative stability assessment method, based on network topology, helps to identify weak load buses that are vulnerable to voltage instability [5].

Voltage stability analysis is performed using several methods: (a) P-V and Q-V curves [6], (b) L-index [7], (c) line stability index (Lmn) [8], and (d) voltage stability index (VSI) [9]. A P-V curve study is commonly used to assess voltage stability in both simple circuits and large interconnected power networks [10]. A voltage stability criterion, based on power transmission concepts in a single line, helps to predict voltage collapse using the line stability index [11]. The VSI is used to evaluate the proximity to voltage breakdown in power systems. It is based on the maximum loading capacity of a bus, determined using a Thevenin equivalent approach, and is applicable to transmission networks [12].

DOI: 10.1201/9781003537878-3

3.2 POWER SYSTEM STABILITY

3.2.1 Basic Concepts and Definitions

Power system stability traditionally refers to rotor angle stability, which involves maintaining the synchronous operation of generators. A power system becomes unstable when voltages drop sharply due to equipment outages, increased load, or other disturbances. To better understand stability issues, power system stability is broadly classified into:

- rotor angle stability (including steady-state and transient stability),
- frequency stability, and
- voltage stability (categorized into short-term and long-term, and primarily load-driven) [13].

3.2.2 Voltage Stability and Rotor Angle Stability

Voltage stability and rotor angle stability are closely related. Transient voltage stability is typically linked with transient rotor angle stability, whereas slower forms of voltage stability are associated with small-signal (small-disturbance) rotor angle stability. Both types of stability are influenced by reactive power control. However, voltage stability is primarily a load-related issue, whereas rotor angle stability is mainly concerned with generator behavior [14].

It very well may be said that, if there is voltage breakdown at a point in a transmission framework remote from burdens, it is an edge solidness issue. On the off chance that the voltage crumples in a heap region, it is presumably principally a voltage precariousness issue.

3.3 METHODOLOGY

3.3.1 Voltage Stability Margin (VSM)

VSM can be controlled by using the following equation [15–16]:

$$VSM = \frac{V_{\text{initial}} - V_{\text{instable}}}{V_{\text{instable}}} \times 100. \tag{3.1}$$

3.3.2 Burden Real Power (P) Margin

Condition (3.2) exhibits the importance of the store P edge [17]:

$$Burden\ P\ Margin = \left(P_{\text{instable}} - P_{\text{initial}}\right). \tag{3.2}$$

3.3.3 Burden Reactive Power (Q) Margin

Condition (3.3) exhibits the importance of the load Q edge:

$$Burden\ Q\ Margin = \left(Q_{\text{instable}} - Q_{\text{initial}}\right), \tag{3.3}$$

where
Q_{instable} = responsive force of burden (MW) at the basic point
Q_{initial} = receptive force of burden (MVar) at the ordinary working conditions.

3.3.4 QV Modal Analysis

To find the most vulnerable transport in the organization, QV modular review is utilized dependent on the Q load. In the following subsections, a few steps of this process are explained.

1. Reduced Jacobian Matrix (Jr)
   The first step in the modal analysis method is to determine the reduced Jacobian matrix (Jr) of the power system. The injected real power (P) and reactive power (Q) at the system buses are represented in Equation (3.4), which forms the Jacobian matrix:

$$\begin{bmatrix} \Delta P \\ \Delta Q \end{bmatrix} = \begin{bmatrix} J_{P\delta} & J_{PV} \\ J_{Q\delta} & J_{QV} \end{bmatrix} \begin{bmatrix} \Delta\delta \\ \Delta V \end{bmatrix}, \tag{3.4}$$

   where
   ΔP = steady change in transport genuine power
   ΔQ = gradual change in transport receptive power
   $\Delta\delta$ = gradual change in the transport voltage point
   ΔV = gradual change in the transport voltage size.

   To obtain the reduced Jacobian matrix (Jr), set $\Delta P = 0$ in Equation (3.4), resulting in Equation (3.5). From Equation (3.5), we can derive Equations (3.6) and (3.7). By substituting Equation (3.6) into Equation (3.7), we obtain Equation (3.8):

$$\begin{bmatrix} 0 \\ \Delta Q \end{bmatrix} = \begin{bmatrix} J_{P\delta} & J_{PV} \\ J_{Q\delta} & J_{QV} \end{bmatrix} \begin{bmatrix} \Delta\delta \\ \Delta V \end{bmatrix} \tag{3.5}$$

$$\Delta\delta = -J_{P\delta}^{-1} J_{PV}\ \Delta V \tag{3.6}$$

$$\Delta Q = J_{Q\delta}\Delta\delta + J_{QV}\Delta V \tag{3.7}$$

$$\Delta Q = \Delta V\left[J_{QV} - J_{Q\delta}J_{P\delta}{}^{-1}J_{PV}\right] \text{ or } \Delta Q = Jr\ \Delta V. \tag{3.8}$$

The relationship between the small variations in bus voltage and reactive power (Q) load is shown in Equation (3.9), which is derived by rearranging Equation (3.8):

$$\Delta V = Jr^{-1}\ \Delta Q. \tag{3.9}$$

2. Assurance of the Most Critical Mode
   The power framework is steadily indicated by positive eigenvalues. The most critical (weakest) mode of the power system is determined by the smallest eigenvalue of the reduced Jacobian matrix (Jr):

$$Jr = \xi \Delta \eta. \tag{3.10}$$

3. Transport Participation Factor
   Condition (3.11) is utilized to work out the transport support factor. The transport cooperation factor is in a lattice structure The number of buses is represented by the rows of the matrix, whereas the modes of the power network are represented by the columns of the matrix:

$$Pki = \xi i\ \eta i. \tag{3.11}$$

3.3.5 PV Modal Analysis

To identify the most vulnerable bus in the network, a QV modal analysis is used, similar to the QV modal method. The only difference in this approach is based on the real power (P) load. To obtain the reduced Jacobian matrix (Jr) for PV modal analysis, set $\Delta Q = 0$ in Equation (3.4), as shown in Equation (3.12):

$$\begin{bmatrix} \Delta P \\ 0 \end{bmatrix} = \begin{bmatrix} J_{P\delta} & J_{PV} \\ J_{Q\delta} & J_{QV} \end{bmatrix} \begin{bmatrix} \Delta\delta \\ \Delta V \end{bmatrix} \tag{3.12}$$

$$\Delta P = \Delta\delta\left[J_{P\delta} - J_{PV}J_{QV}{}^{-1}J_{Q\delta}\right] \text{ or } \Delta P = Jr\ \Delta\delta. \tag{3.13}$$

The relationship between the small variations in bus voltage and reactive power (Q) load is shown in Equation (3.14), which is formed by rearranging Equation (3.13):

$$\Delta\delta = Jr^{-1}\ \Delta P. \tag{3.14}$$

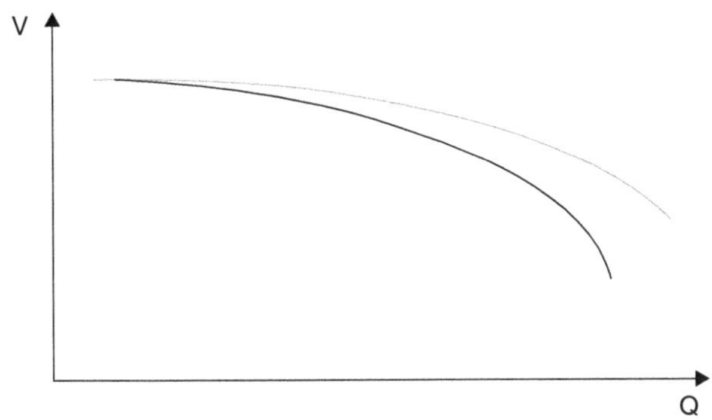

Figure 3.1 QV curve

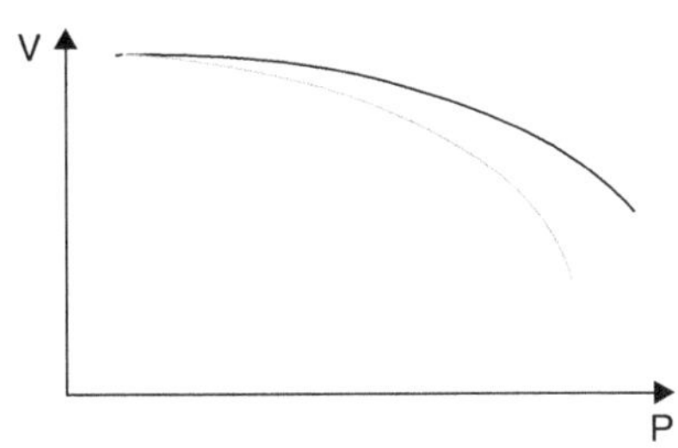

Figure 3.2 PV curve

To determine the most critical mode and the bus participation factor, the same methods explained in Sections 3.2 and 3.3 are used, respectively, for PV modal analysis.

3.4 SIMULATION RESULTS

Figures 3.1 and 3.2 indicate the PV curve and QV curves. Bus 26 is selected for stability analysis because it is the weakest bus, as shown in Figure 3.3. Figures 3.4 and 3.5 indicate the voltage instability PV curve and QV curves for 50% loading. Figurer 3.6 and 3.7 indicate the voltage instability PV and QV curves for 100% loading. Figures 3.8 and 3.9 indicate the voltage instability PV and QV curves for 150% loading. Figures 3.10 and 3.11 indicate the voltage instability PV and QV curves for 200% loading.

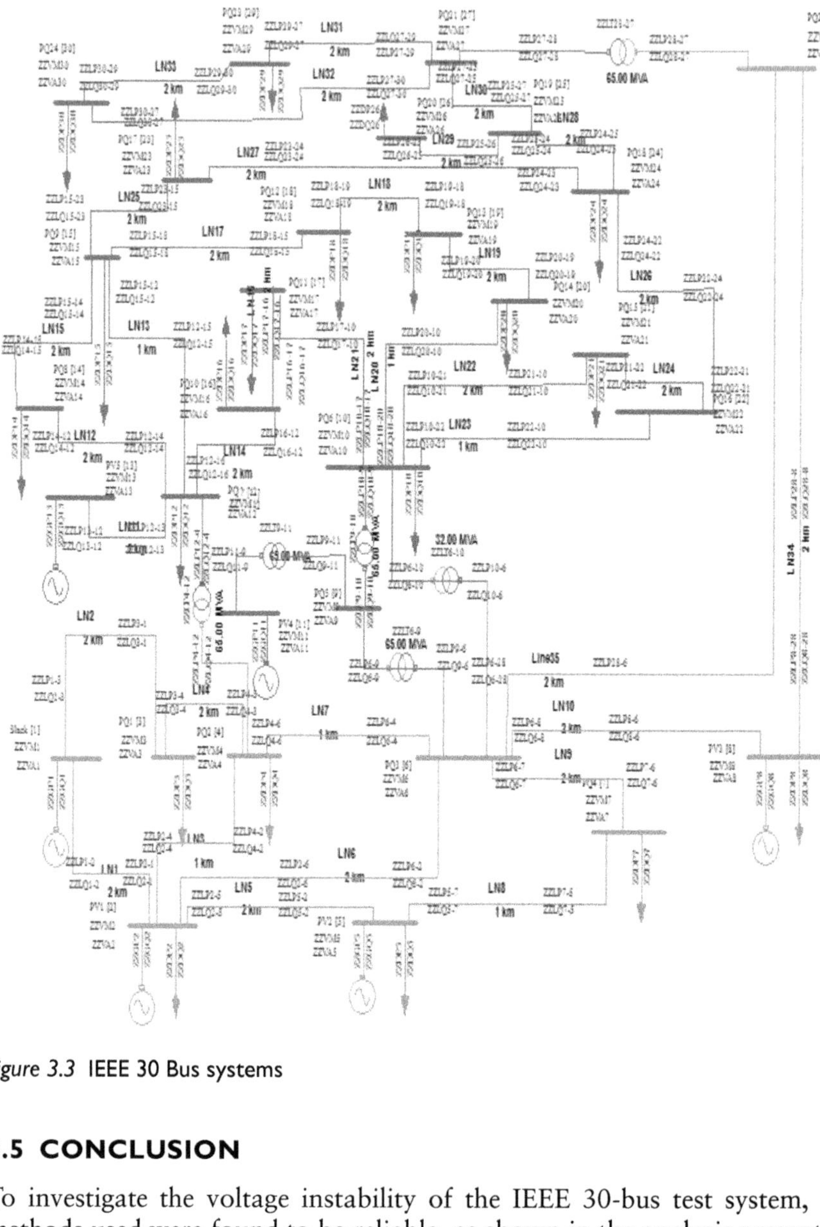

Figure 3.3 IEEE 30 Bus systems

3.5 CONCLUSION

To investigate the voltage instability of the IEEE 30-bus test system, all methods used were found to be reliable, as shown in the analysis presented in this chapter. Both P-V and Q-V analysis methods were applied to identify the weakest bus. The results showed that Bus 26 is the most vulnerable (weakest) bus, as predicted by both Q-V and P-V modal analysis.

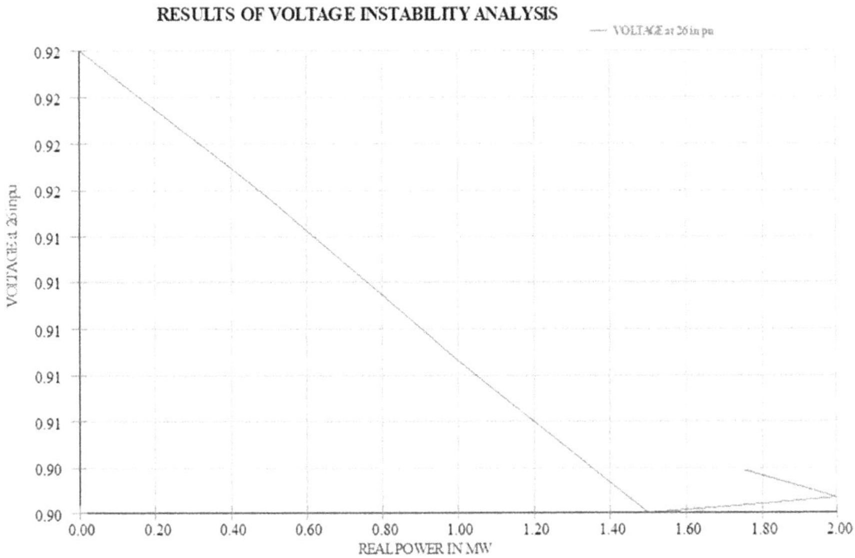

Figure 3.4 Voltage instability PV curve for 50% loading

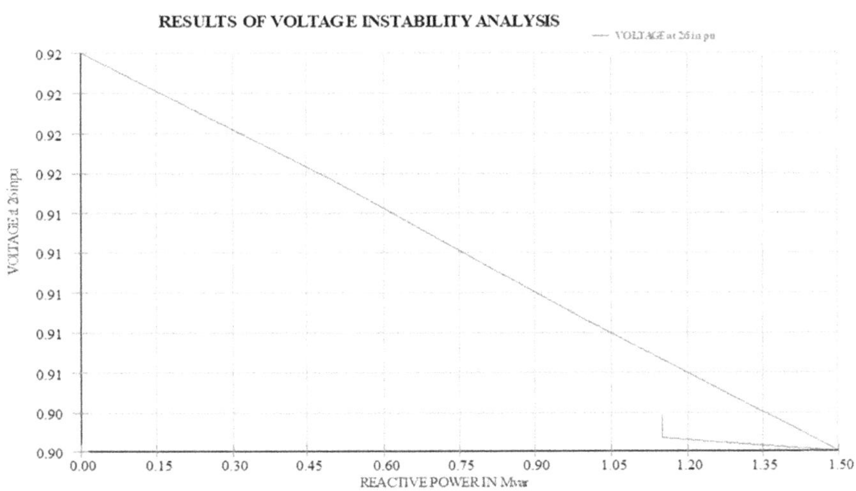

Figure 3.5 Voltage instability QV curve for 50% loading

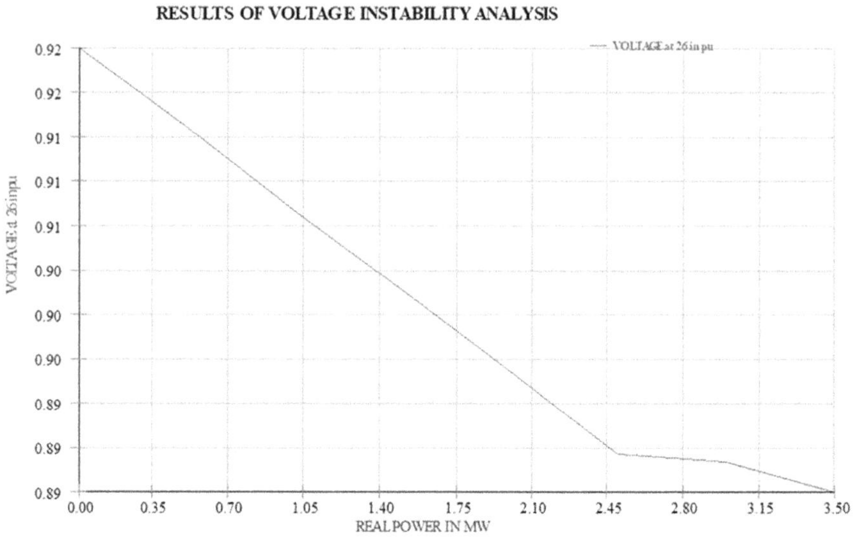

Figure 3.6 Voltage instability PV curve for 100% loading

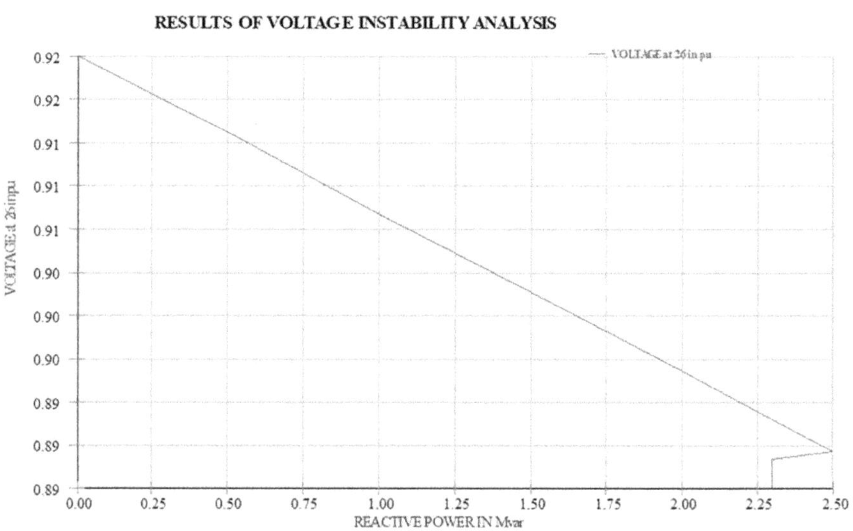

Figure 3.7 Voltage instability QV curve for 100% loading

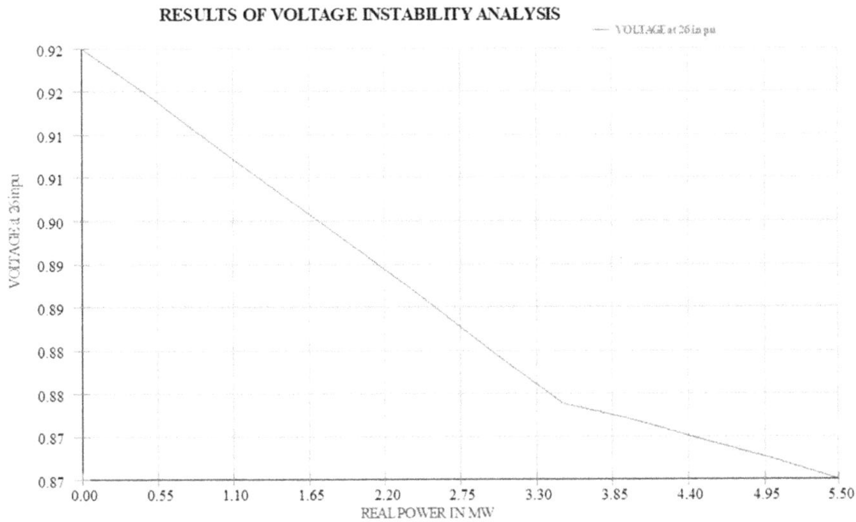

Figure 3.8 Voltage instability PV curve for 150% loading

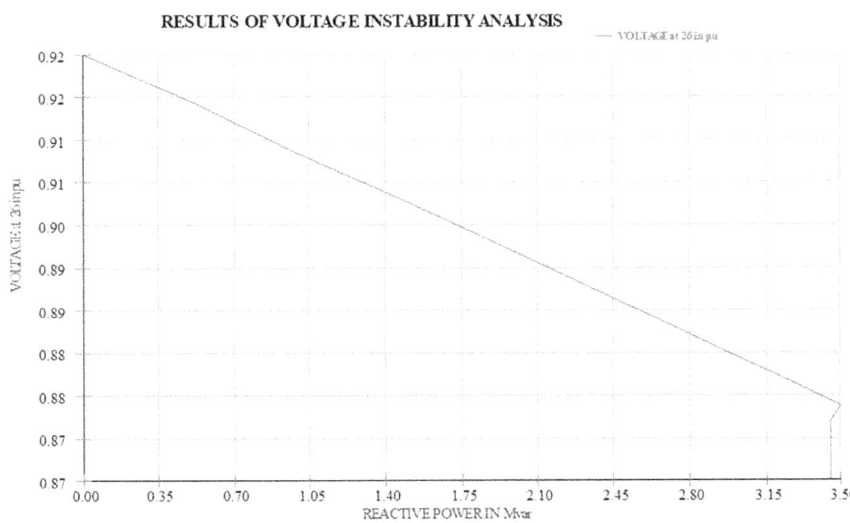

Figure 3.9 Voltage instability QV curve for 150% loading

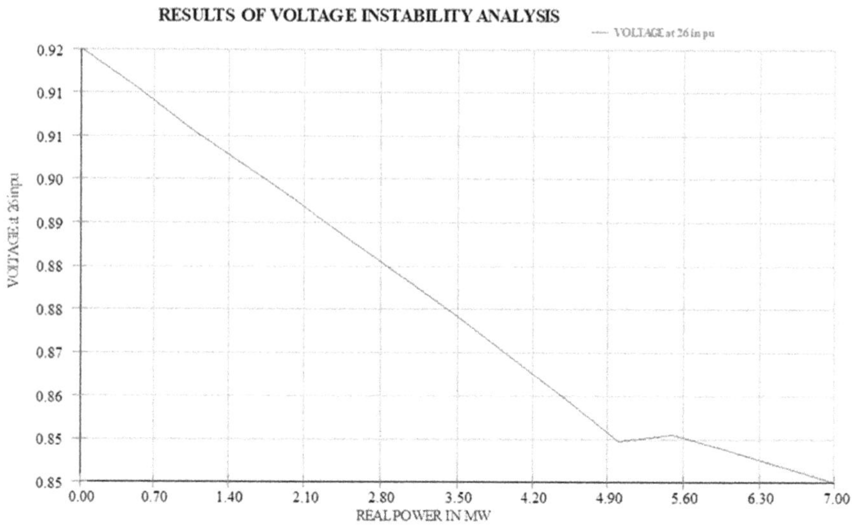

Figure 3.10 Voltage instability PV curve for 200% loading

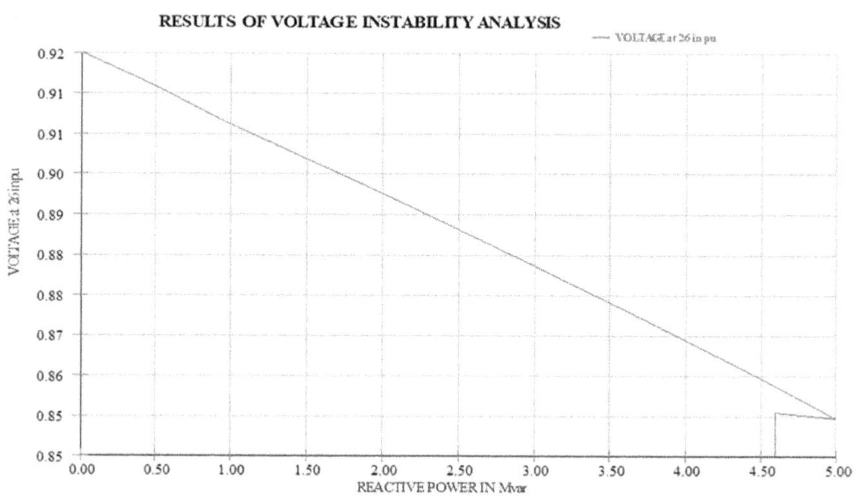

Figure 3.11 Voltage instability QV curve for 200% loading

Table 3.1 Voltage stability margin, and real and reactive power for different loading

% Loading	*Real Power (MW)*	*Reactive Power (MVAr)*	*Voltage in p.u.*	*%VSM or % Error*
50	1.75	1.15	.906	2.2
100	3.5	2.3	.890	3.7
125	4.375	2.875	.877	5.4
150	5.25	3.45	.870	6.3
175	6.125	4.025	.855	7.6
200	7	4.6	.850	8.2

REFERENCES

[1] R. Manivasagam and R. Prabakaran, "Power quality improvement by UPQC using ANFIS-based hysteresis controller," *International Journal of Operational Research*, vol. 37, no. 2, pp. 174–197, 2020. https://doi.org/10.1504/IJOR.2020.105368.

[2] M. Youjie, L.V. Shaofeng, X. Zhou and Z. Gao, "*Review Analysis of Voltage Stability in Power System*," ICMA.IEEE, 2017.

[3] R. Manivasagam and R. Raghavi, "Modeling of a grid connected new energy vehicle charging station," *International Journal of Applied Engineering Research*, vol. 10, pp. 15870–15875, 2015.

[4] A. Tbaileh, C. Mishra, and K. Thomas, "*PV Impacts on Dynamic Voltage Stability*," IEEE, 2017.

[5] R. Manivasagam, P. Parthasarathy, and R. Anbumozhi, "Robust analysis of T-S fuzzy controller for nonlinear system using H-infinity," *Advances in Intelligent Systems and Computing*, vol. 949, pp. 643–651, 2019. Available at: https://doi.org/10.1007/978-981-13-8196-6_56.

[6] I.G. Adebayo, A.A. Jimoh and A.A. Yusuff, "An alternative method for voltage stability assessment in power system," *2017 IEEE AFRICON*, Cape Town, South Africa, 2017, pp. 1084–1088, doi: 10.1109/AFRCON.2017.8095633.

[7] V. Parmar, and M. Shah, "Load flow and voltage instability analysis using MI power software," *IJESI*, vol. 1, no. 6, ver. 3, pp. 49–55, 2018.

[8] R. Manivasagam, and D. Aarthi, "Design of UPFC using ten switch converter with switch reduction," *International Journal for Scientific Research & Evelopment*, vol. 5, pp. 490–493, 2017.

[9] R. Manivasagam, "Comparison of various control strategies for UPQC to mitigate PQ issues," *Journal of The Institution of Engineers (India): Series B*, 2020. https://doi.org/10.1007/s40031-020-00502-4.

[10] A. Yazdanpanah-Goharrizi, and R. Asghari, *A Novel Line Stability Index (NLSI) for Voltage Stability Assessment of Power Systems,* WSEAS, ICPS, No.15–17, 2007.

[11] M. Sulaiman, "PV and QV curves approach for voltage instability analysis on mesh-type electrical power networks using DIgSILENT, MAGNT Research Report (ISSN. 1444-8939) ," vol. 3, no. 6, pp. 36–42, 2015. (DOI: dx.doi.org/14.9831/1444-8939.2015/3-6/MAGNT.04)

[12] R. Manivasagam, and V. Dharmalingam, "Power quality problem mitigation by unified power quality conditioner: An adaptive hysteresis control technique," *International Journal of Power Electronics*, vol. 6, pp. 403–425, 2014. Available at: https://doi.org/10.1504/ijpelec.2014.067442.

[13] R. Maharjan, and S. Kamalasadan, "*Voltage Stability Index for Online Voltage StabilityAssessment*," IEEE, 2012.

[14] R. Manivasagam, and D Aarthi, "Design and implementation of UPFC using ten switch converter with switch reduction," *International Journal of Engineering and Computer Science*, vol. 6, no. 6, pp. 21861–21869, 2017.

[15] R. Manivasagam, and S.P. Richard, "Instant ash monitoring using a load cell in the boiler," *Materials Today: Proceedings,* vol. 39, pp. 1754–1761, 2021. https://doi.org/10.1016/j.matpr.2020.06.496.

[16] W. Qin, and W. Zhang, "*Power System Reliability Based on Voltage WeakestBus Identification*," IEEE, 978-1-4577-1002-5/11, 2011.

[17] S.P. Richard, R. Manivasagam, and P. Vigneshwaran, "Three-level topology for shunt active power filters," *IOP Conference Series: Materials Science and Engineering, Volume 937, Recent Trends on Renewable Energy Smart Grid and Electric Vehicle Technologies (RESGEVT 2020)*, Vellore, India, pp.1–16, 9 July 2020. https://doi.org/10.1088/1757-899X/937/1/012002.

Chapter 4

Deployment of Non-Isolated Multiport Converters in Smart Grid Systems

N. Sivakumar, S. Charles Raja, C. Balasundar

4.1 INTRODUCTION

In recent years, the demand for electricity has risen worldwide. In this dire situation and energy disaster, the energy distribution function will considerably alter all electricity usage from different sources. Due to environmental concerns, renewable energy sources (RESs) are the future emergent sources for high demand [1]. The distribution of energy becomes possible with the integration of a power electronic converter into smart grid systems [2]. This distribution function has a cohesive and resilient infrastructure, with distributed energy divisions, such as home appliances, power plants, and electric vehicles (EVs). The system can share energy depending on demand [3]. In past years, the vast RES utility required converter technology to efficiently convert electrical power. It has changed rapidly because of power electronics components.

In this chapter, power electronics multiport converters are connected in smart grid management systems and their control technologies are discussed. In addition, autonomous alternating current (AC) power systems and direct current (DC) power systems are discussed with coordination and power management structures [4]. Grid-to-vehicle (G2V) can be used to charge EV batteries during off-peak times when electricity rates are low and RES production is excessive. Using RES can support all types of energy sources through storing excess energy when there is low demand. Most grids connect domestic loads using a multiport converter based on a multi-winding transformer [5]. This system is designed to transfer energy to the load, storage unit, and utility in the form of flux. It can isolate using the converter port and sources; thus, loads can increase system safety [6]. By adjusting the transformer ratio, the minimum voltage DC in RES can be stepped up easily. Additionally, the power transfer between DC-DC ports is controlled by the phase shift angle. The transformer eliminates the flow of DC/AC leakage currents in the PV panel from the grid, which results in system safety [7].

DOI: 10.1201/9781003537878-4

Vehicle-to-grid (DC2AC) and G2V are two-way energy transfer concepts involving EVs and the smart grid. This concept is not only used by EVs to draw energy from the utility, it also returns energy back into the grid, contributing to grid stability, efficiency, and resilience. Compared with other multiport methods, power is applied to the battery from the grid source on demand, which increases system suppleness [8].

For EV applications, the proposed multiport DC converter can carry energy to the load from different sources. This converter contains a few components and minimum voltage stress power switches, and enables high efficiency. The proposed converter operates in both buck and buck-boost mode. It is suited for low power applications, like hybrid EVs (HEVs) and renewable energy applications with storage units [9]. Depending on consumer usage, multiport converters require the proper power range for the converter design. Most people want to use only single resources along with utility energy [10]. This chapter discusses an overview of typical generation systems with existing power electronic multiport converters. Then it introduces utility systems with multiport converter-based grid connections and control technologies. Then the control flow of electrical energy depicting self-governing utility systems is considered with the corresponding process. Next, DC power systems are explained [11]. Contrastingly, residential users are the main users for future connections who can access RES like PV and wind in grid power. For this purpose, there is a change of structure and an improvement of converters to relieve the new models. This has fast power flow control, power density standards, and data communications with a converter controller, high flexibility, and efficiency [12]. The proposed converter topology can be integrated with different sources and energy storage units. It mainly utilizes three methods to connect from different sources to transfer loads. Figure 4.1 shows the overall system structure of the multiport power electronics converter for smart grid systems. The system is connected to the actual grid to stimulate DC lines through a transformer and rectifier devices. The three-phase AC voltage with a basic power system load and other sources are connected [13]. By contrast, energy is distributed via the DC lines without being dissipated. Power flow is controlled by converters. This power is distributed by active rectifier devices, which also regulate the DC system voltage.

4.2 METHODOLOGY

Power electronics converters have resulted in various progressions in this field. Additionally, energy converters have a central zone for all applications and exclusive features. Essentially, power electronics converters are categorized into isolated and non-isolated converters, as shown in Figure 4.2. To avoid

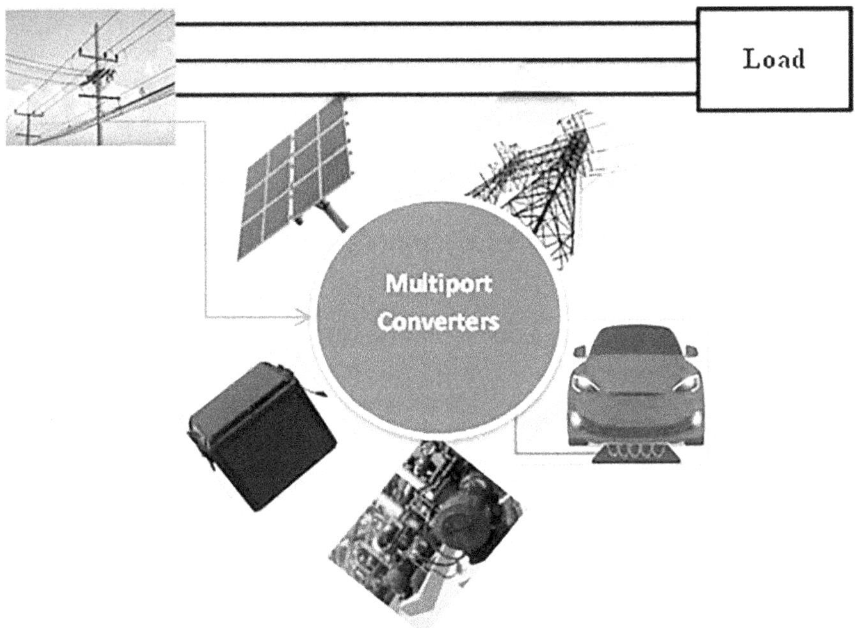

Figure 4.1 Conventional system structure

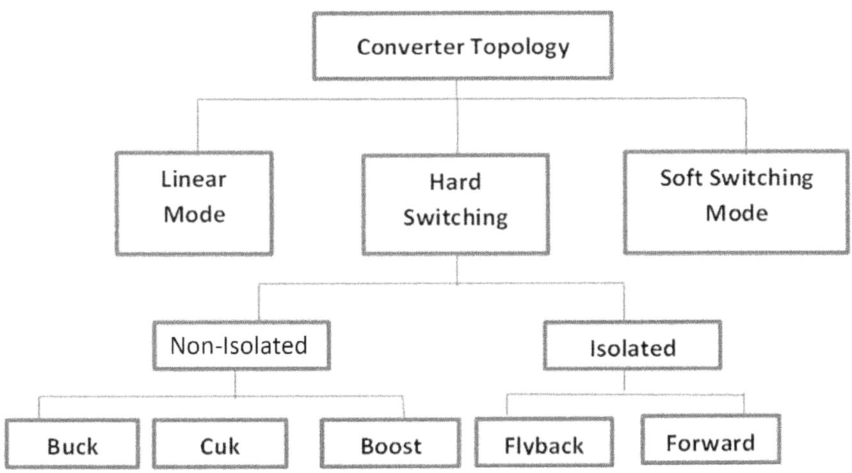

Figure 4.2 Converter topologies

some struggles between sources to load the converter, isolation is preferred. In this method, a high-frequency transformer (HFT) can act as an electrical barrier. It is employed for high-voltage applications. When there is no electrical barrier, the system is called a non-isolated converter, such as Ćuk, SEPIC, boost, and buck-boost.

4.2.1 Proposed Multiport Power Electronics Converters

This chapter on power conversion methods covers the connection of two or more converters with switching elements. Different energy sources are connected by a multiport power electronics converter. Figure 4.3 shows many input sources connected by a common DC bus. Separate DC-DC conversion is also active for every source. All sources are controlled separately by a control unit. A unidirectional converter is connected with the load, whereas a bidirectional multiport converter is connected with the storage unit. This multiport converter performs the vital role of fitting in all energy source device boundaries to provide load power. Different topologies of converters have been proposed that all types of end-user can access easily. Hence, the design of the converter should be suitable for applications either connected to the grid or not.

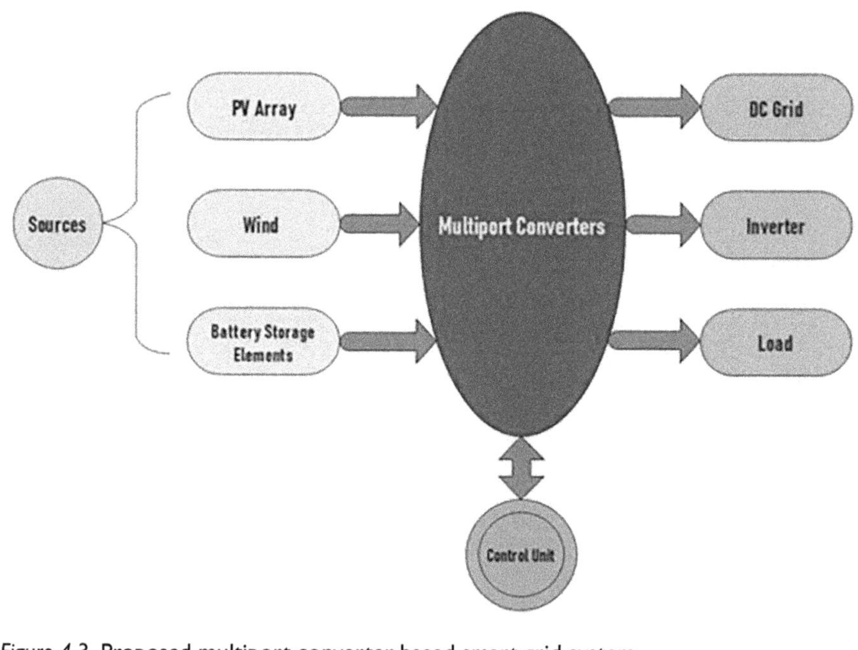

Figure 4.3 Proposed multiport converter-based smart grid system

Two types of power electronics converters exist that are designed to accomplish different power sources and loads. The proposed system incorporates various power sources, which plays the entire location as a single power converter. The load may connect with high/low DC output voltage. A simple design with minimum components and small conversion stages is suitable for EVs, battery energy storage, and HEVs. This chapter covers both isolated and non-isolated converters. They involve many phases of power conversion, as per requirements. The evolution of both converters appears in the literature. It's nothing but energy transfer between two or more electrical circuits without any electrical connections. In such a case safety and isolation are critical, as in RES, PV, and batteries connected to the grid. Converters have high demands due to their high-end performance in power rating systems. Their adaptability is demonstrated in steering the power flow between many ports.

4.3 ISOLATED CONVERTER

An isolated converter performs with an HFT. Additionally, in an isolated converter, the low DC voltage side is electrically separated from the high DC voltage side to minimize shock risk. This isolation enables operation at high voltages while avoiding the use of electronic devices with high voltage and current ratings. However, this topology requires a transformer, which increases the overall system cost and size. But the non-isolated converter has few components and reduces the active switch counts in the system. It can be used when there is no requirement for isolation between the source and the load. The flyback converter is the best example of an isolated converter and is shown in Figure 4.4.

4.3.1 Flyback Converter

This converter refers to an isolated converter. During the ON time, the inductance may store energy and deliver it to the output through the secondary winding while the device is in reset mode. During the OFF time, the stored inductive energy is transferred to the output, and also supplied from the source. In the converter, energy certainly does not flow straight from the source to the load, as for the buck-boost converter. For this reason, the converter is defined as an isolated buck-boost converter. Also known as a magnetic field transformer, it transfers energy into one winding without any delay and there is no deliberate energy storage in the core. Figure 4.5 shows that the irresistible flyback converter stores energy from the primary coil in the core during the ON time and transfers the energy to the output via the secondary coil during a retuned or demagnetization interval. This converter is also not a coupled inductor. This converter allows bidirectional power

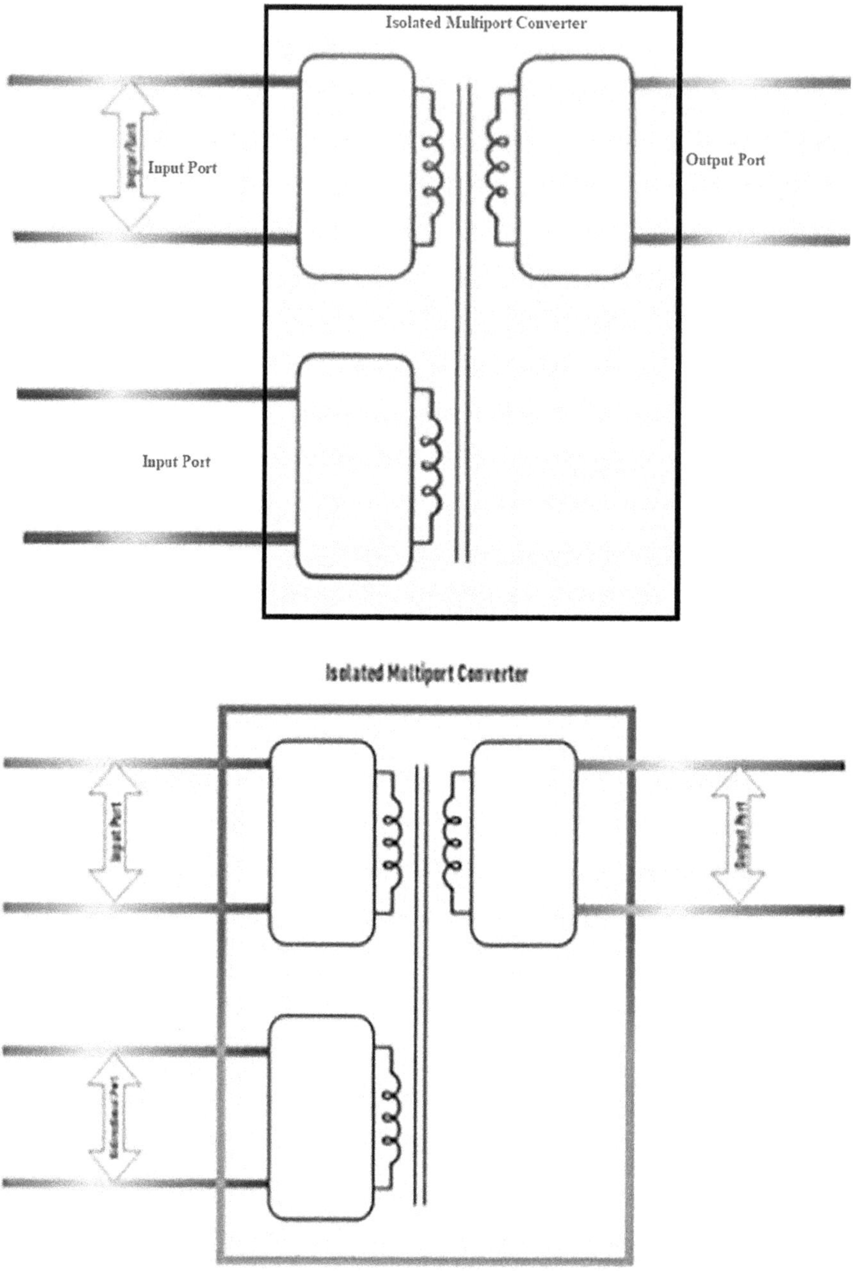

Figure 4.4 Structure of isolated converters

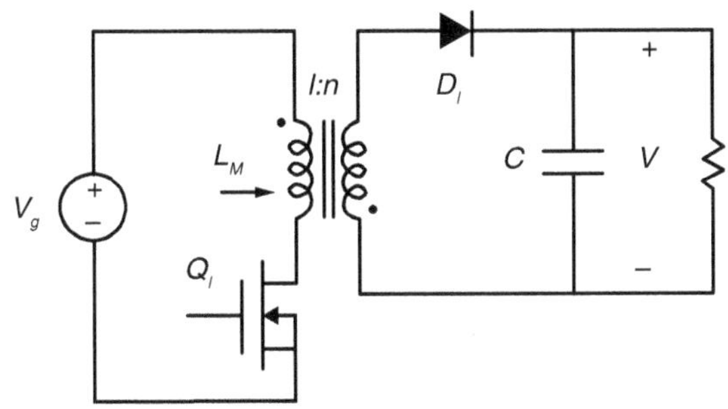

Figure 4.5 Circuit diagram of the flyback converter

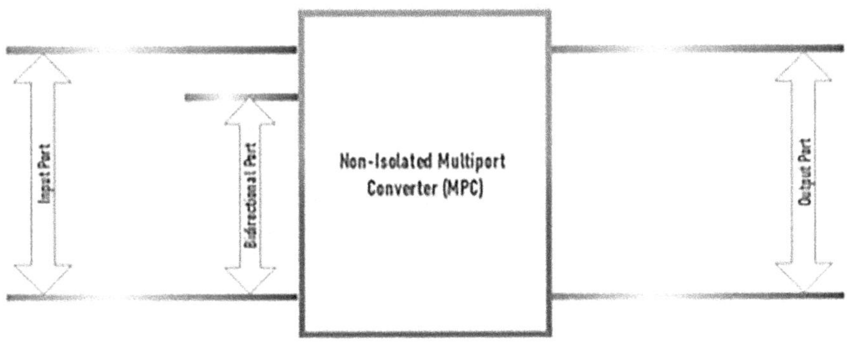

Figure 4.6 Structure of the bidirectional non-isolated converter

flow across the ports; it produces a high-power density and lower ripple current.

4.4 NON-ISOLATED CONVERTER

As shown in Figure 4.6, the non-isolated converter produces high efficiency and reduces the active switches counts in smart grid systems. It's used in real-time applications, such as an EV charging system and RES, where isolation is not mandatory. These converters are flexible power electronic devices, which are critical in connections with other power sources and loads. They are mainly used in EV and RES applications where the panels and batteries are isolated for safety and reliability. This creates a major impact in integrating multiple power resources and applications, while ensuring electrical isolation for safety, efficiency, and reliability in various applications, particularly in renewable energy systems and the electric vehicle infrastructure. The buck-boost bidirectional converter is the best example of a non-isolated converter.

4.4.1 Buck-Boost Converter

In this topology, the ripples in the currents are very low when the converter is operated in continuous conduction mode (CCM). This converter has high efficiency when it's operated in high-load applications. This converter operates in two modes that are controlled with a pulse-width modulation (PWM) controller, which is shown in Figure 4.7.

4.5 RESULTS AND DISCUSSIONS

To validate the performances of the proposed multiport converters, they were statistically developed using MATLAB/Simulink software. As can be observed from the simulation results, the distribution transformer load from AC-DC and DC-AC for various penetration levels obtains corresponding topologies. In AC-DC, the output voltage is high depending on the region and charging specification. The output current for AC-DC charging is determined by the charging station and the capabilities of the EV's onboard charger. AC charging stations typically provide output currents ranging from a few amps to over 30 amps, depending on the charging rate and EV compatibility. DC fast chargers can deliver much higher output currents, often exceeding 100 amps, to enable the rapid charging of EV batteries.

4.5.1 Simulation Results for the Multiport Converter

The isolated grid-connected inverter operates in discontinuous conduction mode (DCM), which is integrated with the flyback converter topology. The converter has a dynamic response and the system also seems to be less complex. The efficiency of flyback converters can be enhanced by their applications. They achieve minimum ripples and obtain the higher efficiencies shown in Figure 4.8.

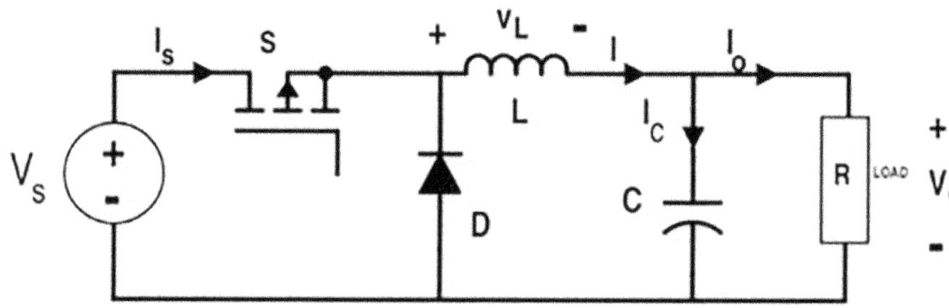

Figure 4.7 Circuit diagram of buck-boost converters

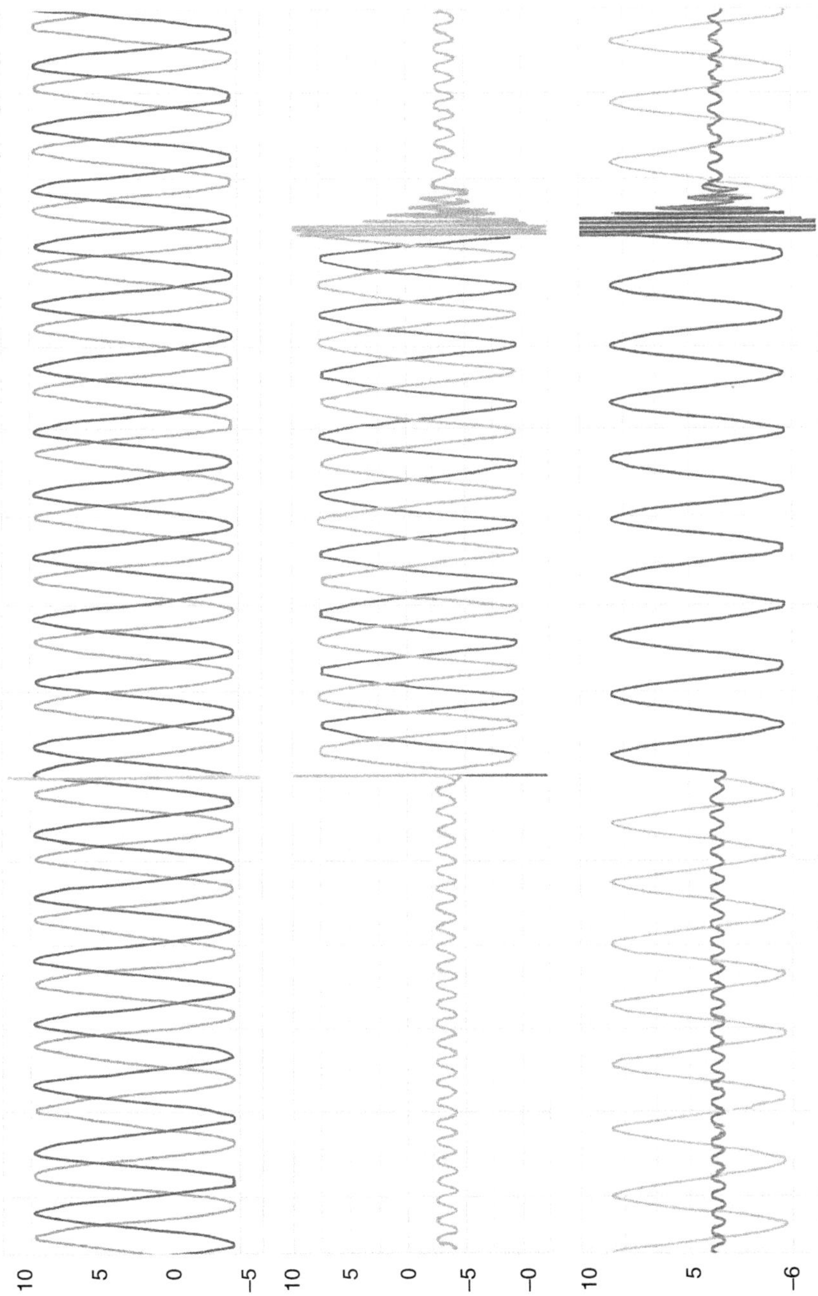

Figure 4.8 Simulation results using the flyback converter

The buck-boost converter simulation results also have greater efficiency with high state of charge (SoC) when the converter is operated with the leading heavy load applications. Depending on the applications, multiport-based smart grid systems operate with PWM in two modes, such as buck and boost. This depends on the changes in source voltage consistent with the output voltage. For buck-boost converter performance designed for batteries, the battery voltage increased above 230 V output voltage from 50 V with 60% of SoC. Figure 4.9 shows that the proposed converter has a high operating efficiency for different output powers.

4.6 CONCLUSION

In this chapter, multiport power electronics converters are implemented for controlling the overall power distribution in a smart grid-based system. The commercial base load profile needs to control the power flow and energy stability. The main objective in this study is to analyze the performance of two different converters based on the load. In a smart grid system, distributed energy system (DES) devices can minimize the total impact of power interruption and the power utility by providing electricity to loads in peak time. In this work, non-isolated multiport converters in smart grid connected systems and other distributer systems are considered, which are increasing the permeation of solar devices and EV batteries. The structure of the multiport converter allows voltage enhancement while producing low ripple content. This method customizes energy loading, which is obtained from the inductance coil to induce the voltage gain. The current stress may reduce on the active switch modules. This method minimizes the commutation losses and enables them to obtain a low current. For this technique, using buck-boost converters was analyzed extensively. Both the compared isolated and non-isolated converters are sufficient for the solar PV system, which was implemented and verified. Regarding demand timing, this chapter supports its necessity for specific applications.

ACKNOWLEDGEMENTS

The authors would like to thank the research excellence strand of the Savitha Project for supporting this work under the Thiagarajar Research Fellowship (TRF) scheme (File.no: TRF/Jul2022/08).

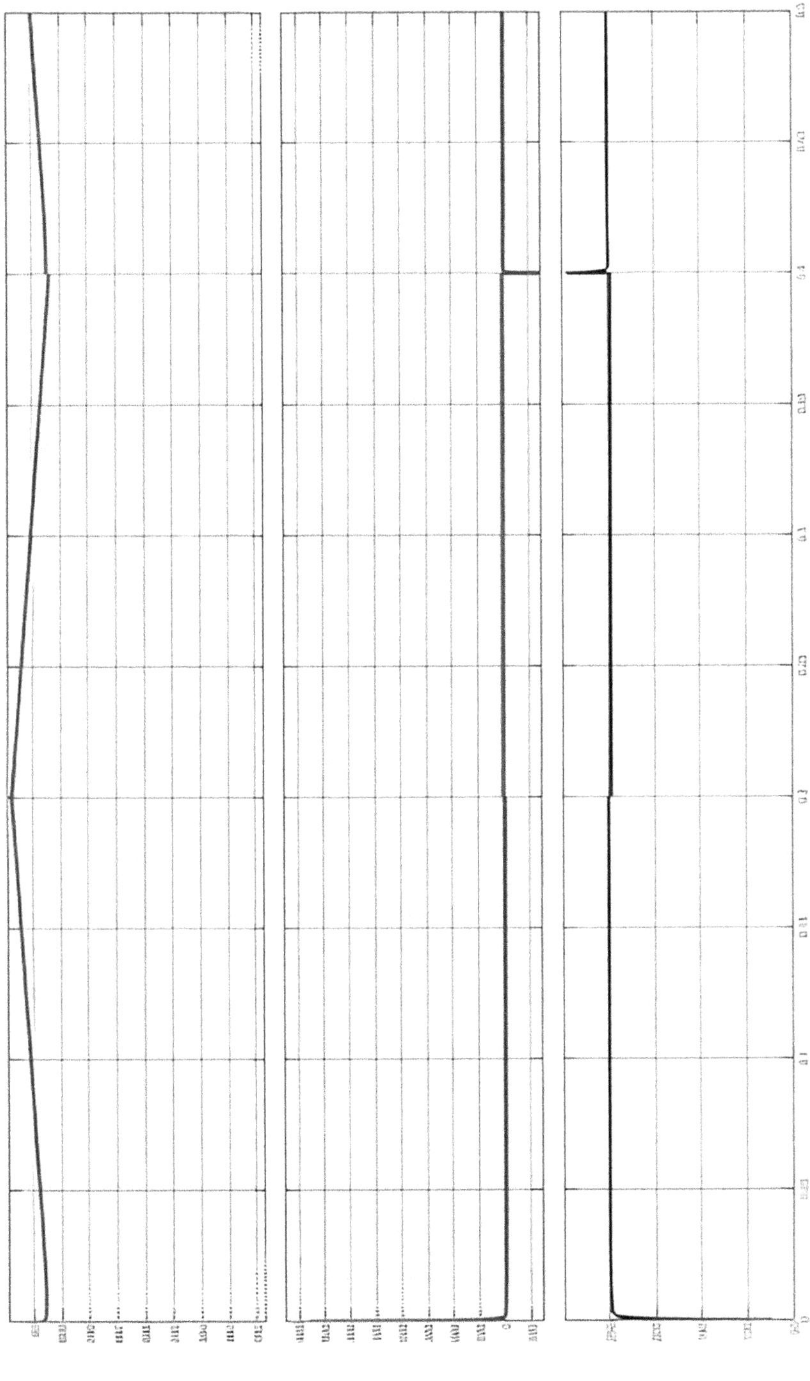

Figure 4.9 Simulation results for batteries using a buck-boost converter

REFERENCES

[1] Galkin, I. A., et al. "Interface converters for residential battery energy storage systems: Practices, difficulties and prospects." *Energies*, vol. 14, no. 12, p. 3365. 2021.

[2] Shayeghi, H., S. Pourjafar, and S. Majid Hashemzadeh. "A switching capacitor based multi-port bidirectional DC–DC converter." *IET Power Electronics*, vol. 14, no. 9, pp. 1622–1636, 2021.

[3] Jafari, M., et al. "A multi-port converter based renewable energy system for residential consumers of smart grid." *IECON 2015-41st Annual Conference of the IEEE Industrial Electronics Society*. IEEE, 2015

[4] Tarassodi, P., J. Adabi, and M. Rezanejad. "A power management strategy for a grid-connected multi-energy storage resources with a multiport converter." *International Journal of Circuit Theory and Applications,* 2023.

[5] Gevorkov, L., et al. "Modern multiport converter technologies: A systematic review." *Applied Sciences*, vol. 13, no. 4. p. 2579, 2023.

[6] Bubovich, A., P. Suskis, and I. Galkin. "Initial evaluation of multiport power supply system for medical lighting equipment." *2023 IEEE 17th International Conference on Compatibility, Power Electronics and Power Engineering (CPE-POWERENG)*. IEEE, 2023.

[7] Firdous, Z., et al. "Multiport converters based EV charging: Architectures, challenges, performance analysis and emerging trends." *2023 IEEE International Conference on Energy Technologies for Future Grids (ETFG)*. IEEE, 2023.

[8] Jafari, M., and Z. Malekjamshidi. "A topology of DC-DC converter based on multi-winding transformer for grid integration of multiple renewable energy resources." *Inventions*, vol. 5, no. 3, p. 31, 2020.

[9] Nduwamungu, A., et al. "Control strategies and stabilization techniques for DC/DC converters application in DC MGs: Challenges, opportunities, and prospects—A review." *Energies*, vol. 17, no. 3, p. 669, 2024.

[10] Uno, M. and K. Sugiyama, "Switched capacitor converter based multiport converter integrating bidirectional PWM and series-resonant converters for standalone photovoltaic systems", *IEEE Transactions on Power Electronics*, vol. 34, no. 2, pp. 1394–1406, Feb. 2019.

[11] Jiang, W. and B. Fahimi, "Multiport power electronic interface-concept modeling and design", *IEEE Transactions on Power Electronics*, vol. 26, no. 7, pp. 1890–1990, 2011.

[12] Lai, C. M., Y. J. Lin, M. H. Hsieh and J. T. Li, "A newly-designed multiport bidirectional power converter with battery/supercapacitor for hybrid electric/fuel-cell vehicle system", *2016 IEEE Transportation Electrification Conference and Expo Asia-Pacific ITEC Asia-Pacific 2016*, pp. 163–166, Jul. 2016.

[13] Zhao, C., S. D. Round and J. W. Kolar, "An isolated three-port bidirectional dc-dc converter with decoupled power flow management", *IEEE Transactions on Power Electronics*, vol. 23, no. 5, pp. 2443–2453, 2008.

Chapter 5

Adaptive Demand Response Management: Introduction and Case Study at a Higher Learning Institution Building

A. C. Vishnu Dharssini, S. Charles Raja, T. Karthick, J. Jeslin Drusila Nesamalar

5.1 INTRODUCTION

Contemporary power systems grapple with various challenges, such as drastically growing electricity demand, the extensive proliferation of distributed renewable sources, the widespread adoption of numerous Internet of Things (IoT) devices, and the rise of threats to cyber-physical systems. These factors contribute to a significant gap and imbalance between energy generation and consumption, underscoring the imperative role of sustainable power policies, strategies, and regulations to enhance energy services. The overarching goal is to ensure the stability and reliability of the system.

The notable surge in energy consumption requires consumers to transition from traditional energy management strategies to more modern and advanced techniques. Conventional systems, reliant on manual operation and exhibiting limited flexibility, struggle to align with the escalating demand. Addressing these challenges requires the exploration of cutting-edge technology and sustainable solutions to ensure the secure and dependable operation of the power system amid significant strain.

A survey indicates that approximately 40% of global annual energy consumption is attributed to buildings [1]. Recognizing buildings as major contributors, modeling the energy demand of a building with a comprehensive understanding of demand patterns and available sources emerges as a crucial strategy. This approach is essential for defining both the quantity and quality of a building's energy performance, achieved through the analysis of time-constrained patterns, identification of clusters with common profiles, or the forecast of demand patterns [2].

The optimal management of building energy through building energy management systems (BEMSs) is greatly facilitated by the implementation of smart meters. These devices play a crucial role in collecting comprehensive numerical data for various system parameters, both electrical and non-electrical, across the entire system [3]. In the context of smart grids, these multidimensional data are sourced from entities such as the control center, grid operators, and equipment providers, all of which participate in

DOI: 10.1201/9781003537878-5

the regulatory process. Smart meters also facilitate the implementation of tailored control strategies for specific users by integrating user power regulation with feedback from smart contracts. Moreover, they continuously monitor and update records of system conditions periodically in the cloud [4]. The stored data follows a request-response model, meaning it can be retrieved upon request, initiated from the user end. To enable the extraction of meaningful insights from this data, it undergoes preprocessing before being fed into algorithms. Consequently, artificial intelligence emerges as a modern and effective tool, with the choice of algorithm being entirely dependent on the specific application and user preferences (UPs) [5].

With the advent of smart grids, utilities are exploring methods to encourage small-scale renewable energy generation, integration, and ways to align electricity demand with supply during peak usage hours. This ability to manage usage is known as demand-side management (DSM), which has been proven to be a highly promising strategy for reducing both utility energy production and customer energy consumption [6]. DSM schemes encompass initiatives such as demand response (DR) programs, energy efficiency programs, and energy conservation programs, all aimed at providing advantages to consumers. Key strategies within DSM include load shifting, peak clipping, valley filling, conservation, flexible load shaping, and load building, all of which are introduced as effective energy-saving programs [7]. The dynamic approach of DSM aims to reduce energy bills by relying on the time-of-use (ToU) principle, achieved through proper appliance scheduling within buildings based on user dependency and comfort preferences [6].

The study documented in [8] introduces a strategic approach for residential building energy management aimed at addressing DR problems. This method is developed through optimization using probability-constrained programming, integrating the particle swarm optimization technique with the two-point estimation method. Historically, predominant studies concerning home energy management systems have relied on centralized optimization approaches [9]. Typically, these methods assume accurate uncertainty predictions, allowing for optimal performance. However, this assumption is often unrealistic in practical scenarios, as the optimization model possesses complete environmental information, eliminating all errors in predictions.

Furthermore, the inclusion of numerous binary or integer variables in certain methods can result in computational expenses that are prohibitively high. By contrast, reinforcement learning (RL), an evolving branch of machine learning, demonstrates impressive decision-making capabilities in environments with limited initial information. The integration of RL into decision-making processes offers significant advantages. First, RL determines optimal actions through interaction with the environment, thus eliminating the need for initial knowledge acquisition, which can be difficult to implement in real-world scenarios. Second, RL can acclimate to diverse application domains by undergoing both offline training and online implementation, autonomously handling inherent uncertainties. Third, RL proves to be more accessible for real-world implementation when compared to traditional optimization methods, as it can attain optimal outcomes from a lookup table, leading to enhanced computational efficiency.

Recent literature has demonstrated increasing interest in employing RL for resolving energy management challenges. Consequently, there is a growing need for a unified framework to facilitate the automation of building energy management. To this end, the principles outlined in the study are as follows:

- to promote overall building energy management by adopting strategic surveys and measures at the customer end
- to present a UP-based ADR framework for ensuring satisfactory utilization and proper scheduling of each load to avoid the risk of reaching the maximum demand limit and peak demand
- to originate time-based workspace prioritization followed by device-based categorization of loads as primitive stages of the proposed framework
- finally, to confer and validate the outcomes of ADR in terms of the cost involved by adopting the "time of use" concept-based energy tariff.

5.2 OVERVIEW AND BACKGROUND OF THE RESEARCH

The main objective of this research is to implement a qualitative adaptive DR technique for the purpose of reducing the energy tariff associated with a commercial building. The primary goal is to prevent exceeding the maximum demand limit and incurring penalties. The case study involves the Thiagarajar College of Engineering (TCE) campus, a higher educational institution situated in the Madurai district of Tamil Nadu, India, which encompasses various departments conducting diverse laboratory sessions that vary in accordance with the day's schedule and allocated hours. Scheduling is contingent on the availability of staff and laboratories for accommodation, yet the current framework neglects energy considerations. This pilot study utilizes smart meter data from TCE to propose a systematic scheduling approach for appliances in the laboratory, incorporating the adaptive DR (ADR) strategy. With smart energy meters, consumption data are periodically recorded and stored in the cloud, retrievable as needed from the centralized host server system. Following procedural data analysis, a suitable ADR technique is recommended, as depicted in Figure 5.1.

The study pertains to the Electrical and Electronics Engineering (EEE) department of TCE. This department offers undergraduate, postgraduate (PG), and Ph.D. programs, catering to over 500 students. Additionally, it accommodates more than 80 research scholars, over 20 faculty members,

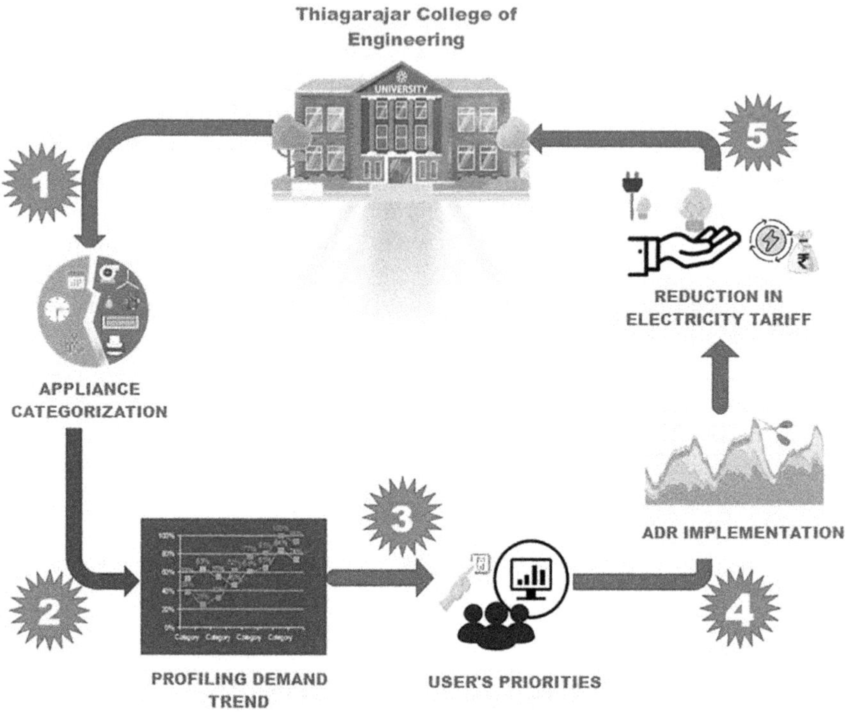

Figure 5.1 Overview of the ADR analysis

and a supporting staff of over 20 individuals. The proliferation of technology and the extensive infrastructure of educational institutions contribute to a substantial demand for energy. Given that laboratories constitute a significant portion of energy consumption, the EEE department's eight well-equipped laboratories adhere to a standard operational schedule from 9:00 a.m. to 4:40 p.m. These laboratories include the Applied Electronics Laboratory, Project/Simulation Laboratory, Embedded Laboratory, Power Electronics Laboratory, P. G. Laboratory, Control System Laboratory, and Computer Laboratory. Typically, laboratory activities are divided into two time slots: from 11:15 a.m. to 1:15 p.m. and from 2:15 p.m. to 4:15 p.m. The priorities in operating loads in relation to workspaces are outlined in Table 5.1.

To gain comprehensive insights into energy consumption, approximately 60 smart energy meters from Elmeasure are installed throughout the main block of TCE. This facilitates real-time monitoring, controls the escalation of demand, and alerts consumers about excessive energy utilization. These meters are interlinked and connect to a regional concentrator. Collectively,

Table 5.1 Relative time slot and workspace-based priorities

Time slot	*Schedules*	*Priority*
11:15 a.m. to 1:15:00 a.m. & 2.15 p.m. to 4.15 p.m.	**Laboratory and lecture sessions**	**P1**
9:00 a.m. to 11:00 a.m.	**Only lecture sessions**	**P2**
1:15 p.m. to 2:15 a.m. & 4:15 p.m. to 6:00 p.m.	**Break time and extracurricular activities**	**P3**

the meters gather data on 266 electrical parameters from all feeders at each timestamp, which are then stored in the cloud. The aggregated data are retrievable from a centralized host server system using Elmeasure's ELNet software, provided in a comma-separated value file format.

5.3 POWER CONSUMPTION TREND

To visualize the net energy pattern of the department during a typical working day, the individual consumption levels of each laboratory are extracted and plotted. This demonstrates that the consumption pattern exhibits a range, beginning to rise around 8:00 a.m., reaching its peak around noon, maintaining that peak until approximately 2:00 p.m., and experiencing a dip during the lunch break. Subsequently, the energy trend resumes an upward trajectory, characterized by moderate consumption, before gradually declining and reaching its lowest point around 5:00 p.m., as depicted in Figure 5.2.

The schedule for each laboratory is subject to variations based on the sequential order of days, in accordance with the time plan of TCE. Several established patterns of laboratory utilization are outlined as follows:

- Project and PG laboratories activities remain consistently operational from 9:00 a.m. to 4:40 p.m., designated primarily based on the research and PG scholars' utilization ratios.
- The embedded laboratory undergoes active sessions exclusively during the odd semesters of the academic year, whereas the Power Electronics laboratory is active solely during the even semesters.
- Similarly, experiments are conducted in the Machines Laboratory with DC machines during odd semesters and AC machines during even semesters.

Each laboratory adheres to a distinct pattern, with the Machines Laboratory exhibiting a notable peak demand of 15.8 kW. Subsequently, the Simulation Laboratory registers a higher demand, whereas the remaining laboratories exhibit demand below 2.7 kW. On assessing overall net consumption, it

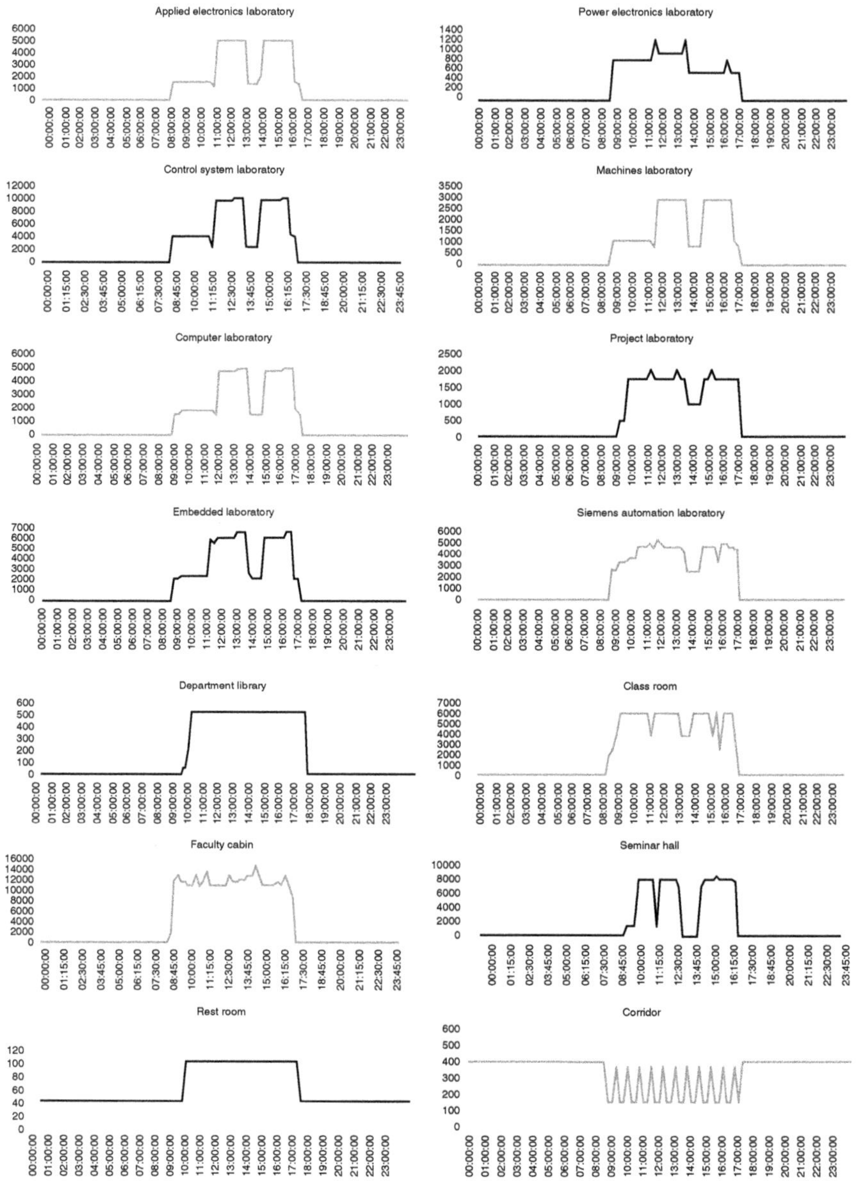

Figure 5.2 Smart meter-retrieved energy trends

becomes evident that the range surpasses the established maximum demand limit for the department, thereby incurring penalties in the electricity tariff.

5.4 PROPOSED ADAPTIVE DEMAND RESPONSE TECHNIQUE

The proposed approach of ADR implementation in a higher education building is entirely considered using existing usage trends, source availability, and cost expenditure for energy. The incorporated strategy ensures reliable energy utilization with minimal outlay using an RL algorithm. The multi-agent Q learning algorithm extracts the traces of energy utilization using past meter records fed into it and acts as an ADR switching unit. Loads are shifted accordingly with the precise decision of the adopted RL algorithm. It is also designed to send pop-up alert notifications on the operating status of appliances and suggests switching accordingly with respect to time. The multi-agent Q learning algorithm is initially fed load details for switching, available source data for allocation, and segmented time zones as shoulder, peak, and off-peak with corresponding charges in rupees (Rs.) per kWh consumption. After learning, the algorithm sends a response as a suggestion for incorporating ADR and source allocation along with cost computation. The trained algorithm suggests shifting appliances operating in each time slot based on the categorization shown in Table 5.2.

Appliances are placed into a set of three categories: intractable, power tractable, and time tractable. Appliances that are essential for academic activities are placed in the intractable category; similarly, for devices that can either be deferred to another time slot or tuned/adjusted to have lower consumption ranges. In this regard, the multi-agent Q learning model is adopted with pre-defined state and action sets for each appliance, as shown in Table 5.3.

Following this, a ToU-based tariff structure is implemented, leading to a substantial reduction in energy tariffs without compromising overall net consumption. Under this structure, the day is divided into 24-hour segments, distinguishing between peak, off-peak, and shoulder hours. The cost associated with the standard usage of devices during normal hours is

Table 5.2 Categorization of loads

Intractable appliances	*Power tractable appliances*	*Time tractable appliances*
Lights	Air conditioners	Water dispensers
Fans	Electric vehicle (EV) chargers	Uninterruptible power system (UPS)
Printers		Exhausters
Projectors		
Camera and routers		

Table 5.3 State set and action set for the multi-agent Q learning algorithm

Appliances/agents	*State set*	*Action set*
Lights	$\{(\varphi_t^{P'}, \varphi_{t+1}^{P'}, \ldots, \varphi_T^{P'})\}$	1 - "on", 0 –"off"
Fans	$\{(\varphi_t^{P'}, \varphi_{t+1}^{P'}, \ldots, \varphi_T^{P'})\}$	1 - "on", 0 –"off"
Printers	$\{(\varphi_t^{P'}, \varphi_{t+1}^{P'}, \ldots, \varphi_T^{P'})\}$	1 - "on", 0 –"off"
Projectors	$\{(\varphi_t^{P'}, \varphi_{t+1}^{P'}, \ldots, \varphi_T^{P'})\}$	1 - "on", 0 –"off"
Camera and routers	$\{(\varphi_t^{P'}, \varphi_{t+1}^{P'}, \ldots, \varphi_T^{P'})\}$	1 - "on"
Air conditioners	$\{(\varphi_t^{P'}, \varphi_{t+1}^{P'}, \ldots, \varphi_T^{P'})\}$	{1.4, 1.5, ..., 2.3} - "power ratings in kW"
EV chargers	$\{(\varphi_t^{P'}, \varphi_{t+1}^{P'}, \ldots, \varphi_T^{P'})\}$	{5.6, 5.7, ..., 7} - "charging rates"
Water dispensers	$\{(\varphi_t^{P'}, \varphi_{t+1}^{P'}, \ldots, \varphi_T^{P'})\}$	1 - "on", 0 –"off"/scheduled with respect to priorities
UPS	$\{(\varphi_t^{P'}, \varphi_{t+1}^{P'}, \ldots, \varphi_T^{P'})\}$	1 - "on", 0 –"off"/scheduled with respect to priorities
Exhausters	$\{(\varphi_t^{P'}, \varphi_{t+1}^{P'}, \ldots, \varphi_T^{P'})\}$	1 - "on", 0 –"off"/scheduled with respect to priorities

φ – electricity tariff, P'– power consumption, and [t-T] – time slots.

Table 5.4 ToU-based tariff structure

ToU tariffs for commercial user end	*Tag*	*Pertinent charges (Rs./unit)*
06:00 a.m. to 10:00 a.m. & 5:00 p.m. to 9:00 p.m.	Peak	20% extra charge of Rs. 7.65
10:00 a.m. to 5:00 p.m.	Shoulder	Rs. 7.65
9:00 p.m. to 06:00 a.m.	Off-Peak	5% reduction of Rs. 7.65

Rs. 7.65 per unit. However, during peak hours, an additional 20% surcharge is applied, while the off-peak tariff experiences a 5% reduction. It is essential to note that this cost calculation is specifically applicable to energy procurements from the Tamil Nadu Electricity Board, aligning with the categorized time zones outlined in the demand charge catalog presented in Table 5.4.

5.5 CONSUMPTION AND COST ANALYSIS

The proposed prudent approach has been identified as instrumental in enhancing the energy performance of the building from the customer's perspective. This is achieved through the utilization of details pertaining to the net loads connected and energy consumption patterns pertaining to smart meters. The mean annualized pattern, extracted from analytics of smart meter data, illustrates energy trends before the adoption of automated DR. Concurrently, the amalgamation of pre-defined UPs and the academic

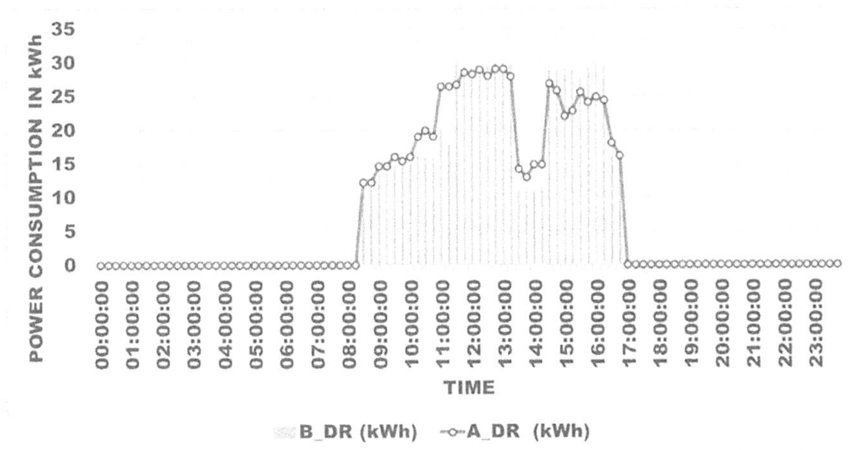

Figure 5.3 Net consumption trends before and after ADR incorporation

timetable of TCE has resulted in the creation of an RL multi-agent Q learning algorithm. This algorithm is crafted to make well-informed decisions about load shifting during designated hours in alignment with the established schedule.

The comparative graphical representation in Figure 5.3 illustrates the energy trend achieved through appropriate decision-making. The traced-back demand curve is depicted through column plots of the hourly consumption trend before ADR. By contrast, the line with the marker plot signifies the estimated trend after the implementation of the proposed ADR strategy.

According to a cost analysis, the cost involved after ADR techniques is found to be low compared to that before ADR estimation, as shown in Figure 5.4. The figure shows the average consumption cost estimation for a typical working day in rupees. As the study is for a higher educational institution, only limited appliances have the possibility to shift. Hence, a slight variation in the energy tariff was observed for the considered test system.

5.6 CONCLUSION

This chapter introduces a systematic approach for improving commercial building energy management systems that completely relies on a Q-learning algorithm. The methodology leverages real-world data to analyze existing energy trends. Subsequently, this insight integrates a multi-agent Q-learning algorithm for decision-making. Additionally, its processes aim at scheduling various appliances with consideration for customer preferences. By

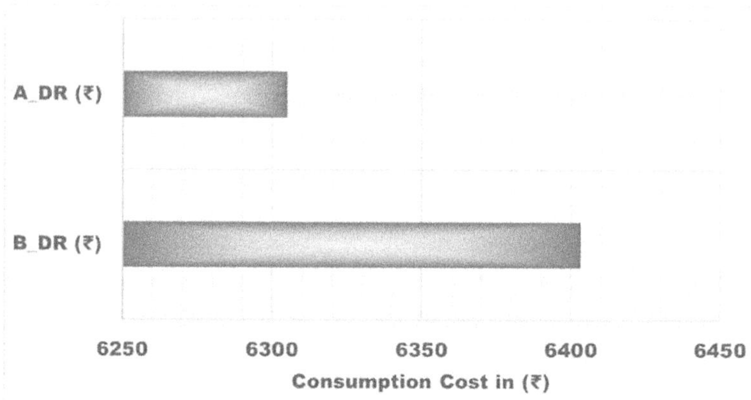

Figure 5.4 Cost of consumption per day before and after ADR incorporation

implementing the proposed ADR method, the objective of minimizing electricity bills through day-ahead energy consumption scheduling is achieved. Studying the test system demonstrates that the proposed ADR method is not specific in terms of reducing electricity costs and ensuring promising performance, but enhances computational efficacy. Future enhancements to the BEMS will include the consideration of uncertainty predictive approaches to further improve overall performance. Moreover, the utilization of multi-agent Q-learning facilitates effective decision-making processes for the adopted ADR technique. However, it is evident from the tests that the model is better suited for residential communities with a diverse range of appliances rather than commercial industries.

ACKNOWLEDGEMENT

The authors express their sincere gratitude to the management of TCE for providing extensive facilities and financial support. They also acknowledge the support received from the research excellence component of the Savitha Project and the Thiagarajar Research Fellowship (TRF) scheme (File No: TRF/Jul2022/02).

REFERENCES

[1] X. Yang, G. Wang, H. He, J. Lu, and Y. Zhang, "Automated demand response framework in ELNs: Decentralized scheduling and smart contract," *IEEE Transactions on Systems, Man, and Cybernetics: Systems*, vol. 50, no. 1, pp. 58–72, Jan. 2020, https://doi.org/10.1109/TSMC.2019.2903485.

[2] M. C. Bozchalui, S. A. Hashmi, H. Hassen, C. A. Canizares, and K. Bhattacharya, "Optimal operation of residential energy hubs in smart grids," *IEEE Transactions on Smart Grid*, vol. 3, no. 4, pp. 1755–1766, Dec. 2012, https://doi.org/10.1109/ISGT.2015.7131906.
[3] F. A. Qayyum, M. Naeem, A. S. Khwaja, A. Anpalagan, L. Guan, and B. Venkatesh, "Appliance scheduling optimization in the smart home networks," *IEEE Access*, vol. 3, pp. 2176–2190, 2015, https://doi.org/10.1109/ACCESS.2015.2496117.
[4] M. A. A. Pedrasa, T. D. Spooner, and I. F. MacGill, "Scheduling of demand side resources using binary particle swarm optimization," *IEEE Transactions on Power Systems (TPWRS)*, vol. 24, pp. 1173–1181, 2009.
[5] K. Tamilarasu, C. R. Sathiasamuel, J. D. N. Joseph, R. Madurai Elavarasan, and L. Mihet-Popa, "Reinforced demand side management for educational institution with incorporation of user's comfort," *Energies*, vol. 14, no. 10, pp. 2855, 2021, https://doi.org/10.3390/en14102855.
[6] T. Karthick, S. Charles Raja, J. Jeslin Drusila Nesamalar, and K. Chandrasekaran, "Design of IoT based smart compact energy meter for monitoring and controlling the usage of energy and power quality issues with demand side management for a commercial building," *Sustainable Energy, Grids and Networks*, vol. 26, pp. 100454, 2021, ISSN 2352–4677, https://doi.org/10.1016/j.segan.2021.100454.
[7] Y. Huang, L. Wang, W. Guo, Q. Kang, and Q. Wu, "Chance constrained optimization in a home energy management system," *IEEE Transactions on Smart Grid*, vol. 9, no. 1, pp. 252–260, 2016.
[8] J. R. Vázquez-Canteli, and Z. Nagy, "Reinforcement learning for demand response: A review of algorithms and modeling techniques," *Applied Energy*, vol. 235, pp. 1072–1089, 2019.
[9] F. Ruelens, B. J. Claessens, S. Vandael, B. De Schutter, R. Babuška, and R. Belmans, "Residential demand response of thermostatically controlled loads using batch reinforcement learning," *IEEE Transactions on Smart Grid*, vol. 8, no. 5, pp. 2149–2159, 2016.

Chapter 6

Multi-Objective Optimal Power Flow Algorithm for Optimal Performance in a Distribution System

K. T. M. U. Hemapala, Rajapaksha Mudiyanselage Tharangani Lakmali, O.V. Gnana Swathika

6.1 INTRODUCTION

6.1.1 Background Study

The electricity requirement is steadily increasing. Therefore, power generation capacities have to be increased to fulfil countries' demands using technical, economical, and environmentally feasible solutions. Recently, the penetration of renewable (solar, wind, geothermal, ocean, and fuel cell) and nonrenewable (microturbines, combined cycles, internal combustion engines, and combustion turbines) distributed generation (DG) has developed to fulfill electrical energy consumption needs [1].

DG can be defined as small-scale generation used to achieve the ever-increasing electrical demand at or near to the premises of utility customers and interconnected with the distribution system.

The technical benefits of DG integration include voltage support, reduction of energy loss, system reliability improvement, and release of system capacity. The economical benefit of DG comprises a hedge against high electricity prices. When improper DG is integrated into the distribution system, the performance of the distribution network changes, increasing system losses, increasing DG cost, and causing voltage and stability problems. Therefore, DG has an impact on the operating characteristics of the system regarding the voltage profile, reliability-stability, and electrical losses. Hence, it is necessary to properly evaluate and determine the optimal size and location of DG units in the distribution system. The use of optimized DG planning helps to achieve the maximum techno-economical benefits of DG integration.

6.1.2 Sri Lankan Distribution Network with DG

Sri Lanka is a small island with a peak demand of around 3000 MW. In Sri Lanka, electricity generation is mainly covered by conventional energy sources such as hydro, coal, and petroleum. Plants are vested by the Ceylon Electricity Board and some private companies.

 DOI: 10.1201/9781003537878-6

According to the Sustainable Energy Authority, Sri Lanka is aspiring to become a carbon-neutral country by 2050. To achieve this, the Ministry of Power and Energy, Ceylon Electricity Board, Sri Lanka Sustainable Energy Authority, and Public Utilities Commission of Sri Lanka are working jointly and encouraging private developers to invest in renewable energy-based DG. The future capacity development of renewable energy technologies in Sri Lanka is indicated in Figure 6.1.

Due to geo-climate conditions in Sri Lanka, the main renewable energy resources for electricity generation are hydro, solar, biomass, and wind. The above generating units are located at resource focus plants that are interconnected with the national grid as DG units. Additionally, some number of diesel combustion engines placed in different locations are considered as DG. The present status of non- conventional renewable energy (NCRE) in Sri Lanka is listed in Table 6.1.

6.1.2.1 Small Hydro Power

A few kW to 10 MW small-hydro power plants are integrated into the distribution network in Sri Lanka. Hydro power is the most attractive renewable source and has highly expanded in the country. It converts the potential energy in water into electricity. Mainly, these types of power plants are located in hill country areas in Sri Lanka. Small-hydro plants are mainly used for stand-alone and base load applications, and rural area electrification. Both active and reactive power are provided in small-hydro generation.

6.1.2.2 Wind Power

A number of wind plants are located around the country, such as the Ambewela, Puttalam, Kilinochchi, and Hambantota areas. Wind turbines convert the kinetic energy of wind to electricity. Wind plants are free of emissions and have no fuel cost requirement. A few hundred W to 10 MW wind power plants are located in Sri Lanka and are mainly used for stand-alone and base load applications. Active and reactive power are both provided by wind turbines according to wind speed.

6.1.2.3 Biomass Power

Biomass is one of the new trending technologies in Sri Lanka used to feed electrical energy into the distribution network. At the plant, biomass materials are transformed to bio-gas. Bio-gas is cleaned and upgraded to natural gas, and becomes bio-methane. Bio-gas is used in gas turbines or piston-driven engines for electricity generation. Biomass electrical power generation can handle active power and reactive power from synchronous

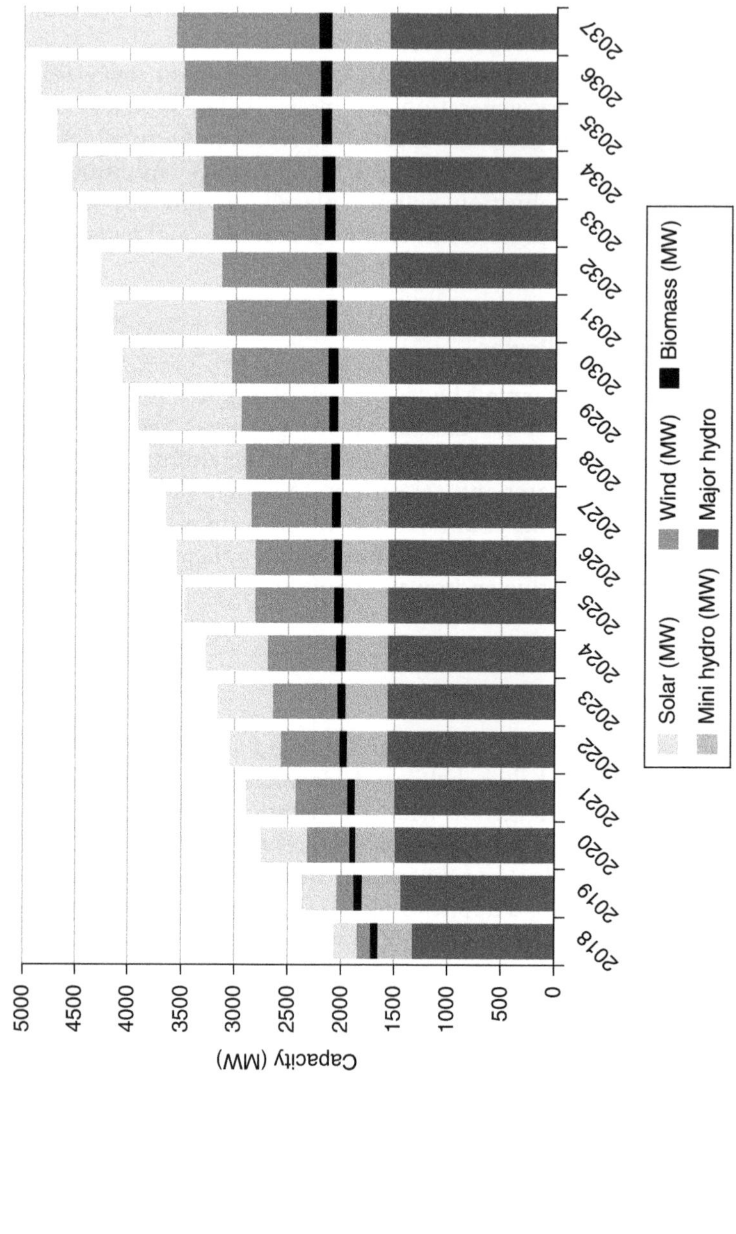

Figure 6.1 Renewable energy capacity development [2]

Table 6.1 Present status for NCRE as at May 31, 2020 [3]

Technologies	*No. of DG Plants*	*Total capacity (MW)*
Small-hydro	181	353.744
Wind	15	128.85
Biomass	9	24.10
Solar	8	51.36

generators. A few kW to 10 MW biomass plants are available in the country to produce electricity for stand-alone and base load applications.

6.1.2.4 Solar Power

Solar power generation is another trending technology present in Sri Lanka. Solar photovoltaic (PV) technologies directly convert sunlight into electrical energy. This also requires very low operating costs and no fuel cost. Solar PV is a unity power factor (PF) generation technology that can generate "active-power." In the Public Utilities Commission of Sri Lanka (PUCSL) and Ceylon Electricity Board (CEB) definition, most solar PV systems tend to have two types: utility-scale solar installations with a capacity usually more than 1 MW to 10 MW, and small-scale ground-mounted and rooftop generation of a few W to 1 MW. Many rooftop solar power generators are located around the country. Utility-scale ground-mounted solar plants are mainly found in the Hambantota, Polonnaruwa, and Vavuniya areas.

6.1.3 Motivation

According to the last two decades' data and the Long-Term Generation Expansion Plan of Sri Lanka, DG connected to the distribution system in Sri Lanka has increased rapidly.

There are no agreeable geological and technical limitations for the location of DG in a distribution network. The main limitations emerge from electrical necessities. Additionally, clear guidelines are never included in the selection of the DG size to be introduced in a feeder in a distribution network. The main aspect used as a guide to select the DG size and location is voltage criteria, that is, to check whether the voltage at the node is within an acceptable range (± 6%) after adding a DG unit.

The inappropriate placement and sizing of DGs would lead to an increment of system losses, and installation and operating costs. It would also occur reliability, stability, and voltage profile problems. The proper placement and sizing of DGs could enhance network potential in terms of the voltage profile and minimize system losses while improving the quality of the power supply.

Worldwide, many researchers have developed mathematical approaches to DG allocation in distribution networks. However, there is no highly efficient, proper, and fast method for solving this problem with appropriate variables and constraints related to Sri Lanka. Our research focuses on fulfilling this requirement. Additionally, the study will be beneficial to the upcoming distribution network and microgrid applications.

6.1.4 Research Objectives

The goal of this chapter is to develop a multi-objective optimal power flow (OPF) algorithm to achieve optimal performance in the distribution system by best placement and sizing DGs.

The specific objectives of the research are highlighted as follows:

1. to propose an optimal power loss reduction algorithm for the distribution system by applying different constraints of DGs
2. to determine the economical cost of DGs
3. to compute and analyze the results of the proposed multi-objective mathematical formulation using MATLAB
4. to apply the developed algorithm in a practical scenario.

6.1.5 Proposed Methodology

The proposed methodology is as follows:

1. select suitable OPF methods for hybrid optimization by analyzing existing methods
2. identify objective functions, DG variables, and constraints
3. develop a mathematical model of a multi-objective algorithm in the MATLAB interface
4. simulate and analyze the results using an IEEE bus system
5. enhance and validate the algorithm by considering the results
6. apply the mathematical model to selected distribution feeders in Sri Lanka for optimal DG planning.

6.2 LITERATURE REVIEW

6.2.1 Formulate the Optimal Placement and Sizing of DGs

In the following subsections, mathematical formulations are provided that have been included in the literature to attain optimal DG planning. This includes appropriate objective functions, DG variables, load variables, and constraints.

6.2.1.1 Single-Objective Functions

The objective functions for the optimal placement and sizing of DGs are listed below. A mathematical model has been developed from the single-objective functions or multi-objective functions below.

Minimization of the power losses of the system
DG placement in the power system induces impacts on electrical power system losses. Power losses are identified as two types: real and reactive power system losses. The real-power loss occurs due to the resistance of the line conductors. The value of the real-power loss is indicated by the overall efficiency of the power system. The reactive loss occurs due to reactive elements in the power system. Reactive power flow helps to maintain a fixed voltage level for the system properly.

Certain DG technologies are able to deliver active power, such as "solar PV, microturbines and fuel cells." The optimal placement and sizing of DGs are capable of reducing active power loss in the network significantly. A number of DG technologies are adept at delivering reactive power, for example, "synchronous compensators." Therefore, the optimal placement and sizing of these DG units are capable of reducing the reactive power loss of the power system considerably. Additionally, synchronous generators are able to provide active and reactive powers, and the DG units that are optimally placed and sized help to achieve the minimization of active and reactive power losses significantly [1], [4].

Minimization of total costs in the power system
The investment and operating costs play a significant role in determining DG capacity. The consideration of the investment cost contains the cost for installing DG technology at specific locations. The operational cost has two components: the cost of real power supplied from the substation and the cost of real power supplied by the installed DG. With the optimal placement and sizing of DG, the total costs of the system can be minimized properly. Additionally, the maintenance cost of DG can be minimized. The planning of cost minimization provides the quick screening of different investment possibilities of DG [1], [5].

Minimization of the voltage deviation
DGs cause voltage profile variation in the power system. The voltage profile is identified as a change of voltage in the distribution system as the load changes. Further voltage levels drop along the length of the feeders. Additionally, the improper placement and sizing of DG in the feeder may create a voltage rise above the appropriate voltage limit. Moreover, the minimization of voltage deviation reduces excessive automated tap changing on distribution transformers in the distribution system [5].

Minimization of average interruption

Optimal DG planning enhances the reliability of the distribution system. The main reliability indices are the "systems average-interruption duration index" (SAIDI) and "systems average-interruption frequency index"(SAIFI). By minimizing these interruptions, customers can obtain a reliable power supply with DG installation [5].

Maximization of profit

Considering the DG owner's point of view, profit maximization is a more important factor. The optimal sizing of DG has a considerable effect on profit [5].

6.2.1.2 Methodology for the Multi-Objective Formulation

The technical and economical merits can be achieved by optimal DG planning with multiple objectives. This is a complex combinatorial optimization issue due to the concurrent optimization of multiple objectives. Multi-objective optimization is generally solved using one of following methodologies [1]:

1. Transformation into single-objective functions with weights
   Selected objective functions are multiplied by a factor called the "weighting factor" and an optimal solution can be obtained from a single multi-objective function.
2. Using the goal multi-objective index
   The multi-objective formulation is converted into a single-objective function by providing a goal programming method.

6.2.1.3 DG Variables

1. Number of DG units
   The number of DGs units that can be introduced into a distribution system can be taken as a free variable parameter. Therefore, a predefined single or multiple DG can be considered during modeling.
2. Size and location
   The size and location of DGs depend on primary fuel availability with geological conditions, investment cost, and profit for DG and the existing distribution system. In this study, this is optimized in accordance with given conditions and limits.
3. Primary energy
   The primary energy of DG is illustrated in DG technology and can be used for power generation. Technologically, DG varies with commercial availability and geological conditions. Solar, wind, mini-hydro, biomass, and reciprocating engines are currently used as the primary energy of DG in Sri Lanka. Additionally, some of the newest technology is still being tested.

4. DG type
   The type of DG depends on DG technology. To facilitate optimal DG modeling, DG types are categorized into four types based on real-power and reactive-power delivering capacities [1], [6]:
   - Type 1: DGs capable of injecting active power only
   - Type 2: DGs capable of injecting real and reactive powers
   - Type 3: DGs capable of injecting reactive power only
   - Type 4: DGs capable of injecting real power, but they consume reactive power.

6.2.1.4 Load Variables

An accurate load profile is an important factor for DG planning. However, it is very difficult to obtain a proper load profile in the distribution system due to continuous load varying conditions. To minimize the load varying error, the following are considered for load modeling according to a variety of loads [1]:

1. one-load level
2. multi-load level
3. time varying load
4. probabilistic load
5. fuzzy loads.

6.2.1.5 Constraint

The constraints of the design are the most important parameters for DG modeling. Objective functions should satisfy specified constraints, otherwise a malfunction will occur in the performance of the distribution system. Constraints are mainly identified as two types: equity constraints and inequity constraints [1], [4].

1. Equity constraints
   The "power balance" constraint of the power flow equation is considered in the equity constraint. The constraint helps to keep the balanced system in the distribution network.
2. Inequity constraints
   The following constraints are considered as this category. For optimal DG modeling, the parameters in the inequity constraints should be varied within prescribed limits to maintain system stability and to maintain DG with geological conditions:
   - bus voltage or voltage drop limits
   - total harmonic voltage distortion limit
   - line/transformer overloading/capacity limit

- real and reactive power generation limits
- maximum number of DG units;
- DG type
- budget limit.

6.2.2 Mathematical Methods for Optimal DG Planning

DG is a solution for higher demand in the present world. The optimal planning of DG is important and most researchers have been involved in optimal DG planning in DG systems using OPF methods. The classification of OPF methods is shown in Figure 6.2 [1], [5].

Mainly, two types of approaches exist: mathematical methods and heuristic methods.

The analytical approach is one of the OPF methods in optimal DG planning. The approach is applied for a radial feeder with uniformly distributed loads; the technique may not be effective on nonuniformly distributed loads. Using this method, the "2/3 - rule" introduces the installation of DG at "2/3 capacity" of incoming generation at "2/3 of the length of the line." The optimized results of the method are lower than those of other methods. Additionally, the method has not been used directly due to complexity and specific characteristics in the distribution system [1].

Different numerical methods are used in the optimal sizing and placement of DGs. In [7], a numerical approach was introduced using the gradient method to lower losses and improve the voltage profiles in electrical microgrids with DG. It was shown that about 12% of loss reduction is achieved by applying this method for a fixed DG size. Furthermore, a Newton–Raphson (NR)-based approach for the optimal placement of

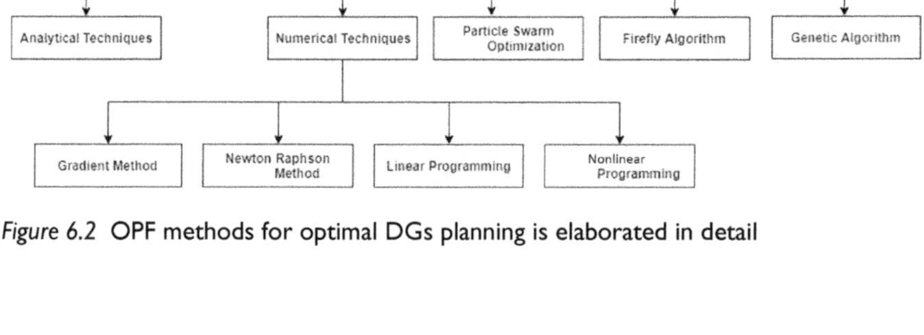

Figure 6.2 OPF methods for optimal DGs planning is elaborated in detail

DG in an electrical distribution system was proposed in [8]. It was able to reduce power loss by about 15% and fixed the 20 MW solar PV installed in this paper.

In a linear programming method, the objective problem should be characterized as a linear function and constraints are only linear inequalities. For nonlinear problems, the problem should contain discrete and continuous variables, and the constraints should be in a nonlinear form. These approaches are not suitable for objective functions that have many decision variables and a long computation time [9].

Heuristic methods are computed-oriented techniques for optimization problems. The particle swarm optimization (PSO) method is one of the heuristic methods used in optimal DG planning. The technique was used in [10] to determine the optimal location and size of solar PV (SPV) in a microgrid. This was modeled using the IEEE bus system and around 42% loss reduction was obtained.

In [11], a loss minimization approach for microgrids was introduced using the analytical firefly algorithm and it identified a 8.67% power loss reduction in the optimal placement and sizing of DGs. The firefly algorithm has a high probability to be trapped in locally optimal solutions; this is one of the main drawbacks of the approach.

A genetic algorithm (GA)-based optimal DG planning solution was proposed in [12]. It identified around 47% power loss reduction with a single DG placement. The computation time for a GA is higher than that for other methods and convergence problems can occur due to unguided mutation.

Higher efficiency can be identified from hybrid optimization models by combining two different approaches. A comparison of the features of two main OPF techniques is listed in Table 6.2.

6.2.3 Discussion on Problem Formulation and Mathematical Methods

A suitable input, output parameters, and OPF methods for the proposed model are identified by analyzing the literature review and analyzing suitable conditions in Sri Lanka. The parameters are selected as explained below.

The multi-objective formulation is used to develop the model by transforming the objectives into a single multi-objective function with weights. The following single-objective functions are considered to develop a techno-economical optimization model:

- minimization of the active power losses of the system
- minimization of the reactive power losses of the system
- minimization of the cost of the system.

Table 6.2 Comparison of OPF methods [1], [4], [5]

Technique	*Features*	*Limitations*
Numerical technique (NR)	• Accurate values of losses can be identified for the optimal solution • Can use DG for optimal placement only	• Difficult to control distribution power system equations as an optimization model • Percentage reduction around 15–20%
Heuristic technique (PSO)	• Simple and flexible in solving optimization problems with a large search space and discrete properties • Can search for near optimal DG solutions • Does not require converting power system models into optimization programming models • Can handle multi-objective optimization	• Requires computational time • Percentage reduction around 25–48%

Additionally, the following main equity and inequity constraints are considered in the modeling to achieve proper operation:

- power balanced equation
- bus voltage limits
- DG type
- power generation limits of DG.

Furthermore, Type 1 and Type 2 DGs are analyzed using a mathematical model considering one-load level and multi-load level data.

Another important consideration is the selection of suitable mathematical approaches for the hybrid model. By a qualitative and quantitative analysis of the literature survey, the following methods are selected:

- NR method
- PSO.

The design variables of the model are the optimal location and optimal size of DG.

6.3 CONCLUSION

This chapter introduces background information about the optimal placement and sizing of DG. Additionally, the chapter includes a comprehensive literature review on the formulation of DG planning and mathematical approaches to DG planning.

REFERENCES

[1] P.S. Georgilakis and N.D. Hatziargyriou, "Optimal distributed generation placement in power distribution networks: Models, methods and future research," *IEEE Transactions on Power Systems*, vol. 28, no. 3, pp. 3420–3428, Aug. 2013.

[2] "CEB long term generation expansion plan 2020–2039," Draft, [online] www.pucsl.gov.lk/lcltgep-2020-2039/.

[3] "Present Status of NCRE as at 31st May 2020," [online] https:// ceb.lk/business-with ceb/national-energy/en.

[4] P. Prakash and D.K. Khatod, "Optimal sizing and siting technologies for distributed generation in distribution system: A review," *Renewable and Sustainable Energy Reviews*, vol. 57, Elsevier, pp. 111–130, 2016.

[5] S. Daud, A.F.A. Kadir, M.Y. Lada, and C.K. Gan, "A review: Optimal distributed generation planning and power quality issues," *International Review of Electrical Engineering (I.R.E.E.)*, vol. 11, no. 2, ISSN 1827-6660, 2016.

[6] V. Roja, and M. S. Sujitha, "A Review of optimal DG allocation in distribution system for loss minimization," *IOSR Journal of Electrical and Electronics Engineering*, pp. 15–22, 2016.

[7] F.A.H. Ramirez, E.R. Trujillo, and J.A.H. Mora, "Implementation of an optimal power flow to reduce losses and improve the profiles voltages in electrical microgrids with distributed generation," *IEEE Thirty Fifth General American and Panama Convention (CONCAPAN XXXV)*, 2015.

[8] M. Bouzguenda, A. Samadi, and S.R. Mohamed, "Optimal placement of distributed generation in electrical distribution networks," *IEEE International Conference on Intelligent Technologies in Control, Optimization and Signal Processing*, 2017.

[9] A.S. Hassan, Y. sun, and Z. Wang, "Optimization techniques applied for optimal planning and integration of renewable energy sources based on distributed generation: Recent trends," *Cogent Engineering*, vol. 7, no. 1, pp. 1766394, DOI:10.1080/23311916.2020.1766394, 2020.

[10] S.M. Dawold, L. Xiangning, F.M.F. Flaih, and M.I. Okba, "PSO algorithm for optimal placement of multiple SPV based distributed generators in Microgrid," *IEEE Power and Energy Engineering Conference (APPEEC), IEEE PES Asia-Pacific*, pp. 125–129, 2016.

[11] A. Olaniyan, B. Jimoh, and Y. Jibril "Application of Analytical- Firefly Algorithm for optimal location and sizing of distributed generator in standard IEEE 30- bus distribution network," *International Engineering Conference (IEC), Federal University of Technology, Minna, ResearchGate publication*, 2015.

[12] M. Kashyap, A. Mittal, and S. Kansal, "Optimal placement of distributed generation using Genetic Algorithm Approach," www.researchgate.net/ publication/326721495, 2019.

Chapter 7

Techno-Economical Optimal Solution for a Sri Lankan Distribution System

K. T. M. U. Hemapala, Rajapaksha Mudiyanselage Tharangani Lakmali, O.V. Gnana Swathika

7.1 OPTIMIZATION METHODOLOGY

7.1.1 Overview

A new approach is presented in this chapter for the optimal sizing and suitable placement of distributed generation (DG) in a distribution system by considering multi-objective optimization with selected constraints [1–3]. A summary of the input, output, and optimal power flow (OPF) method for the proposed model is illustrated in Figure 7.1.

7.1.2 Newton–Raphson Load Flow Method for Optimal DG Allocation

The Newton–Raphson (NR) method is defined as an iterative approach that is applied to solve a group of nonlinear equations from a Taylor's series expansion and the terms are finite to the first approximation. The origin and formulation of the NR approach dates back to the late 1960s from previous work by Isaac Newton and Joseph Raphson.

One of the objectives of this study is to minimize power losses in a power system. To determine the power loss and voltage magnitude at each bus, the NR technique is used for load flow analysis. From this method, the voltage magnitudes and angles of each bus can be identified. Hence, the power injections at all the buses and the power flow through the system can be calculated and power losses can be obtained. The main steps of the power loss calculation from the NR method are described in Section 7.1.2.1 and a flowchart of the NR method is shown in Figure 7.2.

7.1.2.1 Mathematical Approach of the NR Method

Step 1: Formulate the YBUS from line resistance and reactance.

Step 2: Classify the type of each bus in the power system, and identify the known and unknown parameters, as shown in Table 7.1.

DOI: 10.1201/9781003537878-7

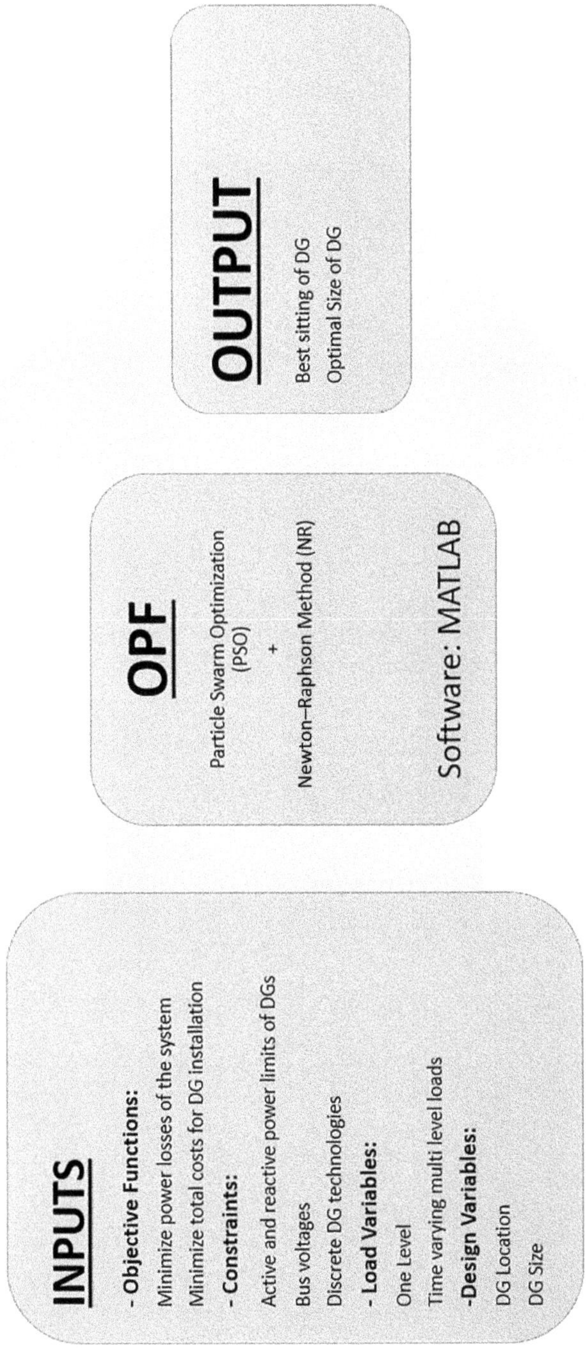

Figure 7.1 Summary of the proposed model

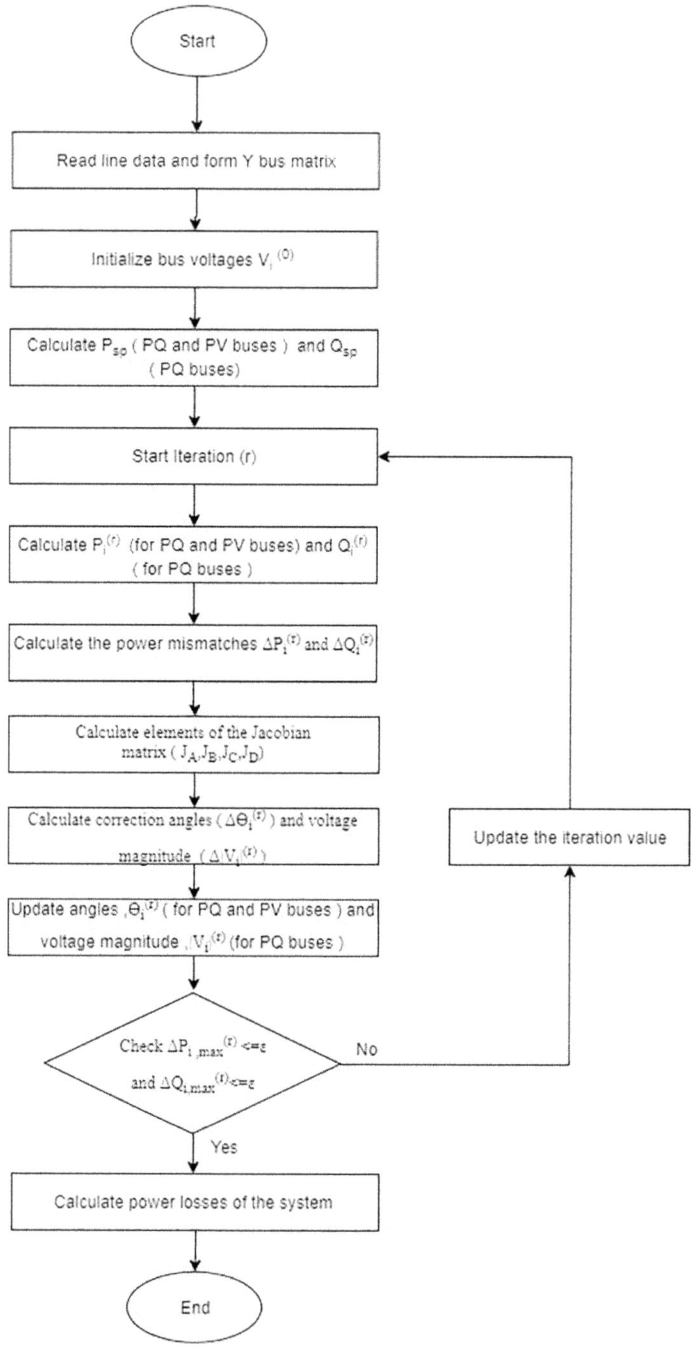

Figure 7.2 Flowchart of the NR method

Table 7.1 Bus classification for load flow analysis

Classification	*Known parameters*	*Unknown parameters*
Slack bus (swing bus)	$\|V\|, \delta$	P, Q
PV bus (generator bus)	$P, \|V\|$	Q, δ
PQ bus (load bus)	P, Q	$\|V\|, \delta$

Step 3: Choose the initial values of the voltage magnitudes and angles for all buses except the slack bus:

$V_i = |V_{i,sp}| \angle 0^0$ (at all PV buses)
$V_i = 1\angle 0^0$ (at all PQ buses)

Step 4: Compute P_i and Q_i for each PQ bus and P_i for each PV bus:

$$P_{i,cal} = \sum_{j=1}^{N} V_i V_j (G_{ij}\cos(\theta_i - \theta_j) + B_{ij}\sin(\theta_i - \theta_j)) \quad (7.1)$$

$$Q_{i,cal} = \sum_{j=1}^{N} V_i V_j (G_{ij}\sin(\theta_i - \theta_j) - B_{ij}\cos(\theta_i - \theta_j)). \quad (7.2)$$

where
$P_{i,cal}$ = real power injected at the i-th bus
$Q_{i,cal}$ = reactive power injected at the i-th bus
V_i and V_j = magnitudes of voltage at nodes i and j, respectively
θ_i and θ_j = voltage angles at the i-th bus and j-th bus with respect to the reference angle, respectively
G_{ij} = real component or conductance of bus admittance matrixY between nodes i and j
B_{ij} = imaginary component or acceptance of bus admittance matrixY between nodes i and j

Step 5: Calculate the power mismatches:

$$\Delta P_i^r = P_{i,sp}^r - P_{i,cal}^r \quad (7.3)$$

$$\Delta Q_i^r = Q_{i,sp}^r - Q_{i,cal}^r \quad (7.4)$$

Step 6: Calculate the elements in the Jacobian matrix:

$$J = \begin{bmatrix} \frac{\delta \Delta P}{\delta \theta} & \frac{\delta \Delta P}{\delta |V|} \\ \frac{\delta \Delta Q}{\delta \theta} & \frac{\delta \Delta Q}{\delta |V|} \end{bmatrix} \quad (7.5)$$

Step 7: Obtain the value of $\Delta\delta$ and $\Delta | V^r |$:

$$\begin{bmatrix} \Delta\delta^r \\ \dfrac{\Delta|V^r|}{|V|} \end{bmatrix} = [J^r]^{-1} \begin{bmatrix} \Delta P^r \\ \Delta Q^r \end{bmatrix} \tag{7.6}$$

Step 8: Update the magnitude and phase angle of voltage at all load buses:

$$|V_i^{r+1}| = |V_i^r| + \Delta|V_i^r| \tag{7.7}$$

$$|\delta_i^{r+1}| = |\delta_i^r| + \Delta|\delta_i^r| \tag{7.8}$$

Step 9: Go to Step 3 and iterate until the power mismatches are within an acceptable tolerance:

$$\Delta P_i^r < \varepsilon_p$$
$$\Delta Q_i^r < \varepsilon_Q.$$

7.1.3 Particle Swarm Optimization (PSO) Technique for Optimal DG Sizing and Allocation

The PSO algorithm is a modern heuristic algorithm in optimization. It was invented by Kennedy and Eberhart in 1995. This algorithm is an iterative search technique based on the swarm's social behavior for birds flocking and fish schooling in the world to find food. The swarm moves in only one direction to moves to the best food in groups.

In PSO, particles move throughout the search space as an objective function to find the best solution. The moment of a particle is based on its own experience and neighbors' experience. Basic elements used in PSO are listed below:

- particle/individual swarm: technical term for each bird/fish;
- particle population: technical term for a flock of particles (group of swarms);
- search space: set of possible solutions for all variables of a given function that is limited by the boundary conditions of the variables by considering the upper bound and lower bound for each variable;
- local best/personal best (P_{best}): best position of a particle among its all positions visited so far; and
- global best (G_{best}): position where the best fitness is achieved among all particles visited so far in the solution space.

The main steps for the PSO algorithm are described in Section 7.1.3.1 and a flowchart of the PSO algorithm is shown in Figure 7.3 [4–5].

7.1.3.1 Mathematical Approach of PSO

Step 1: Develop an objective function:

$$f(x) = f(x_1, x_2, \dots x_n) \tag{7.9}$$

At the beginning of PSO, the objective function needs to be identified clearly. x_i represents the optimizing variables and n is the number of variables. $f(x)$ is the objective function that needs to be maximized or minimized.

Step 2: Initialize the PSO parameters.
It is important to correctly select the PSO parameters for a given problem; the selection has a considerable impact on PSO performance. Therefore, the following parameters need to be initialized:

- Number of variables
  The number of variables depends on the objective function. Additionally, the upper and lower boundaries of the variables need to be initialized.
- Size of the population
  The number of particles in the swarm is called the population size. A moderate population size must be selected according to the problem because a large population may decrease the number of iterations while increasing the complexity of one iteration, whereas a small population may require more iterations to achieve convergence.
- Number of iterations
  The number of iterations is affected by the convergence of the problem. A small number of iterations may cause the termination of the algorithm without acquiring better convergence, whereas a large number of iterations increases the computational time. The selection of the number of iterations also depends on the nature of the problem.
- Acceleration coefficients (c_1 and c_2)
  The acceleration coefficients are used to change the velocity of particles toward their personal and global best together. Generally, all agents in PSO are considered to become attracted toward the average of P_{best} and G_{best}, then $c_1 = c_2$.

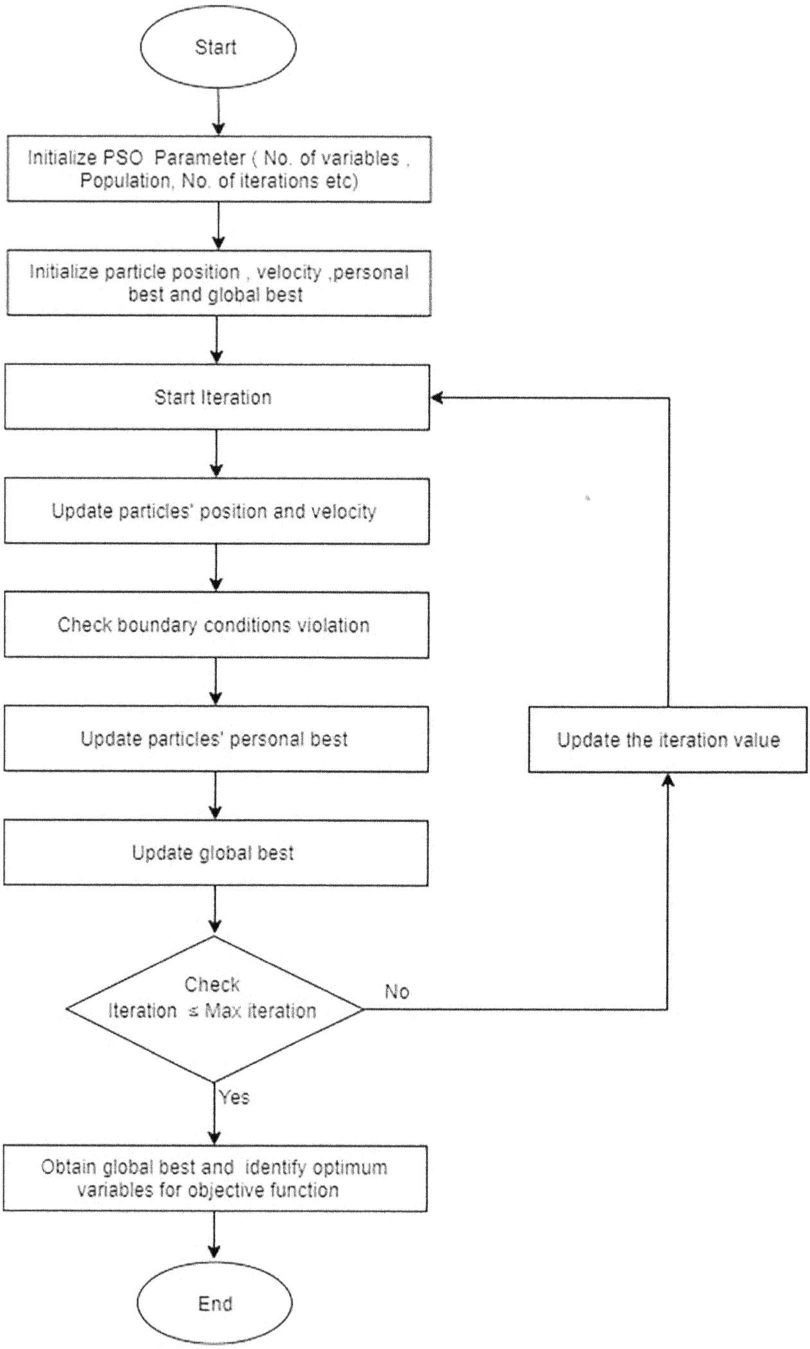

Figure 7.3 Flowchart of the PSO method

- Inertia weights
  The inertia weight (ω) helps to develop a balance between global exploration and local exploration, which means it controls the influence of the previous history of the velocities on the current one. ω_{max} and ω_{min} are initialized in this step and ω is calculated using the following equation for each iteration:

$$\omega = \omega_{max} - \frac{\omega_{max} - \omega_{min}}{i_{max}} \times i, \tag{7.10}$$

where
ω_{max} and ω_{min} = initial and final inertia factor weights, respectively
i = current iteration number
i_{max} = maximum iteration number.

Step 3: Initialize the particle position and velocity, and obtain the personal best and global best from the objective function.
The initial particle population (x_0) is predefined randomly using the boundary values of the variables and the initial velocity (v_0) is defined by considering x_0. The initial values for P_{best} and G_{best} can be identified from x_0 and v_0.

Step 4: Start the iteration.
Step 5: Update the position and velocity of each particle.
To identify the best variable values, each particle in the population tries to change its position to a better position by considering the below factors:

- current position
- current velocity
- distance between the current position and P_{best}
- distance between the current position and G_{best}.

The equations for the updated PSO velocity and position are

$$v_i^{k+1} = \omega v_i^k + c_1 r_1 (P_{best.i} - x_i^k) + c_2 r_2 (G_{best.i} - x_i^k) \tag{7.11}$$

$$x_i^{k+1} = x_i^k + v_i^{k+1}, \tag{7.12}$$

where
c_1 and c_2 = weighting coefficients
r_1 and r_2 = randomly generated numbers from 0 to 1
ω = weighting factor
v_i^k = current velocity of the i-th particle at the k-th iteration
v_i^{k+1} = modified velocity of particle i

x_i^k = current position of the i-th particle at the k-th iteration
x_i^{k+1} = modified position of particle i
$P_{best.i}$ = personal best of particle i
$G_{best.i}$ = global best of the group.

Step 6: Check the criteria of the boundary violation.
Step 7: Evaluate the objective function and update P_{best} and the objective value.
Step 8: Update G_{best} and the best objective value.
Step 9: Check the stopping criteria and repeat from Step 4 until the stopping criteria are satisfied.

7.1.4 Problem Formulation

7.1.4.1 Selection of Objective Functions and Constraints

To obtain a techno-economical optimal solution, the following objective functions and constraints are identified based on the Sri Lankan distribution system [5–6]:

Objective 1: Minimize the sum of the active power losses (F_1):

$$P_L = \sum_{k=1}^{N} G_{kj}\left(V_k^2 + V_j^2 - 2V_kV_j\cos\left(\theta_k - \theta_j\right)\right). \tag{7.13}$$

Objective 2: Minimize the sum of the reactive power losses (F_2):

$$Q_L = \sum_{k=1}^{N} -B_{kj}\left(V_k^2 + V_j^2 - 2V_kV_j\cos\left(\theta_k - \theta_j\right)\right). \tag{7.14}$$

Objective 3: Minimize the total costs of active power produced by the DG unit (F_3):

$$\begin{gathered} C = C_{DG,i} + C_{substation} \\ C_{DG,i} = 1.3 \times Fixed\ Costs + Variable\ Costs \\ Fixed\ Costs = Capital\ Costs \times\ Maximum\ Capacity\ of\ DG \\ Variable\ Cost = \left(Fuel\ Cost + Operation\ \&Maintenance\ Cost\right) \times P_{DG,i}. \\ C_{substation} = Price \times P_{substation} \end{gathered} \tag{7.15}$$

Constraints:
There are two types of constraints that are selected for DG modeling: equality constraints and inequality constraint:

- Equity constraints
  The power balance constraint of the power flow equation is considered as the equity constraint for modeling:

$$P_{Substation} + \sum_{k=1}^{N_{DG}} P_{DG} = P_{Demand} + P_{Losses} \tag{7.16}$$

$$Q_{Substation} + \sum_{k=1}^{N_{DG}} Q_{DG} = Q_{Demand} + Q_{Losses}. \tag{7.17}$$

- Inequity constraints
  The voltage limits of the bus and power generation limits of DG are considered in DG modeling by considering the proper operation of the power distribution network:
  Voltage limit of the bus:

  $$V_{\min} \leq V_i \leq V_{\max}.$$

  Power generation limits of DGs:

  $$P_{DG}^{\min} \leq P_{DGi} \leq P_{DG}^{\max}$$
  $$Q_{DG}^{\min} \leq Q_{DGi} \leq Q_{DG}^{\max}.$$

7.1.4.2 Multi-Objective Planning

The weighting factor method is used to convert the above multi-objectives into a single multi-objective problem using the linear summation of the factors:

Objective function = Minimize {(Active power loss), (Reactive power loss), (DG cost)}
Objective function = Minimize $\{(W_1 * F_1) + (W_2 * F_2) + (W_3 * F_3)\}$,

where W_1, W_2, and W_3 are weighting factors. The values of the weighting factors are chosen based on the importance level of each objective. In order to obtain a desirable set of non-dominated solutions for DG variables, the proposed algorithm is run multiple times with different weighting factors. For a specific set of weighting values, only the optimal solution for DG is obtained. When selecting the values of the weighting factors, the following two considerations are made:

$$W_1 + W_2 + W_3 = 1$$

$$0 \leq W_i \leq 1; \quad i = 1, 2, 3.$$

In each case study for the proposed model, the following scenarios for the objective functions are used by considering different weighting factor values. Type 1 Scenarios are considered only for active power injected DGs, and Type 2 Scenarios are considered for both active and reactive power injected DGs.

Type 1:
Scenario 1: base or reference scenario (DGs are not integrated into the system)
Scenario 2: $W_1 = 0.7$, $W_3 = 0.3$
Scenario 3: $W_1 = 0.5$, $W_3 = 0.5$
Scenario 4: $W_1 = 0.3$, $W_3 = 0.7$
Scenario 5: $W_1 = 1$, $W_3 = 0$
Scenario 6: $W_1 = 0$, $W_3 = 1$.

Type 2:
Scenario 1: base or reference scenario (DGs are not integrated into the system)
Scenario 2: $W_1 = 0.5$, $W_2 = 0.3$, and $W_3 = 0.2$
Scenario 3: $W_1 = 0.4$, $W_2 = 0.4$, and $W_3 = 0.2$
Scenario 4: $W_1 = 0.3$, $W_2 = 0.3$, and $W_3 = 0.4$
Scenario 5: $W_1 = 1$, $W_2 = 0$, and $W_3 = 0$
Scenario 6: $W_1 = 0$, $W_2 = 1$, and $W_3 = 0$
Scenario 7: $W_1 = 0$, $W_2 = 0$, and $W_3 = 1$.

7.1.5 Proposed Algorithm

The flowchart for the proposed mathematical model is shown in Figure 7.4. It is combined with PSO and NR methods to determine the optimal size and optimal location of DGs.

7.1.5.1 Initial Parameters for the Proposed Algorithm

The initial parameters for the proposed mathematical model are as follows:

- Number of variables
  The proposed model is applied by considering two types of DGs in the distribution network. Therefore, two variables are considered throughout the study, such as the active power of DG and reactive power of DG.
- Population size
  According to the nature of the problem, the population is taken as 250.

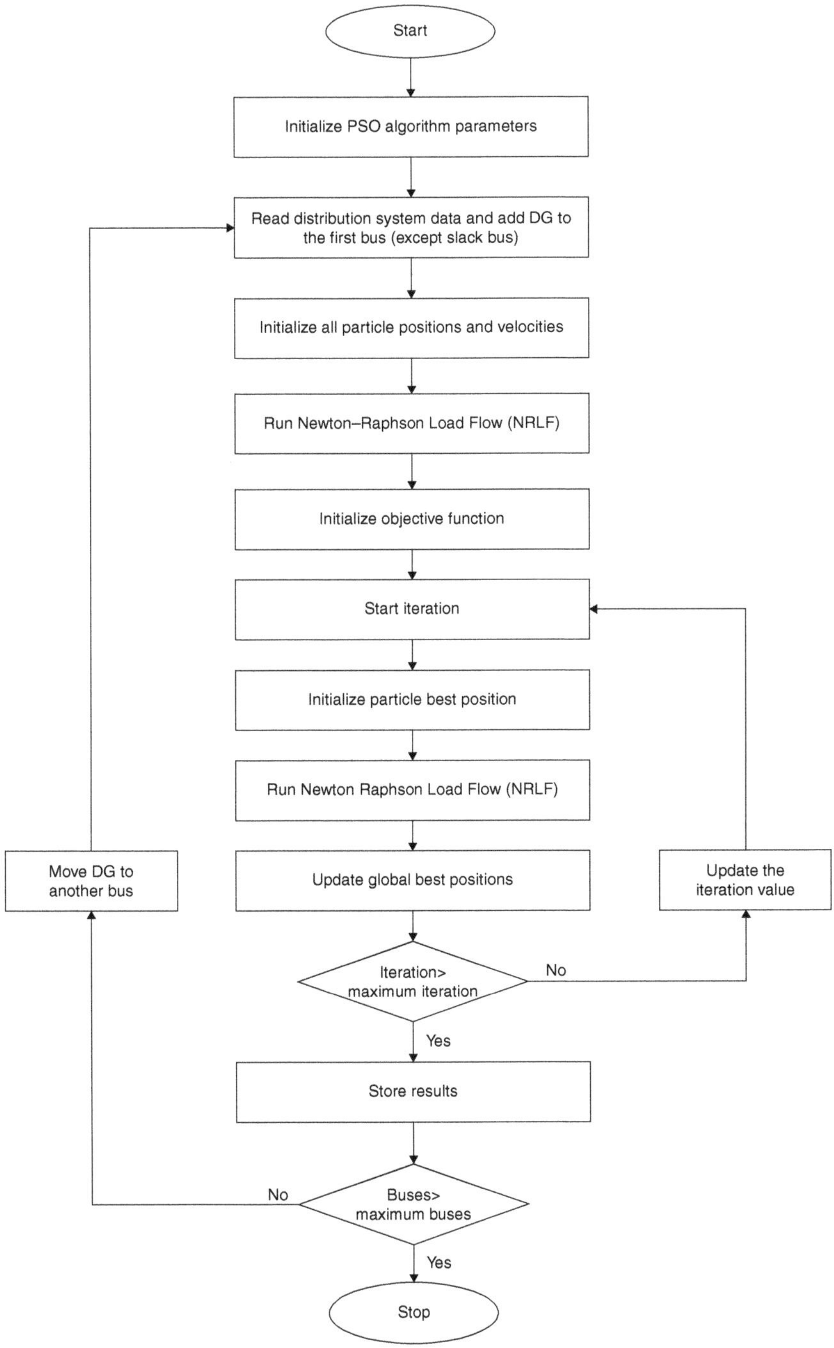

Figure 7.4 Proposed mathematical model

- Number of iterations
  The number of iterations also depends on the nature of the problem. Therefore, it is selected as 10.
- Acceleration coefficients (c_1 and c_2)
  Based on the PSO literature [7], the acceleration coefficients are selected as $c_1 = c_2 = 2$.
- Inertia weights
  Based on the PSO literature,

  $\omega_{max} = 0.9$ and $\omega_{min} = 0.4$.

7.1.6 Distributed Generator Model

Before the DG type is applied to the proposed model, it is necessary to identify the hours of availability and percentage of output of DG. The nature of the main DG types that can apply to Sri Lankan distribution systems are described as follows, and solar photovoltaic (PV) and biomass are selected in my modeling.

7.1.6.1 Solar Power Generation

Solar PV is a fast-growing renewable energy technology in Sri Lanka. New solar energy generation projects are increasing day by day. Therefore, solar PV generation is selected to apply and validate the proposed mathematical model.

Solar is a non-dispatchable generation resource. The probabilistic nature of solar irradiance is identified from the beta-probability density function from past solar data. The typical output variation of solar PV is illustrated in Figure 7.5. The output mainly depends on solar irradiance.

7.1.6.2 Wind Power Generation

Wind power is another DG technology that is commonly available in Sri Lanka. Wind power production mainly depends on the wind speed at the plant. The kinetic energy of wind is converted into electrical energy using wind turbines. Wind power generation can occur throughout the day and varies with the wind speed. For example, the typical wind speed variation in Kalpitiya Sri Lanka is shown in Figure 7.6.

7.1.6.3 Biomass Power Generation

Biomass is another type of DG that is available in Sri Lanka. According to the generation capacity of the biomass generator, constant power output

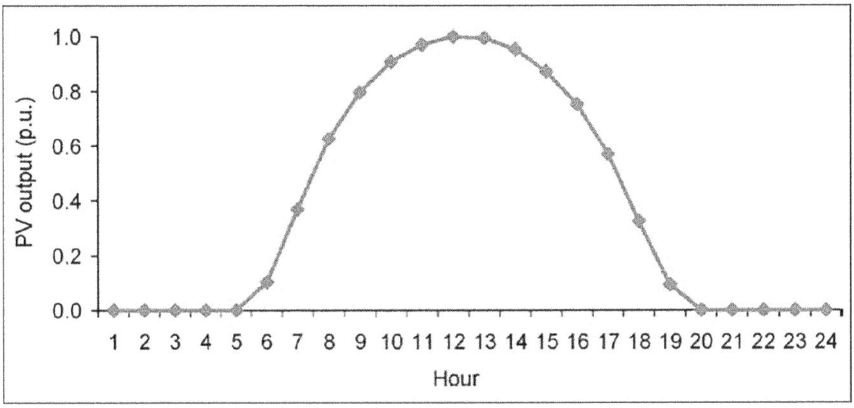

Figure 7.5 Solar output variation

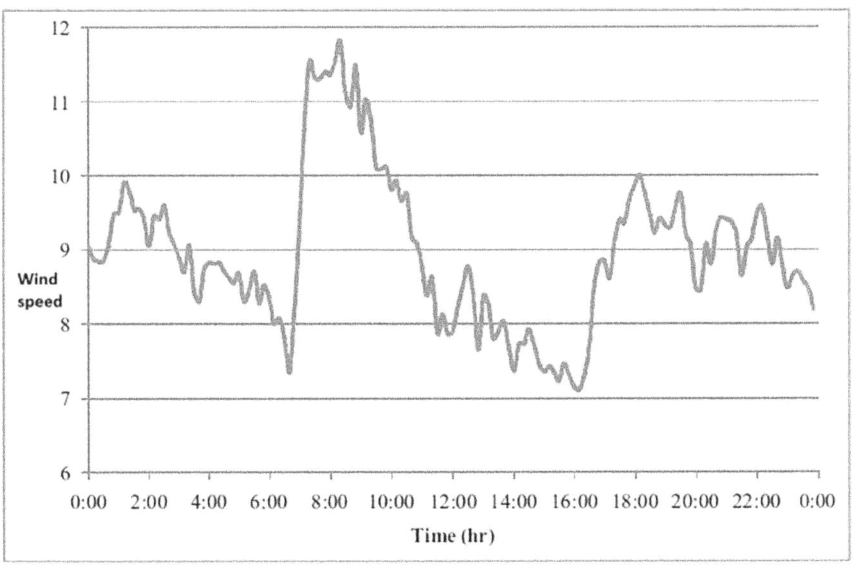

Figure 7.6 Wind speed variation

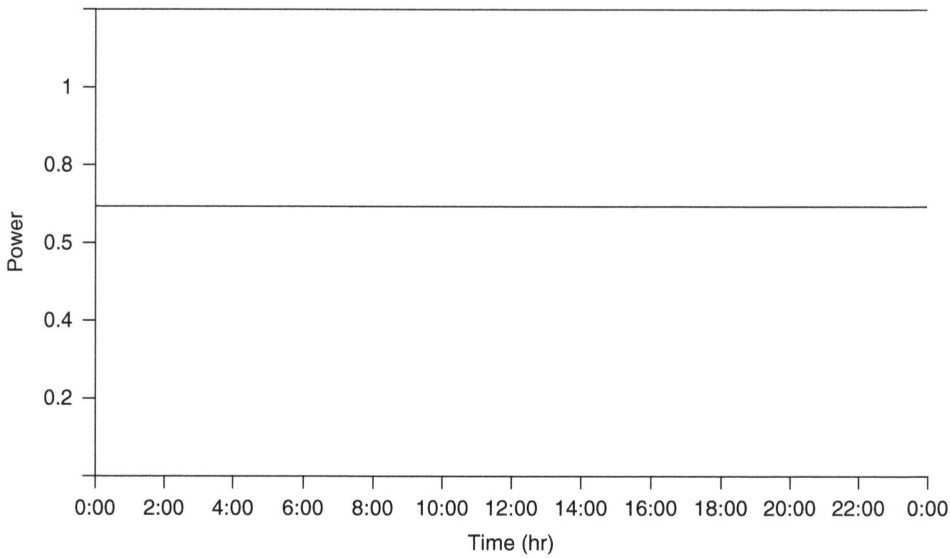

Figure 7.7 Power output variation of biomass DG

can be obtained throughout the day. An example of the power output variation against time is shown in Figure 7.7.

7.1.7 Load Model

Loads in the distribution network vary throughout the day due to different user applications. The average load values for the following time intervals are considered as a load model for the application of the proposed algorithm:

- 12:00 a.m. (midnight) – 3:00 a.m.
- 3:00 a.m. – 6:00 a.m.
- 6:00 a.m. – 9:00 a.m.
- 9:00 a.m. – 12:00 p.m. (noon)
- 12:00 p.m. (noon) – 3:00 p.m.
- 3:00 p.m. – 6:00 p.m.
- 6:00 p.m. – 9:00 p.m.
- 9:00 p.m. – 12:00 a.m. (midnight).

7.2 CONCLUSION

This chapter describes a proposed methodology for optimal DG. The chapter contains the input, output, and constraint data for modeling, and the proposed mathematical approach.

REFERENCES

[1] P.S. Georgilakis, and N.D. Hatziargyriou, "Optimal distributed generation placement in power distribution networks: models, methods and future research", *IEEE Transactions on Power Systems*, vol. 28, no. 3, pp. 3420–3428, Aug. 2013.
[2] "CEB long term generation expansion plan 2020- 2039", Draft, [online] www.pucsl.gov.lk/lcltgep-2020-2039/.
[3] "Present Status of NCRE as at 31st May 2020", [online]https://ceb.lk/business-with ceb/national-energy/en.
[4] P. Prakash, and D.K. Khatod, "Optimal sizing and siting technologies for distributed generation in distribution system: A review", *Renewable and Sustainable Energy Reviews*, vol. 57, Elsevier, pp. 111–130, 2016.
[5] S. Daud, A.F.A. Kadir, M.Y. Lada, and C.K. Gan, "A review: Optimal distributed generation planning and power quality issues", *International Review of Electrical Engineering (I.R.E.E.)*, vol. 11, no. 2, ISSN 1827-6660, 2016.
[6] V. Roja, and M.S. Sujitha, "A Review of optimal DG allocation in distribution system for loss minimization", *IOSR Journal of Electrical and Electronics Engineering*, pp. 15–22, 2016.
[7] D.B. Prakash, and C. Lakshminarayana, "Multiple DG placements in distribution system for power loss reduction using PSO algorithm", *Global Colloquium in Recent Advancement and Effectual Researches in Engineering, Science and Technology*, 2016.

Chapter 8

Case Study – Kalubowila Feeder: Optimal DG Planning for Distribution Feeders in Sri Lanka

K. T. M. U. Hemapala, Rajapaksha Mudiyanselage Tharangani Lakmali, O.V. Gnana Swathika

8.1 OPTIMAL DISTRIBUTED GENERATION (DG) PLANNING FOR DISTRIBUTION FEEDERS IN SRI LANKA

8.1.1 Case Study – Kalubowila Feeder

The Kalubowila feeder is selected for the optimal solar photovoltaic (PV) installation as a case study. The feeder is connected to the Ethulkotte grid substation. The main details of the feeder are shown in Table 8.1 and geographical details of the feeder are shown in Figure 8.1.

The locations and currently available transformer details of the feeder are in Appendix A.1.

8.1.1.1 Feeder Details

The line resistance and reactance details of the feeder are shown in Appendix A.2. The conductor types of the feeder used for the simulation are listed below:

- 11 kV 150 mm2 AAC conductor: Rk = 0.20457 (Ω/kmk); kX = 0.33056 (Ω/kmk); C = 11.2894 (nF/km)
- 11 kV 60 mm2 AAC conductor: kRk = 0.50498 (Ω/kmk); kX = 0.36253 (Ω/kkm); C = 10.3054 (nF/km).

The load data of the feeder for selected time interval periods are presented in Appendix A.3. According to solar power generation, optimal solar PV planning is considered for the 6:00 a.m. to 6:00 p.m. time period. Load variation for 6:00 a.m. to 6:00 p.m. at different nodes are shown in Figure 8.2.

 DOI: 10.1201/9781003537878-8

Table 8.1 Details of the Kalubowila feeder

Parameter	*Value*
Voltage level	11 kV
Main line length	2.752 km
Branch conductor length	3.083 km
No. of nodes	70
No. of transformers	51
Feeder demand	2.02 MW (min) 3.07 MW (max)

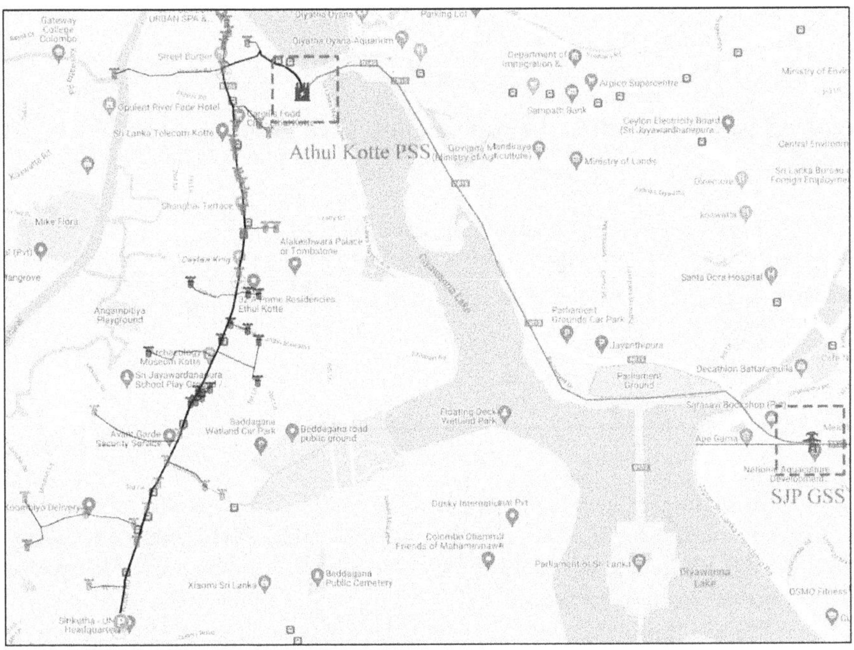

Figure 8.1 Geological view of the Kalubowila feeder

8.1.1.2 Constraint Data

- According to the distribution code of Sri Lanka [1–5], the voltage limits of each bus are considered as ±6%.
- In the Public Utilities Commission of Sri Lanka (PUCSL) and Ceylon Electricity Board (CEB) definitions, most solar PV systems tend to have two types: utility-scale solar installations, with capacity usually more than 1 MW, and small-scale ground-mounted and rooftop generation, with capacity up to 1 MW. Therefore, P_{DG} is considered below the range and 20% of load variation is considered for optimal solar PV planning in accordance to P_{DG} variation.

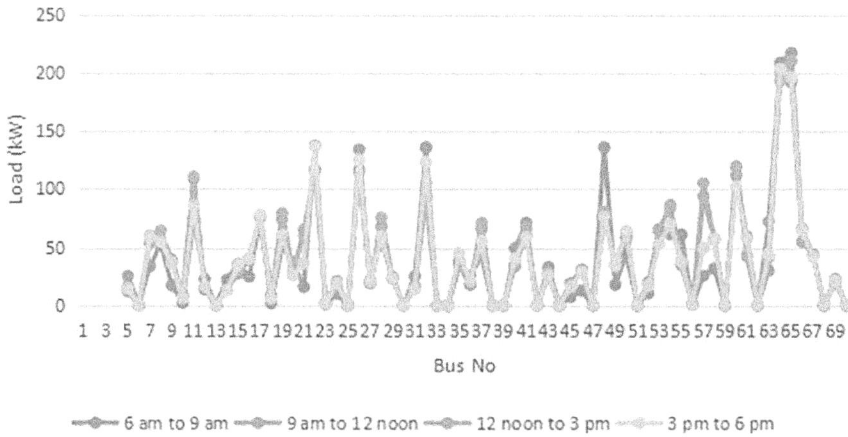

Figure 8.2 Load variation at the feeder nodes from 6 a.m. to 6 p.m.

Table 8.2 Percentage values of the solar PV output

Time interval	*Solar PV generation (%)*
6:00 a.m.–9:00 a.m.	50
9:00 a.m.–12:00 p.m. (noon)	90
12:00 p.m. (noon)–3:00 p.m.	95
3:00 p.m.–6:00 p.m.	65

- According to geological conditions in Sri Lanka, solar generation is considered from 6:00 a.m. to 6:00 p.m. Table 8.2 shows the selected solar PV generation details for the time intervals by considering solar power output variation.
  The weighting factors are selected by analyzing the results as W1 = 0.7 and W3 = 0.3.

8.1.1.3 Results and Analysis – Kalubowila Feeder

The results obtained for the Kalubowila feeder are shown below. Load variations at 6:00 a.m.–6:00 p.m. are considered for the Kalubowila feeder due to solar PV integration. Objective function variation, power loss variation, DG cost variation, optimal DG size variation, and bus voltage variation with respect to a selected DG for different time intervals are shown in Figures 8.3–8.7, respectively.

From an analysis of the observed results, summaries for optimal DG planning in different time intervals are tabulated in Tables 8.3 and 8.4.

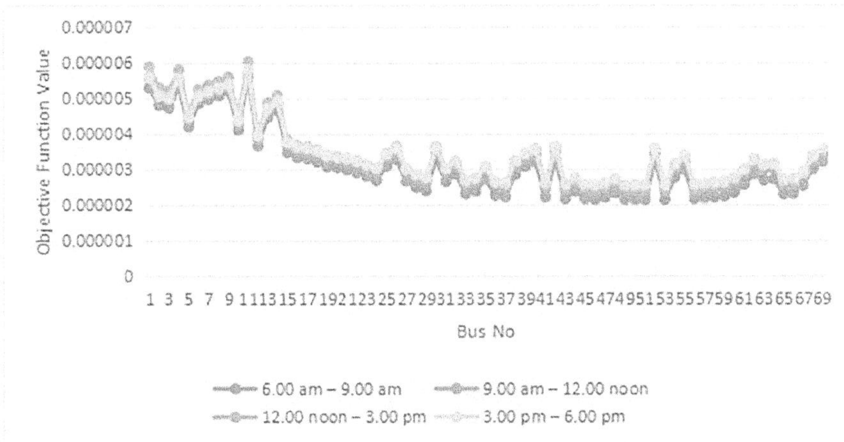

Figure 8.3 Objective function variation in Bus 1 to Bus 70

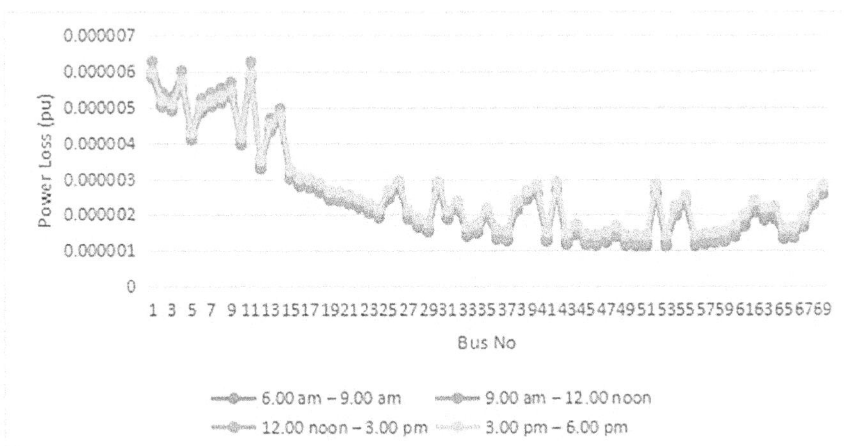

Figure 8.4 Power loss variation in Bus 1 to Bus 70

Additionally, a graphical representation of the summary of the results for the Kalubowila feeder is shown in Figure 8.8.

In these results, different DG sizes are needed to fulfill optimal DG requirements in four time intervals for given constraints. However, the optimal DG location is the same for all four cases. From a consideration of all the above results, a selected DG of 9:00 a.m.–12:00 p.m. (noon) is proposed as the optimal DG for the Kalubowila feeder considering its loss and cost reduction. The geographical view of the proposed DG location is indicated in Figure 8.9.

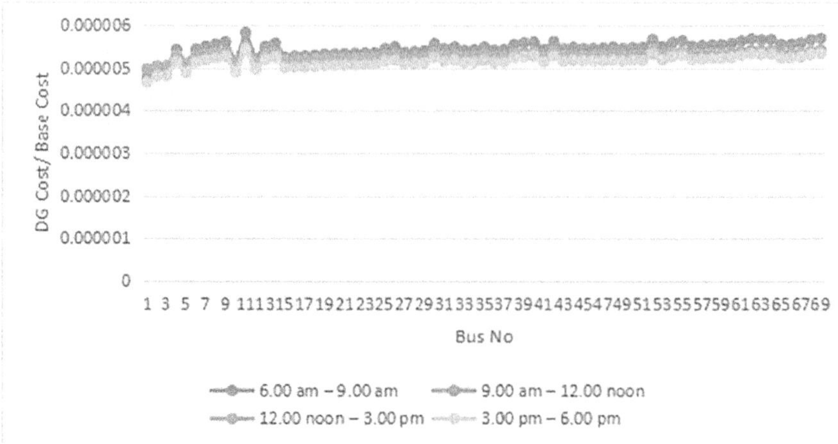

Figure 8.5 DG cost variation in Bus 1 to Bus 70

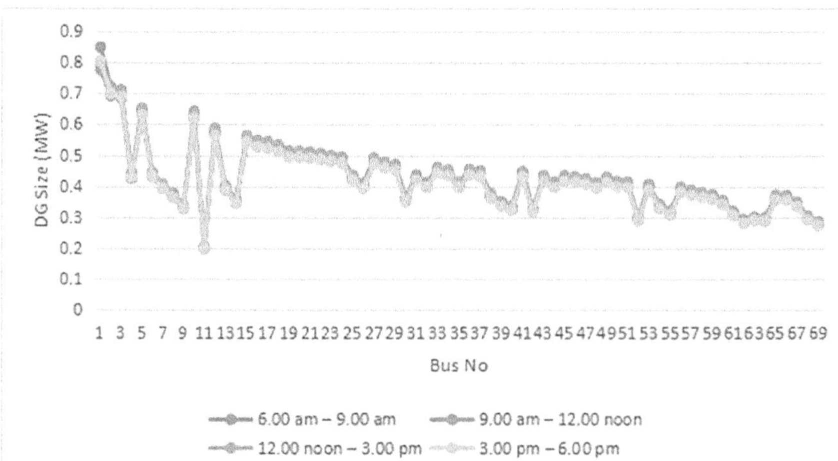

Figure 8.6 Optimal DG size variation in Bus 1 to Bus 70

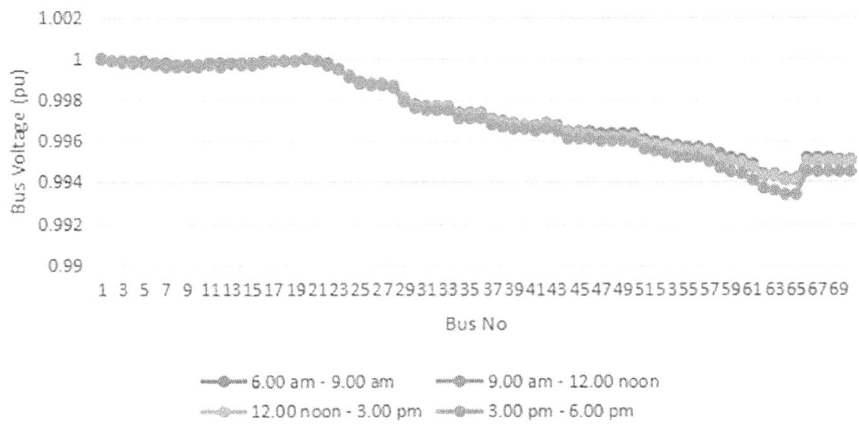

Figure 8.7 Bus voltage variation in Bus 1 to Bus 70

Table 8.3 Optimal DG planning results – Kalubowila feeder

Time interval	*Base case loss (kW)*	*Node*	*P_{Loss} (kW)*	*DG cost (MRs)*	*DG size (kW)*	*Solar PV output (%)*	*Actual DG size (kW)*
6:00 a.m.–9:00 a.m.	0.674325	52	0.113681	134.82	407.078	50	814.156
9:00 a.m.–12:00 p.m. (noon)	0.666977	52	0.111573	113.95	400.687	90	445.208
12:00 p.m. (noon)–3:00 p.m.	0.720877	52	0.122698	113.52	415.377	95	437.239
3:00 p.m.–6:00 p.m.	0.679029	52	0.142169	116.83	396.151	65	609.463

8.1.2 Validation of the Power Flow Results for the HY Kalubowila Feeder

The optimal results obtained from the algorithm are validated using Power System Simulator for Engineering (PSS/E) software. The network diagram, load data, and line data of the Kalubowila feeder on PSS/E are shown in Figures 8.10–8.12.

The optimal DG size for the Kalubowila feeder is obtained as 400.69 kW (actual size with 90% solar irradiance = 445.2 kW) from the proposed algorithm. The DG is applied to randomly selected locations and power loss reduction is obtained. The observed percentage values of power loss reduction are shown in Table 8.5 and the analyzed results are shown in

Table 8.4 Analysis of optimal DG planning results – Kalubowila Feeder

Time interval	*Optimal DG location*	*Actual DG size (kW)*	*Loss reduction (%)*	*Cost reduction (%)*	*Minimum voltage (p.u.)*
6:00 a.m.–9:00 a.m.	52	814.156	83.14	44.93	0.9942
9:00 a.m.–12:00 p.m. (noon)	52	445.208	83.27	52.98	0.9941
12:00 p.m. (noon)–3:00 p.m.	52	437.239	82.38	53.02	0.9941
3:00 p.m.–6:00 p.m.	52	609.463	78.68	51.85	0.9934

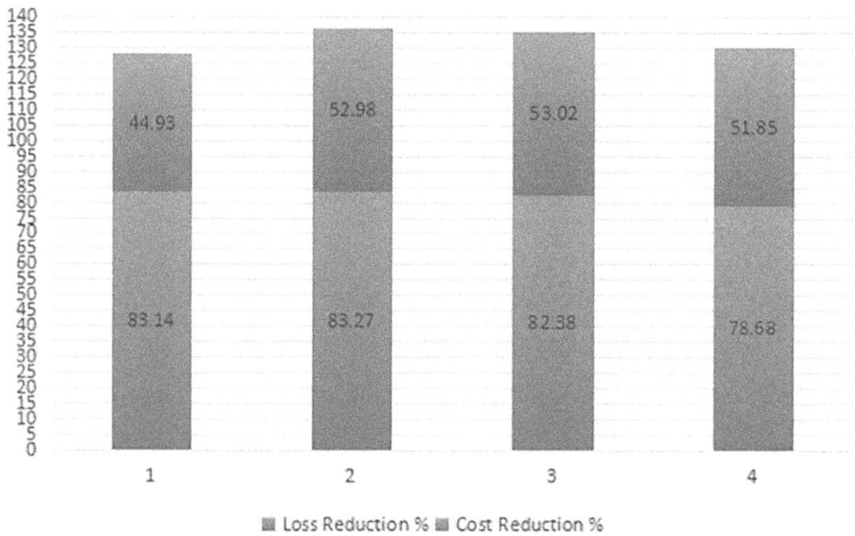

Figure 8.8 Summary of the optimal DG planning results – Kalubowila feeder

Figure 8.9 Geological view of the Proposed DG plant location – Kalubowila Feeder

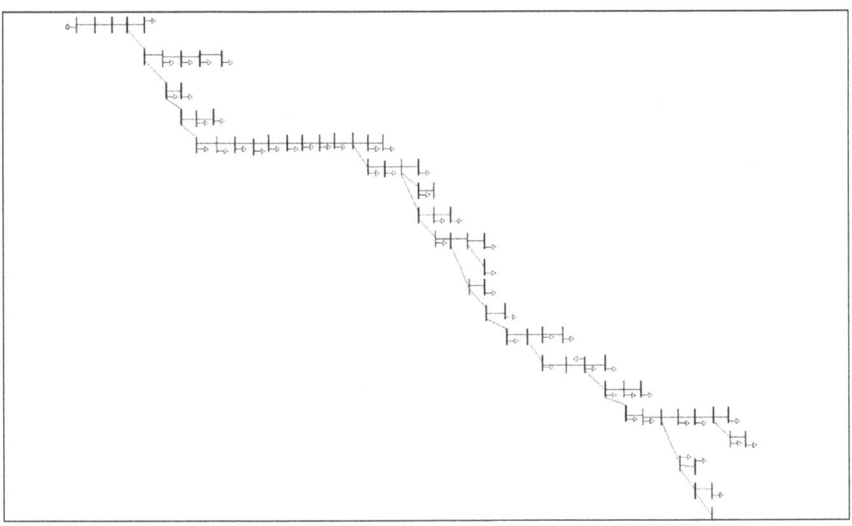

Figure 8.10 Network diagram on PSS/E software – Kalubowila feeder

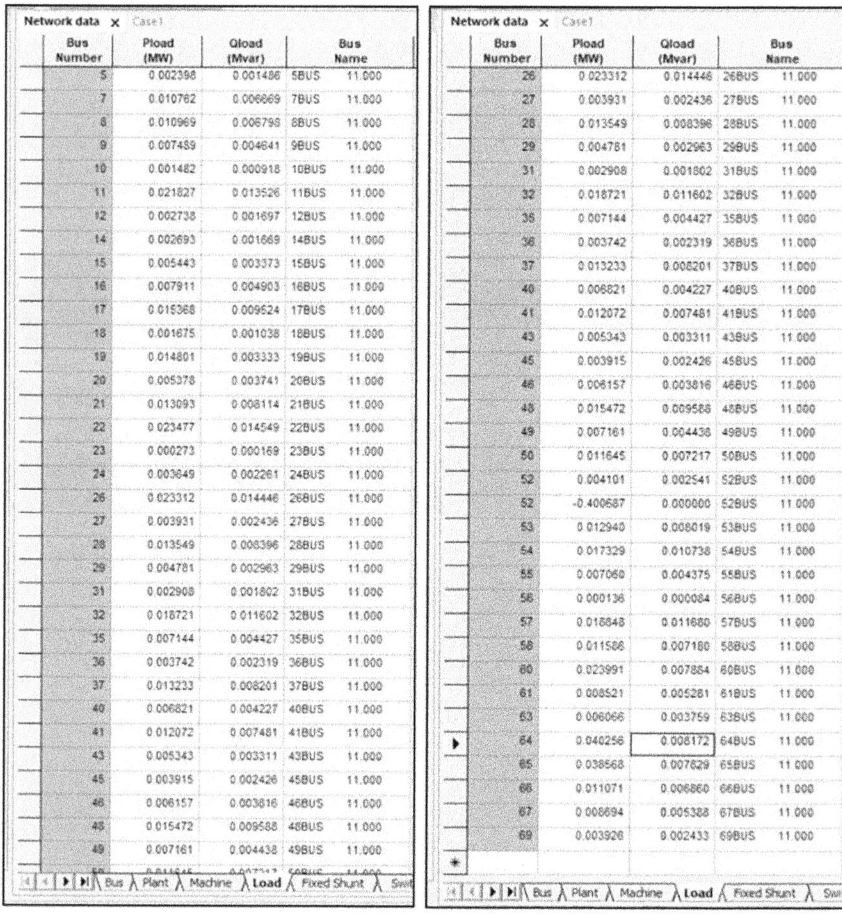

Bus Number	Pload (MW)	Qload (Mvar)	Bus Name
5	0.002398	0.001486	5BUS 11.000
7	0.010762	0.006669	7BUS 11.000
8	0.010969	0.006798	8BUS 11.000
9	0.007489	0.004641	9BUS 11.000
10	0.001482	0.000918	10BUS 11.000
11	0.021827	0.013526	11BUS 11.000
12	0.002738	0.001697	12BUS 11.000
14	0.002693	0.001669	14BUS 11.000
15	0.005443	0.003373	15BUS 11.000
16	0.007911	0.004903	16BUS 11.000
17	0.015368	0.009524	17BUS 11.000
18	0.001675	0.001038	18BUS 11.000
19	0.014801	0.003333	19BUS 11.000
20	0.005378	0.003741	20BUS 11.000
21	0.013093	0.008114	21BUS 11.000
22	0.023477	0.014549	22BUS 11.000
23	0.000273	0.000169	23BUS 11.000
24	0.003649	0.002261	24BUS 11.000
26	0.023312	0.014446	26BUS 11.000
27	0.003931	0.002436	27BUS 11.000
28	0.013549	0.008396	28BUS 11.000
29	0.004781	0.002963	29BUS 11.000
31	0.002908	0.001802	31BUS 11.000
32	0.018721	0.011602	32BUS 11.000
35	0.007144	0.004427	35BUS 11.000
36	0.003742	0.002319	36BUS 11.000
37	0.013233	0.008201	37BUS 11.000
40	0.006821	0.004227	40BUS 11.000
41	0.012072	0.007481	41BUS 11.000
43	0.005343	0.003311	43BUS 11.000
45	0.003915	0.002426	45BUS 11.000
46	0.006157	0.003816	46BUS 11.000
48	0.015472	0.009588	48BUS 11.000
49	0.007161	0.004438	49BUS 11.000

Bus Number	Pload (MW)	Qload (Mvar)	Bus Name
26	0.023312	0.014446	26BUS 11.000
27	0.003931	0.002436	27BUS 11.000
28	0.013549	0.008396	28BUS 11.000
29	0.004781	0.002963	29BUS 11.000
31	0.002908	0.001802	31BUS 11.000
32	0.018721	0.011602	32BUS 11.000
35	0.007144	0.004427	35BUS 11.000
36	0.003742	0.002319	36BUS 11.000
37	0.013233	0.008201	37BUS 11.000
40	0.006821	0.004227	40BUS 11.000
41	0.012072	0.007481	41BUS 11.000
43	0.005343	0.003311	43BUS 11.000
45	0.003915	0.002426	45BUS 11.000
46	0.006157	0.003816	46BUS 11.000
48	0.015472	0.009588	48BUS 11.000
49	0.007161	0.004438	49BUS 11.000
50	0.011645	0.007217	50BUS 11.000
52	0.004101	0.002541	52BUS 11.000
52	-0.400687	0.000000	52BUS 11.000
53	0.012940	0.008019	53BUS 11.000
54	0.017329	0.010738	54BUS 11.000
55	0.007060	0.004375	55BUS 11.000
56	0.000136	0.000084	56BUS 11.000
57	0.018848	0.011680	57BUS 11.000
58	0.011586	0.007180	58BUS 11.000
60	0.023991	0.007884	60BUS 11.000
61	0.008521	0.005281	61BUS 11.000
63	0.006066	0.003759	63BUS 11.000
64	0.040256	0.008172	64BUS 11.000
65	0.038568	0.007629	65BUS 11.000
66	0.011071	0.006860	66BUS 11.000
67	0.008694	0.005388	67BUS 11.000
69	0.003926	0.002433	69BUS 11.000

Figure 8.11 Load data on PSS/E software – Kalubowila feeder

Figure 8.13. According to the results, the maximum power loss reduction obtained is the same as the results from the proposed model and the proposed mathematical model is validated.

8.2 CONCLUSION

This chapter presents the application of the proposed model in a practical scenario. The mathematical model is applied for two selected distribution feeders in Sri Lanka: the Kalubowila feeder and Mallawapitiya feeder 8.

Network data × Case1

From Bus Number	To Bus Number	Line R (pu)	Line X (pu)
1	2	0.018428	0.029778
2	3	0.018090	0.029231
3	4	0.002536	0.004098
4	5	0.026710	0.019175
4	6	0.021471	0.034695
6	7	0.035891	0.025767
6	11	0.004227	0.006830
7	8	0.010851	0.007790
8	9	0.009599	0.006891
9	10	0.014190	0.010187
11	12	0.166936	0.119845
11	13	0.024346	0.039339
13	14	0.053002	0.038051
13	16	0.011835	0.019123
14	15	0.020867	0.014981
16	17	0.009130	0.014752
17	18	0.002536	0.004098
18	19	0.006255	0.010108
19	20	0.011666	0.018850
20	21	0.002029	0.003278
21	22	0.005241	0.008469
22	23	0.007946	0.012840
23	24	0.010313	0.016665
24	25	0.010482	0.016938
25	26	0.031300	0.022471
25	28	0.004058	0.006557
26	27	0.015859	0.011385
28	29	0.024007	0.038793
29	30	0.013187	0.021309
30	31	0.075536	0.054230
30	32	0.020867	0.014981
30	34	0.017076	0.027592
32	33	0.016694	0.011984
34	35	0.005008	0.003595
34	37	0.011496	0.018577
35	36	0.027544	0.019774
37	38	0.007777	0.012567

Bus | Plant | Machine | Load | Fixed

Network data × Case1

From Bus Number	To Bus Number	Line R (pu)	Line X (pu)
34	37	0.011496	0.018577
35	36	0.027544	0.019774
37	38	0.007777	0.012567
38	39	0.061349	0.044043
38	42	0.003889	0.006283
39	40	0.026710	0.019175
39	41	0.042986	0.030860
42	43	0.108091	0.077599
42	44	0.021302	0.034422
44	45	0.016694	0.011984
44	46	0.001691	0.002732
46	47	0.005748	0.009288
47	48	0.005425	0.003895
47	50	0.002367	0.003825
48	49	0.010851	0.007790
50	51	0.021471	0.034695
51	52	0.004565	0.007376
52	53	0.136887	0.098273
52	54	0.014032	0.022675
54	55	0.070113	0.050335
54	57	0.015216	0.024587
55	56	0.029214	0.020973
57	58	0.016568	0.026773
58	59	0.011327	0.018304
59	60	0.011685	0.008389
59	66	0.010313	0.016665
60	61	0.030094	0.048628
61	62	0.070948	0.050934
62	63	0.048829	0.035055
62	64	0.041734	0.029961
64	65	0.006260	0.004494
66	67	0.002087	0.001498
66	68	0.027558	0.044530
68	69	0.065940	0.047339
68	70	0.034997	0.056550

Bus | Plant | Machine | Load | Fixed

Figure 8.12 Line data on PSS/E software – Kalubowila feeder

Table 8.5 P_{Loss} reduction results in PSS/E with optimal DG – Kalubowila

Node	*Power from grid (kW)*	*Power from DG (kW)*	*Load (kW)*	*Power loss (kW)*	*Power loss for base case (kW)*	*Power loss reduction (%)*
5	124.222	400.69	524.43	0.479	0.667	28.12
10	124.236	400.69	524.43	0.483	0.667	27.55
15	124.177	400.69	524.43	0.434	0.667	34.99
20	123.982	400.69	524.43	0.239	0.667	64.15
25	123.929	400.69	524.43	0.187	0.667	71.99
30	123.886	400.69	524.43	0.143	0.667	78.57
35	123.879	400.69	524.43	0.136	0.667	79.61
40	123.988	400.69	524.43	0.245	0.667	63.21
45	123.872	400.69	524.43	0.129	0.667	80.63
47	123.873	400.69	524.43	0.123	0.667	80.52
48	123.876	400.69	524.43	0.131	0.667	80.43
49	123.877	400.69	524.43	0.134	0.667	79.91
50	123.872	400.69	524.43	0.129	0.667	80.67
51	123.863	400.69	524.43	0.120	0.667	81.95
52	123.849	400.69	524.43	0.111	0.667	83.31
53	123.895	400.69	524.43	0.152	0.667	77.20
54	123.933	400.69	524.43	0.191	0.667	71.44
55	123.958	400.69	524.43	0.215	0.667	67.71
56	124.005	400.69	524.43	0.266	0.667	60.64
57	123.919	400.69	524.43	0.177	0.667	73.48
60	123.939	400.69	524.43	0.196	0.667	70.54
65	124.013	400.69	524.43	0.270	0.667	59.50
70	123.979	400.69	524.43	0.237	0.667	64.54

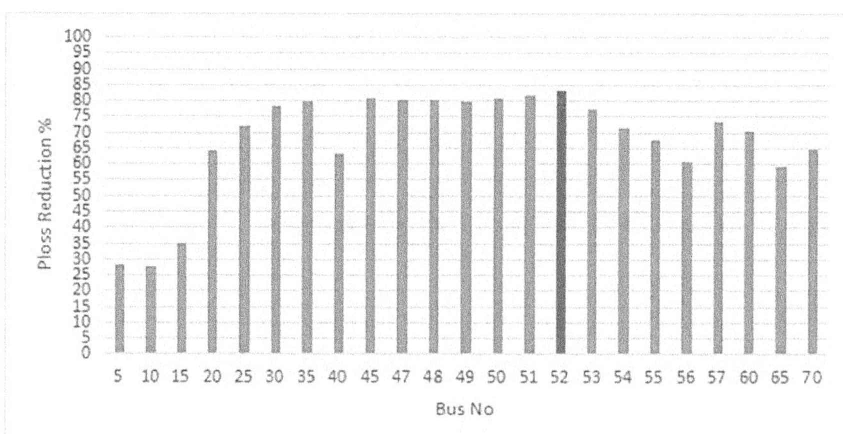

Figure 8.13 P_{Loss} reduction results on PSS/E with optimal DG – Kalubowila feeder

REFERENCES

[1] P.S. Georgilakis and N.D. Hatziargyriou, "Optimal distributed generation placement in power distribution networks: Models, methods and future research", *IEEE Transactions on Power Systems*, vol. 28, no. 3, pp. 3420–3428, Aug. 2013.

[2] "CEB long term generation expansion plan 2020- 2039" Draft, [online] https://www.pucsl.gov.lk/lcltgep-2020-2039/.

[3] "Present Status of NCRE as at 31st May 2020", [online]https://ceb.lk/business-with ceb/national-energy/en.

[4] P. Prakash and D.K. Khatod, "Optimal sizing and siting technologies for distributed generation in distribution system: A review", *Renewable and Sustainable Energy Reviews,* vol. 57, Elsevier, pp. 111–130, 2016.

[5] S. Daud, A.F.A. Kadir, M.Y. Lada and C.K. Gan, "A review: Optimal distributed generation planning and power quality issues", *International Review of Electrical Engineering (I.R.E.E.),* vol. 11, no. 2, ISSN 1827-6660, 2016.

APPENDIX A

A.1 LOCATIONS AND TRANSFORMER DETAILS OF THE KALUBOWILA FEEDER

Node	*T/F Name*	*T/F (kVA)*	*Place*	*Node*	*T/F Name*	*T/F (kVA)*	*Place*
1				**36**	AZ0515	160	Kolobathanthri Mw 2
2				**37**	AZ0217	160	EW Perera MW NR Angampitiya Rd
3				**38**			
4				**39**			
5	AZ1003	100	Wickramasinghe Place	**40**	AZ0219	160	Rampart Road
6				**41**	AZ1047	250	Rulsuru
7	AZ8272	250	Randi Garment	**42**			
8	AZ0530	160	Ethul Kotte Nr Pertrol Shed	**43**	AZ0540	160	Angampitiya Road
9	AZ8721	160	Lanka Orix Information Technology	**44**			
10	AZ8702	160	DR Jayasinghe	**45**	AZ0550	160	RNH Ethul Kotte
11	AZ0202	250	Near Uswatte Mw	**46**	AZ8299	250	Nigata
12	AZ8903	100	KLS Rajapakse	**47**			
13				**48**	AZ8906	400	Prime Lands Ethul Kotte
14	AZ0205	100	Maligawa Rd	**49**	AZ8915	160	SMS Holdings
15	AZ1041	160	Maligawa Road No 18	**50**	AZ1009	160	Bangla JN
16	AZ8701	100	Cargills Ceylon Ltd, Ethul Kotte Rd	**51**			
17	AZ8211	400	Mobitel	**52**	AZ1038	250	SkyLine
18	AZ8210	630	Ceylon Paper Sacks	**53**	AZ0520	160	Mission Road
19	AZ0203	160	Near Jyaweera Mw	**54**	AZ0221	250	Bangla JN
20	AZ0536	100	New Jayaweera Mw	**55**	AZ0447	160	Beddagana Road
21	AZ8260	100	Graphic System	**56**	AZ8266	100	Foot Ball Fedaration

22	AZ8907	400	Capital Trust Residencies	**57**	AZ8706	250	Solis Reception Hall
23	AZ1008	160	Ethul Kotte	**58**	AZ8271	160	Keels Super (Bandul)
24	AZ8286	100	TKN Perera	**59**			
25				**60**	AZ8707	400	Min. of Envir. & National Resources
26	AZ0215	400	Mahindarama Road 1	**61**	AZ0517	160	Epitamulla RD
27	AZ1004	160	Mahindarama Road 2	**62**			
28	AZ0525	160	Ethul Kotte Nr Mati Ambalama	**63**	AZ0228	250	Epitamulla RD
29	AZ8742	100	Dambadiav Vandana	**64**	AZ8231	400	National Water Board
30				**65**	AZ8232	400	National Water Board
31	AZ0539	100	9TH Lane	**66**	AZ0229	160	Kotte Road Nr Epitamulla
32	AZ1031	400	PRIME Homes	**67**	AZ1037	160	Dialog Building
33	AZ0542	100	Alakeshwara Road	**68**			
34				**69**	AZ0524	100	4th Ln Pita KOT
35	AZ0220	160	Kolobathanthri Mw 1	**70**			

A.2 LINE DATA OF THE KALUBOWILA FEEDER

From	To	Length (m)	R (per km)	Rk	R pu	Xw (per km)	X	X pu
1	2	109	0.20457	0.02230	0.018428	0.33056	0.03603	0.029778
2	3	107	0.20457	0.02189	0.018090	0.33056	0.03537	0.029231
3	4	15	0.20457	0.00307	0.002536	0.33056	0.00496	0.004098
4	5	64	0.50498	0.03232	0.026710	0.36253	0.02320	0.019175
4	6	127	0.20457	0.02598	0.021471	0.33056	0.04198	0.034695
6	7	86	0.50498	0.04343	0.035891	0.36253	0.03118	0.025767
7	8	26	0.50498	0.01313	0.010851	0.36253	0.00943	0.007790
8	9	23	0.50498	0.01162	0.009599	0.36253	0.00834	0.006891
9	10	34	0.50498	0.01717	0.014190	0.36253	0.01233	0.010187
6	11	25	0.20457	0.00511	0.004227	0.33056	0.00826	0.006830
11	12	400	0.50498	0.20199	0.166936	0.36253	0.14501	0.119845
11	13	144	0.20457	0.02946	0.024346	0.33056	0.04760	0.039339
13	14	127	0.50498	0.06413	0.053002	0.36253	0.04604	0.038051
14	15	50	0.50498	0.02525	0.020867	0.36253	0.01813	0.014981
13	16	70	0.20457	0.01432	0.011835	0.33056	0.02314	0.019123
16	17	54	0.20457	0.01105	0.009130	0.33056	0.01785	0.014752
17	18	15	0.20457	0.00307	0.002536	0.33056	0.00496	0.004098
18	19	37	0.20457	0.00757	0.006255	0.33056	0.01223	0.010108
19	20	69	0.20457	0.01412	0.011666	0.33056	0.02281	0.018850
20	21	12	0.20457	0.00246	0.002029	0.33056	0.00397	0.003278
21	22	31	0.20457	0.00634	0.005241	0.33056	0.01025	0.008469
22	23	47	0.20457	0.00962	0.007946	0.33056	0.01554	0.012840
23	24	61	0.20457	0.01248	0.010313	0.33056	0.02016	0.016665
24	25	62	0.20457	0.01268	0.010482	0.33056	0.02050	0.016938
25	26	75	0.50498	0.03787	0.031300	0.36253	0.02719	0.022471
26	27	38	0.50498	0.01919	0.015859	0.36253	0.01378	0.011385
25	28	24	0.20457	0.00491	0.004058	0.33056	0.00793	0.006557
28	29	142	0.20457	0.02905	0.024007	0.33056	0.04694	0.038793
29	30	78	0.20457	0.01596	0.013187	0.33056	0.02578	0.021309
30	31	181	0.50498	0.09140	0.075538	0.36253	0.06562	0.054230
30	32	50	0.50498	0.02525	0.020867	0.36253	0.01813	0.014981
32	33	40	0.50498	0.02020	0.016694	0.36253	0.01450	0.011984
30	34	101	0.20457	0.02066	0.017076	0.33056	0.03339	0.027592
34	35	12	0.50498	0.00606	0.005008	0.36253	0.00435	0.003595
35	36	66	0.50498	0.03333	0.027544	0.36253	0.02393	0.019774
34	37	68	0.20457	0.01391	0.011496	0.33056	0.02248	0.018577
37	38	46	0.20457	0.00941	0.007777	0.33056	0.01521	0.012567
38	39	147	0.50498	0.07423	0.061349	0.36253	0.05329	0.044043
39	40	64	0.50498	0.03232	0.026710	0.36253	0.02320	0.019175
39	41	103	0.50498	0.05201	0.042986	0.36253	0.03734	0.030860
38	42	23	0.20457	0.00471	0.003889	0.33056	0.00760	0.006283
42	43	259	0.50498	0.13079	0.108091	0.36253	0.09390	0.077599
42	44	126	0.20457	0.02578	0.021302	0.33056	0.04165	0.034422
44	45	40	0.50498	0.02020	0.016694	0.36253	0.01450	0.011984
44	46	10	0.20457	0.00205	0.001691	0.33056	0.00331	0.002732
46	47	34	0.20457	0.00696	0.005748	0.33056	0.01124	0.009288

From	To	Length (m)	R (per km)	Rk	R pu	Xw (per km)	X	X pu
47	48	13	0.50498	0.00657	0.005425	0.36253	0.00471	0.003895
48	49	26	0.50498	0.01313	0.010851	0.36253	0.00943	0.007790
47	50	14	0.20457	0.00286	0.002367	0.33056	0.00463	0.003825
50	51	127	0.20457	0.02598	0.021471	0.33056	0.04198	0.034695
51	52	27	0.20457	0.00552	0.004565	0.33056	0.00893	0.007376
52	53	328	0.50498	0.16563	0.136887	0.36253	0.11891	0.098273
52	54	83	0.20457	0.01698	0.014032	0.33056	0.02744	0.022675
54	55	168	0.50498	0.08484	0.070113	0.36253	0.06091	0.050335
55	56	70	0.50498	0.03535	0.029214	0.36253	0.02538	0.020973
54	57	90	0.20457	0.01841	0.015216	0.33056	0.02975	0.024587
57	58	98	0.20457	0.02005	0.016568	0.33056	0.03240	0.026773
58	59	67	0.20457	0.01371	0.011327	0.33056	0.02215	0.018304
59	60	28	0.50498	0.01414	0.011685	0.36253	0.01015	0.008389
60	61	178	0.20457	0.03641	0.030094	0.33056	0.05884	0.048628
61	62	170	0.50498	0.08585	0.070948	0.36253	0.06163	0.050934
62	63	117	0.50498	0.05908	0.048829	0.36253	0.04242	0.035055
62	64	100	0.50498	0.05050	0.041734	0.36253	0.03625	0.029961
64	65	15	0.50498	0.00758	0.006260	0.36253	0.00544	0.004494
59	66	61	0.20457	0.01248	0.010313	0.33056	0.02016	0.016665
66	67	5	0.50498	0.00253	0.002087	0.36253	0.00181	0.001498
66	68	163	0.20457	0.03335	0.027558	0.33056	0.05388	0.044530
68	69	158	0.50498	0.07979	0.065940	0.36253	0.05728	0.047339
68	70	207	0.20457	0.04235	0.034997	0.33056	0.06843	0.056550

A.3 LOAD DATA OF THE KALUBOWILA FEEDER

	6 a.m.–9 a.m.		9 a.m.–12 p.m. (noon)		12 p.m. (noon)– 3 p.m.		3 p.m.–6 p.m.	
Noder	P(kW)	Qr (kVAR)	P(akW)	Qr (akVAR)	P (akW)	Qr (akVAR)	P(akW)	Qr (akVAR)
1	0.000	0.000	0.000	0.000	0.000	0.000	0.000	0.000
2	0.000	0.000	0.000	0.000	0.000	0.000	0.000	0.000
3	0.000	0.000	0.000	0.000	0.000	0.000	0.000	0.000
4	0.000	0.000	0.000	0.000	0.000	0.000	0.000	0.000
5	25.263	15.656	11.992	7.431	16.109	9.983	14.426	8.940
6	0.000	0.000	0.000	0.000	0.000	0.000	0.000	0.000
7	33.792	20.941	53.811	33.347	60.090	37.238	58.770	36.420
8	52.700	32.658	54.846	33.988	64.544	39.998	54.696	33.895
9	16.977	10.521	37.444	23.204	39.747	24.631	36.644	22.708
10	2.202	1.364	7.409	4.591	9.782	6.062	5.568	3.450
11	74.345	46.071	109.135	67.631	110.768	68.643	85.979	53.281
12	22.713	14.075	13.692	8.485	15.284	9.472	16.045	9.943
13	0.000	0.000	0.000	0.000	0.000	0.000	0.000	0.000
14	22.331	13.838	13.467	8.345	13.442	8.330	12.552	7.778

(continued)

	6 a.m.–9 a.m.		9 a.m.–12 p.m. (noon)		12 p.m. (noon)–3 p.m.		3 p.m.–6 p.m.	
Noder	P(kW)	Qr (kVAR)	P(akW)	Qr (akVAR)	P (akW)	Qr (akVAR)	P(akW)	Qr (akVAR)
15	35.809	22.191	27.215	16.865	30.152	18.685	35.208	21.819
16	24.731	15.326	39.556	24.513	40.044	24.815	40.293	24.969
17	75.820	46.986	76.840	47.618	77.690	48.144	77.798	48.212
18	1.734	1.075	8.373	5.189	8.859	5.490	5.857	3.629
19	50.782	31.470	74.005	45.861	79.437	49.227	60.666	37.595
20	37.814	23.433	26.888	16.663	30.186	18.706	25.941	16.075
21	15.712	9.736	65.466	40.569	60.050	37.213	36.468	22.599
22	137.806	85.398	117.387	72.745	115.706	71.703	136.868	84.817
23	1.373	0.851	1.364	0.845	1.192	0.739	1.365	0.846
24	9.799	6.072	18.245	11.307	21.433	13.282	19.388	12.015
25	0.000	0.000	0.000	0.000	0.000	0.000	0.000	0.000
26	134.406	83.292	116.559	72.232	114.856	71.176	125.340	77.673
27	21.356	13.234	19.657	12.181	18.806	11.654	20.302	12.581
28	56.937	35.284	67.745	41.982	75.614	46.858	59.509	36.877
29	23.059	14.290	23.907	14.815	24.759	15.343	23.059	14.290
30	0.000	0.000	0.000	0.000	0.000	0.000	0.000	0.000
31	25.263	15.656	14.542	9.012	16.963	10.512	14.426	8.940
32	136.106	84.345	93.607	58.008	105.534	65.400	123.384	76.461
33	0.000	0.000	0.000	0.000	0.000	0.000	0.000	0.000
34	0.000	0.000	0.000	0.000	0.000	0.000	0.000	0.000
35	38.271	23.716	35.721	22.137	34.871	21.609	45.137	27.971
36	25.634	15.886	18.710	11.595	17.861	11.068	21.260	13.175
37	43.023	26.662	66.165	41.003	71.293	44.180	55.804	34.582
38	0.000	0.000	0.000	0.000	0.000	0.000	0.000	0.000
39	0.000	0.000	0.000	0.000	0.000	0.000	0.000	0.000
40	49.507	30.679	34.106	21.136	35.746	22.152	40.821	25.297
41	71.418	44.258	60.361	37.405	66.407	41.152	58.782	36.427
42	0.000	0.000	0.000	0.000	0.000	0.000	0.000	0.000
43	33.409	20.704	26.716	16.556	28.762	17.824	26.303	16.300
44	0.000	0.000	0.000	0.000	0.000	0.000	0.000	0.000
45	7.238	4.486	19.577	12.132	18.857	11.686	15.796	9.789
46	12.784	7.922	30.786	19.078	31.274	19.381	28.996	17.969
47	0.000	0.000	0.000	0.000	0.000	0.000	0.000	0.000
48	136.156	84.376	77.360	47.940	80.769	50.053	76.606	47.473
49	17.981	11.143	35.806	22.189	34.106	21.136	34.860	21.603
50	44.202	27.392	58.226	36.083	63.217	39.176	64.640	40.058
51	0.000	0.000	0.000	0.000	0.000	0.000	0.000	0.000
52	10.328	6.400	20.505	12.707	18.833	11.671	17.129	10.615
53	66.109	40.968	64.699	40.094	64.703	40.097	53.000	32.844
54	61.476	38.097	86.643	53.692	84.302	52.242	69.825	43.270
55	61.328	38.005	35.301	21.876	35.829	22.203	38.429	23.814
56	0.683	0.423	0.679	0.421	0.659	0.408	0.563	0.349
57	25.953	16.083	94.242	58.402	105.627	65.457	49.399	30.612
58	32.174	19.938	57.928	35.898	58.010	35.949	57.729	35.774
59	0.000	0.000	0.000	0.000	0.000	0.000	0.000	0.000
60	112.210	36.872	119.956	39.418	119.108	39.139	102.860	63.742

	6 a.m.–9 a.m.		*9 a.m.–12 p.m. (noon)*		*12 p.m. (noon)–3 p.m.*		*3 p.m.–6 p.m.*	
Noder	*P(kW)*	*Qr (kVAR)*	*P(akW)*	*Qr (akVAR)*	*P (akW)*	*Qr (akVAR)*	*P(akW)*	*Qr (akVAR)*
61	59.606	36.938	42.605	26.403	43.469	26.938	57.835	35.840
62	0.000	0.000	0.000	0.000	0.000	0.000	0.000	0.000
63	72.501	44.929	30.332	18.796	41.142	25.496	44.121	27.342
64	209.187	42.465	201.282	40.860	193.928	39.367	204.111	126.487
65	217.687	44.190	192.840	39.146	210.905	42.814	196.359	121.683
66	66.405	41.151	55.353	34.302	55.381	34.320	64.732	40.114
67	43.216	26.781	43.471	26.939	44.331	27.472	41.759	25.878
68	0.000	0.000	0.000	0.000	0.000	0.000	0.000	0.000
69	23.243	14.404	19.628	12.163	22.171	13.739	21.518	13.335
70	0.000	0.000	0.000	0.000	0.000	0.000	0.000	0.000

Chapter 9

Case Study – Mallawapitiya Feeder 8: Optimal DG Planning for Distribution Feeders in Sri Lanka

K. T. M. U. Hemapala, Rajapaksha Mudiyanselage Tharangani Lakmali, O.V. Gnana Swathika

9.1 CASE STUDY – MALLAWAPITIYA FEEDER 8

Mallawapitiya Feeder 8 is selected as case study 2 for this chapter. The feeder is connected to the Mallawapitiya Grid Substation via the Ibbagamuwa gantry. The feeder details are shown in Table 9.1 and the geological locations of the feeder are illustrated in Figure 9.1.

The locations and currently available transformer details of the selected Mallawapitiya Feeder 8 are provided in Appendix C.1.

9.2 FEEDER DETAILS

The line resistance and reactance details of the feeder are provided in Appendix C.2. The conductor types of the feeder used for the simulation are listed below:

33 kV Racoon conductor: kR = 0.5192 (Ω/kmk); kX = 0.2877 (Ω/kkm); C = 22.36 (nF/km)
33 kV Weasel conductor: kR = 1.2963 (Ω/kmk); kX = 0.3166 (Ω/kkm); C = 24.99 (nF/km).

The load data of the feeder for eight selected time interval periods are provided in Appendix C.3. The load data variations for the eight time intervals at 75 bus locations are shown in Figure 9.2.

9.3 CONSTRAINT DATA

- According to the distribution code of Sri Lanka, the voltage limits of each bus are considered as ±6%.
- Biomass generation is considered for modeling.
- According to the currently available capacities of biomass plants in Sri Lanka, 24 hours availability of the proposed biomass power plant is assumed.

 DOI: 10.1201/9781003537878-9

Table 9.1 Details of Mallawapitiya Feeder 8

Parameter	*Value*
Voltage level	33 kV
Main line length	49.326 km
Branch conductor length	103.402 km
No. of nodes	75
No. of transformers	53
Feeder demand	4.15 MW (min) 4..51 MW (max)

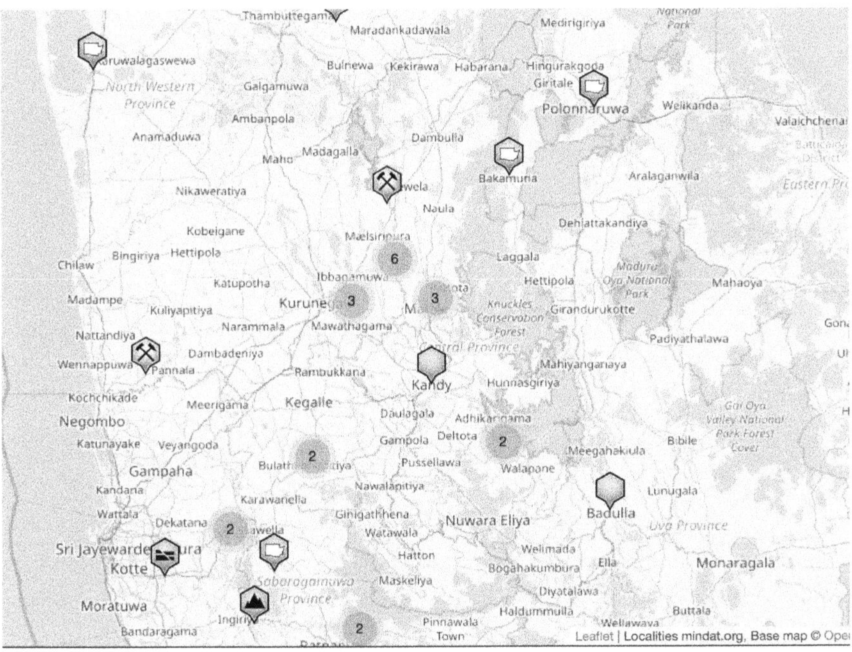

Figure 9.1 Geological view of Mallawapitiya Feeder 8

- The weighting factors are selected by analyzing results as W1 = 0.5, W2 = 0.3, and W3 = 0.2.

9.4 RESULTS AND ANALYSIS – MALLAWAPITIYA FEEDER 8

The obtained results for the objective function variation, active power loss variation, reactive power loss variation, distributed generation (DG) cost variation, optimal DG size (megawatts (MW)) variation, optimal DG size (megavolt-amperes reactive (MVAR)) variation, and bus voltage variation

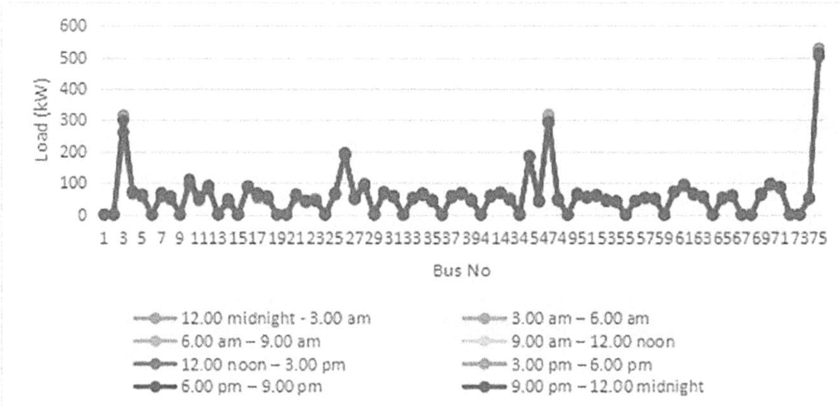

Figure 9.2 Load variation in feeder nodes in a day (eight time intervals)

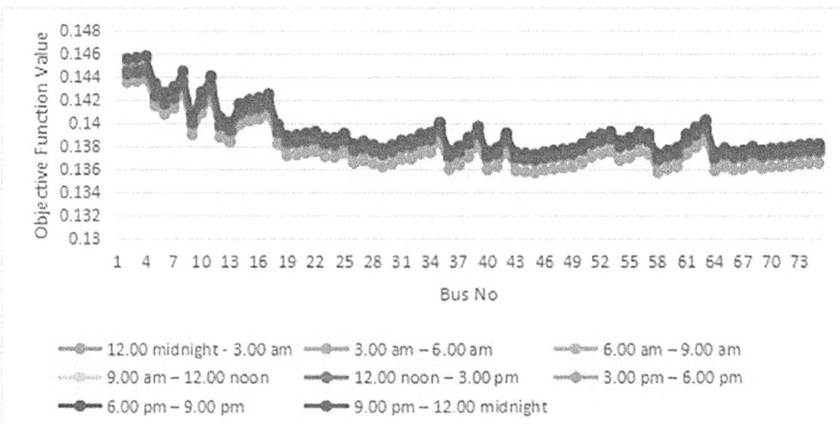

Figure 9.3 Objective function variation in Bus 1 to Bus 75

with respect to the selected DG for eight time intervals are shown in Figures 9.3–9.9, respectively.

An analysis of the observed results yields summaries for optimal DG planning in different time intervals, which are tabulated in Tables 9.2 and 9.3, and Figure 9.10.

According to the above results, different active and reactive power requirements are needed to fulfill optimal DG requirements in the eight time intervals for the given constraints. However, the optimal DG location is the same for all eight cases. DG with the selected active and reactive power at 9.00 a.m.–12.00 p.m. (noon) is proposed as the optimal DG for

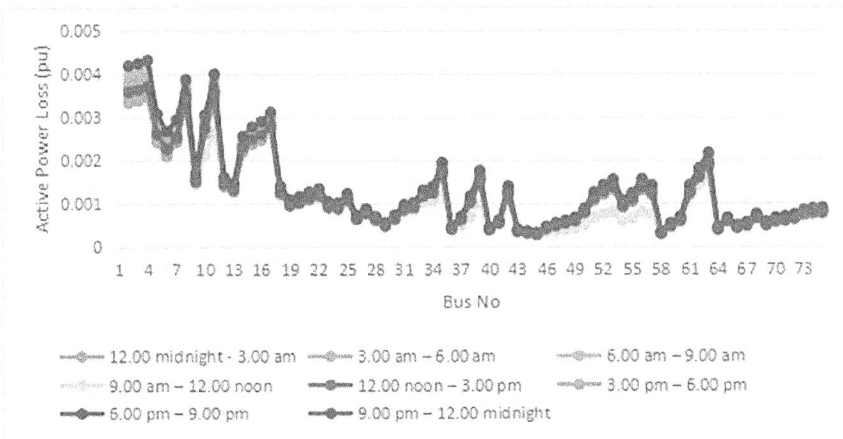

Figure 9.4 Active power loss variation in Bus 1 to Bus 75

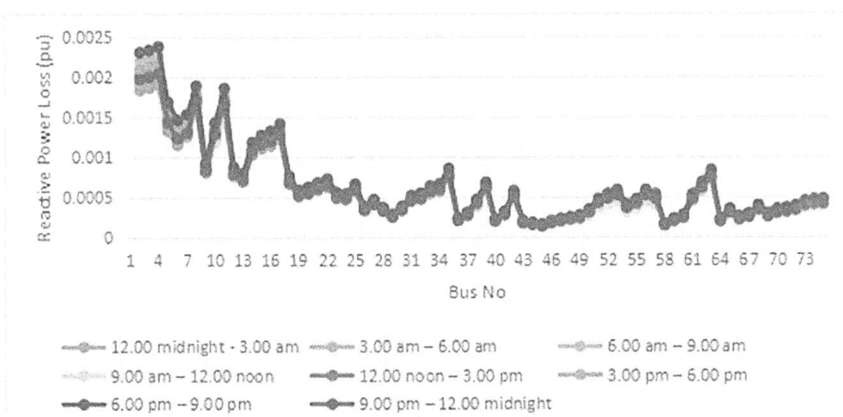

Figure 9.5 Reactive power loss variation in Bus 1 to Bus 75

Mallawapitiya Feeder 8 by considering its higher percentages of power losses and cost reduction compared to others. The geographical view of the proposed DG location is indicated in Figure 9.11.

9.5 VALIDATION OF THE POWER FLOW RESULTS – MALLAWAPITIYA FEEDER 8

The optimal results obtained from the algorithm are validated using Power System Simulator for Engineering (PSS/E) software. The network diagram,

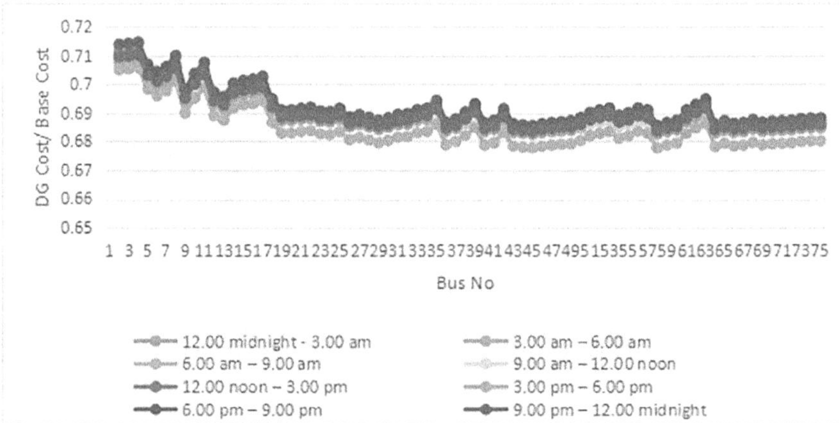

Figure 9.6 DG cost variation in Bus 1 to Bus 75

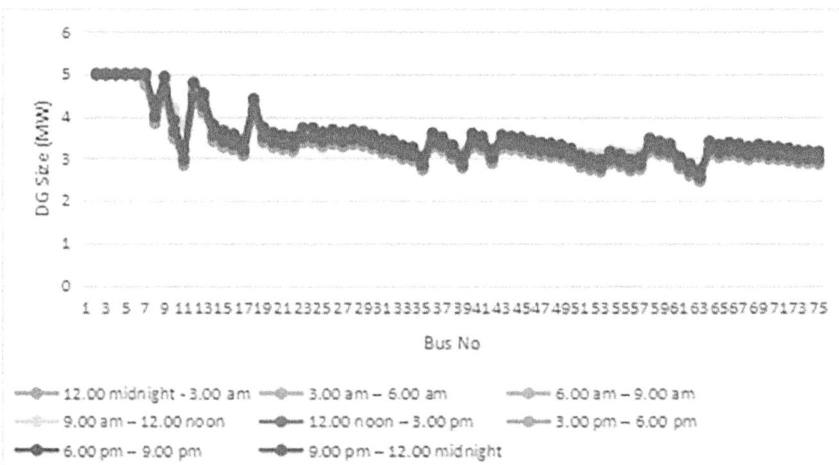

Figure 9.7 Optimal DG size (MW) variation in Bus 1 to Bus 75

load data, and line data of Mallawapitiya Feeder 8 on PSS/E are shown in Figures 9.12–9.14.

The optimal DG size for Mallawapitiya Feeder 8 is obtained as 3.33 MW and 1.596 MVAR from the proposed algorithm. DG is applied to randomly selected locations and power loss reduction is obtained. The observed

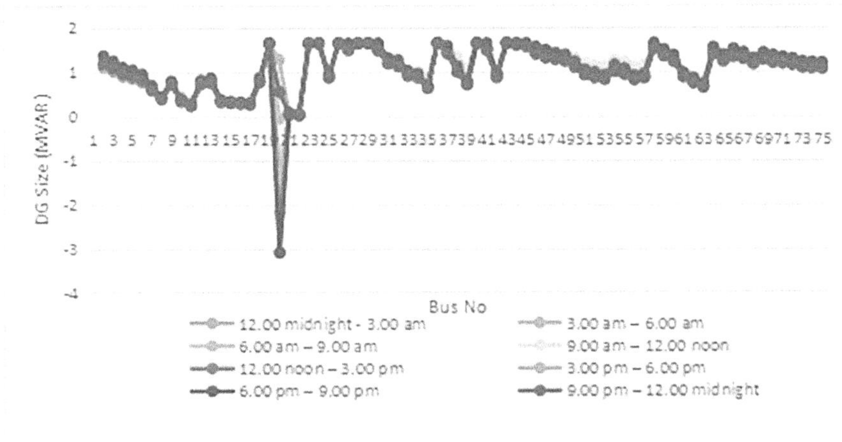

Figure 9.8 Optimal DG size (MVAR) variation in Bus 1 to Bus 75

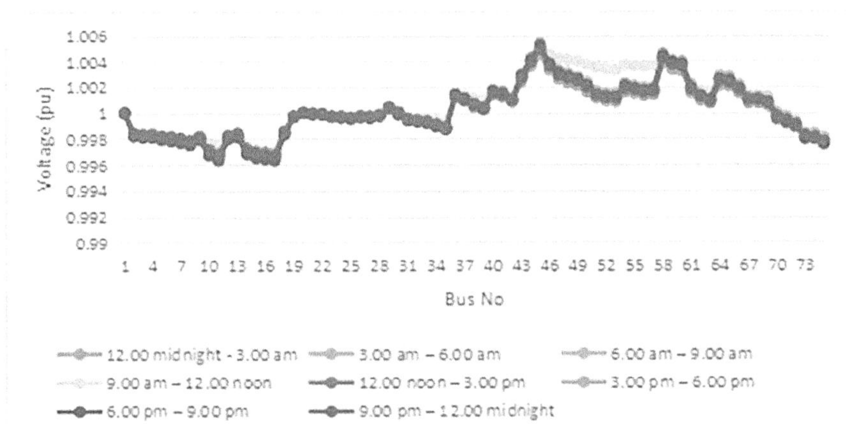

Figure 9.9 Bus voltage variation in Bus 1 to Bus 75

percentage values of the active power loss reduction and reactive power loss reduction are shown in Tables 9.4 and 9.5, respectively. The analyzed results are shown in Figures 9.15 and 9.16. According to the results, the maximum power loss reduction is obtained in the same way as the results from the proposed model and the proposed mathematical model is validated.

Table 9.2 Optimal DG planning results – Mallawapitiya Feeder 8

Time Interval	*Base Case Loss (kW)*	*Base Case Loss (kVAR)*	*Node*	*Plossaa (kW)*	*Qlossaa (kVAR)*	*DGeCost (MRs)*	*DGSize (MW)*	*DGeSize (MVAR)*
12.00 midnight–3.00 am	681.785	376.597	45	26.005	13.283	1659.59	3.174	1.566
3.00 am–6.00 am	705.690	389.821	45	26.552	13.566	1661.25	3.239	1.540
6.00 am–9.00 am	793.620	438.406	45	29.631	15.144	1666.99	3.435	1.608
9.00 am–12.00 noon	744.033	411.949	45	26.120	14.153	1664.42	3.333	1.596
12.00 p.m. (noon)–3.00 p.m.	743.291	410.615	45	27.746	14.190	1664.24	3.325	1.582
3.00 p.m.–6.00 p.m.	763.466	421.736	45	28.429	14.514	1648.09	3.368	1.585
6.00 p.m.–9.00 p.m.	826.731	456.658	45	31.255	15.955	1667.42	3.493	1.643
9.00 p.m.–12.00 p.m. (midnight)	725.947	401.012	45	27.356	13.980	1662.29	3.282	1.567

Table 9.3 Analysis of the optimal DG planning results – Mallawapitiya Feeder 8

Time Interval	*Optimal DG Loc.*	*DGaa Size (MW)*	*DGaSize (MVAR)*	*Plossa Redu. %*	*Qlossa Redu. %*	*Cost Redu. %*	*Minimum Voltage (p.u.)*
12.00 a.m. (midnight)–3.00 a.m.	45	3.174	1.566	96.19	96.47	31.76	0.9969
3.00 a.m.–6.00 a.m.	45	3.239	1.540	96.24	96.52	31.69	0.9968
6.00 a.m.–9.00 a.m.	45	3.435	1.608	96.27	96.53	31.46	0.9965
9.00 a.m.–12.00 p.m. (noon)	45	3.333	1.596	96.49	96.66	31.96	0.9968
12.00 p.m. (noon)–3.00 p.m.	45	3.325	1.582	96.27	96.54	31.57	0.9967
3.00 p.m.–6.00 p.m.	45	3.368	1.585	96.18	96.51	32.23	0.9966
6.00 p.m.–9.00 p.m.	45	3.493	1.643	96.22	96.50	31.44	0.9963
9.00 p.m.–12.00 p.m. (midnight)	45	3.282	1.567	96.23	96.49	31.65	0.9967

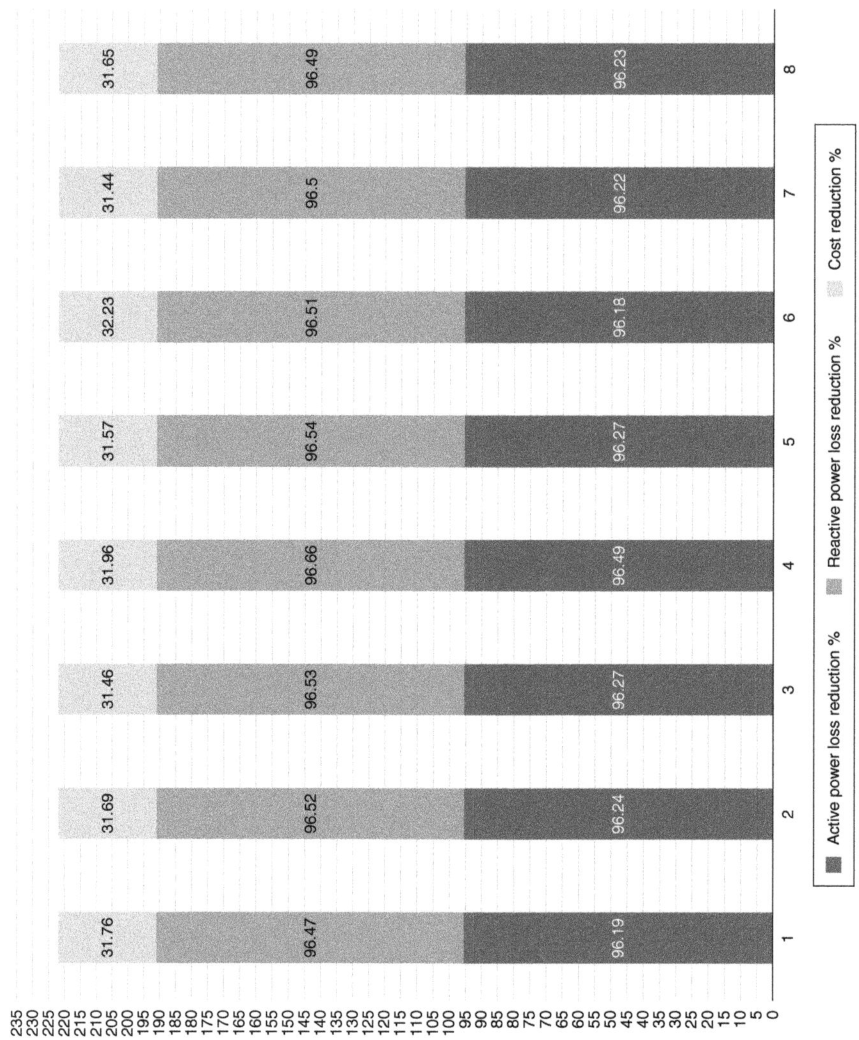

Figure 9.10 Summary of the optimal DG planning results – Mallawapitiya Feeder 8

Figure 9.11 Geological view of the proposed DG plant location – Mallawapitiya Feeder 8

9.6 DG ADDITION IN A DISTRIBUTION SYSTEM IN SRI LANKA

DG integration in the Sri Lankan distribution system has increased rapidly. The following two criteria are considered for DG planning in the Sri Lankan distribution network:

- voltage criteria: within (+6%) or (-6%) of nominal voltage at the point of supply
- loading criteria: up to a maximum of 80% of the transformer capacity.

For DG planning and analysis, the "SynerGEE Electric" software tool is used at the Ceylon Electricity Board (CEB). The procedure of DG planning in SynerGEE Electric software is shown in Figure 9.17.

A comparison of the proposed model and conventional software is illustrated in Table 9.6.

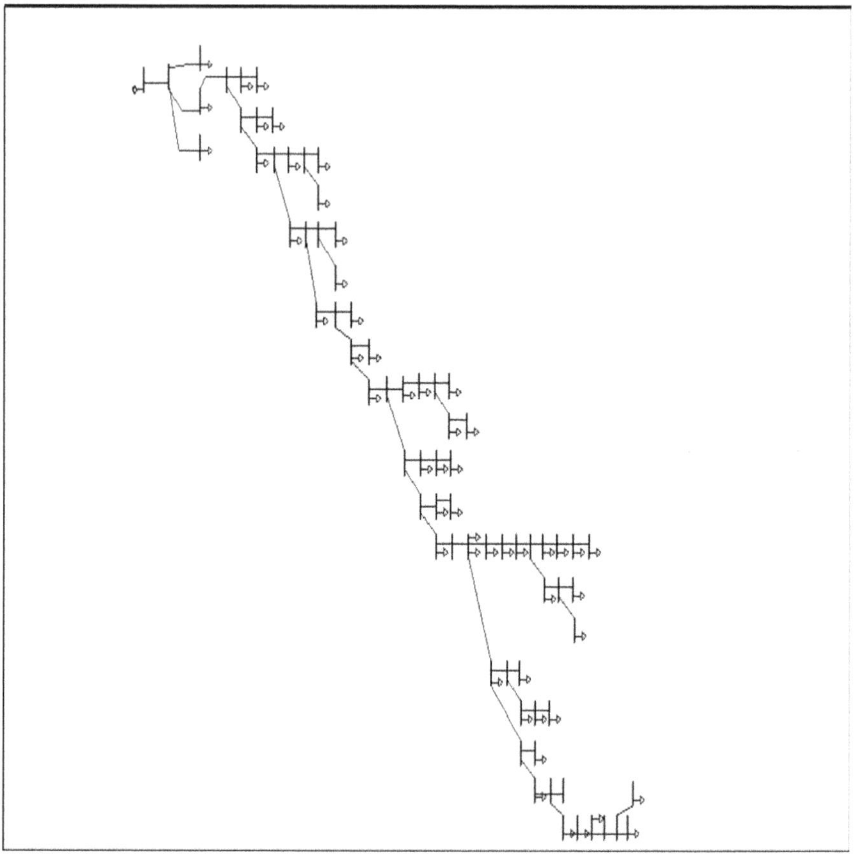

Figure 9.12 Network diagram on PSS/E software – Mallawapitiya Feeder 8

9.7 CONCLUSION

This chapter presents the application of the proposed model in a practical scenario. The mathematical model is applied to Mallawapitiya Feeder 8.

Diagram1.sld Network data ×

Bus Number	Pload (MW)	Qload (Mvar)	Bus Name
3	0.312480	0.193644	3BUS 33.000
4	0.065480	0.040578	4BUS 33.000
5	0.055870	0.034623	5BUS 33.000
7	0.059710	0.037002	7BUS 33.000
8	0.052170	0.032330	8BUS 33.000
10	0.095450	0.059150	10BUS 33.00
11	0.046140	0.028593	11BUS 33.00
12	0.085710	0.053114	12BUS 33.00
14	0.040120	0.024862	14BUS 33.00
16	0.086120	0.053369	16BUS 33.00
17	0.059150	0.036655	17BUS 33.00
18	0.050140	0.031072	18BUS 33.00
21	0.059140	0.036649	21BUS 33.00
22	0.040150	0.024881	22BUS 33.00
23	0.049780	0.030849	23BUS 33.00
25	0.064230	0.039803	25BUS 33.00
26	0.189140	0.117210	26BUS 33.00
27	0.051480	0.031902	27BUS 33.00
28	0.090450	0.056052	28BUS 33.00
30	0.066580	0.041260	30BUS 33.00
31	0.057100	0.035385	31BUS 33.00
33	0.050140	0.031072	33BUS 33.00
34	0.062040	0.038446	34BUS 33.00
35	0.041040	0.025432	35BUS 33.00
37	0.053150	0.032937	37BUS 33.00
38	0.063140	0.039128	38BUS 33.00
39	0.045780	0.028370	39BUS 33.00
41	0.057410	0.035577	41BUS 33.00
42	0.067400	0.041768	42BUS 33.00
43	0.051010	0.031611	43BUS 33.00
45	0.182010	0.112792	45BUS 33.00

Bus | Plant | Machine | **Load** | Fixed

Figure 9.13 Load data on PSS/E Software – Mallawapitiya Feeder 8

Diagram1.sld | Network data ×

Bus Number	Pload (MW)	Qload (Mvar)	Bus Name	
38	0.063140	0.039128	38BUS	33.00
39	0.045780	0.028370	39BUS	33.00
41	0.057410	0.035577	41BUS	33.00
42	0.067400	0.041768	42BUS	33.00
43	0.051010	0.031611	43BUS	33.00
45	0.182010	0.112792	45BUS	33.00
45	-3.330000	1.596000	45BUS	33.00
46	0.040950	0.025377	46BUS	33.00
47	0.312010	0.193353	47BUS	33.00
48	0.047880	0.029671	48BUS	33.00
50	0.063140	0.039128	50BUS	33.00
51	0.053780	0.033327	51BUS	33.00
52	0.058810	0.036445	52BUS	33.00
53	0.043240	0.026796	53BUS	33.00
54	0.040890	0.025340	54BUS	33.00
56	0.042980	0.026635	56BUS	33.00
57	0.053040	0.032869	57BUS	33.00
58	0.048730	0.030198	58BUS	33.00
60	0.072150	0.044711	60BUS	33.00
61	0.092010	0.057019	61BUS	33.00
62	0.063140	0.039128	62BUS	33.00
63	0.054420	0.033724	63BUS	33.00
65	0.050140	0.031072	65BUS	33.00
66	0.059140	0.036649	66BUS	33.00
69	0.062140	0.038508	69BUS	33.00
70	0.094280	0.058425	70BUS	33.00
71	0.083470	0.051726	71BUS	33.00
74	0.049450	0.030644	74BUS	33.00
75	0.525470	0.325634	75BUS	33.00

Bus | Plant | Machine | **Load** | Fixed

Figure 9.13 (Continued)

From Bus Number	To Bus Number	Line R (pu)	Line X (pu)
1	2	0.245111	0.135821
2	3	0.020564	0.011395
2	4	0.047937	0.026563
2	5	0.103232	0.057203
5	6	0.045906	0.025438
6	7	0.110783	0.027057
6	9	0.154057	0.085367
7	8	0.335026	0.081824
9	10	0.599645	0.146454
9	12	0.029569	0.016385
10	11	0.642944	0.157029
12	13	0.053228	0.029495
13	14	0.565461	0.138105
13	18	0.030034	0.016642
14	15	0.141683	0.034604
15	16	0.081098	0.019807
15	17	0.236347	0.057724
18	19	0.184606	0.102294
19	20	0.034882	0.019329
19	23	0.028818	0.015969
20	21	0.068099	0.037735
20	22	0.123485	0.068426
23	24	0.018910	0.010478
24	25	0.125127	0.069336
24	26	0.180368	0.099946
26	27	0.083055	0.046022
26	28	0.022909	0.012695
28	29	0.129269	0.071631
29	30	0.116566	0.064592
29	36	0.110625	0.061300
30	31	0.167909	0.093042

Figure 9.14 Line data on PSS/E software – Mallawapitiya Feeder 8

Diagram1.sld | Network data ×

From Bus Number	To Bus Number	Line R (pu)	Line X (pu)
30	31	0.167909	0.093042
31	32	0.035012	0.019401
32	33	0.209880	0.051260
32	34	0.283338	0.069201
34	35	0.486189	0.118744
36	37	0.147625	0.036055
36	40	0.024761	0.013721
37	38	0.342168	0.083569
38	39	0.492546	0.120296
40	41	0.099557	0.055167
40	43	0.092261	0.051124
41	42	0.569413	0.139070
43	44	0.092738	0.051388
44	45	0.078664	0.043589
45	46	0.155111	0.037883
45	58	0.047154	0.026129
46	47	0.088985	0.021733
47	48	0.056780	0.013868
48	49	0.052768	0.012888
49	50	0.186367	0.045517
49	54	0.329146	0.080388
50	51	0.331930	0.081068
51	52	0.163795	0.040004
52	53	0.121448	0.029662
54	55	0.146656	0.035818
55	56	0.299997	0.073269
55	57	0.168958	0.041265
58	59	0.145856	0.035623
58	64	0.143383	0.079451
59	60	0.104698	0.025571
59	61	0.788652	0.192615

Bus | Plant | Machine | Load | Fixed

Figure 9.14 (Continued)

Diagram1.sld Network data

From Bus Number	To Bus Number	Line R (pu)	Line X (pu)
45	58	0.047154	0.026129
46	47	0.088985	0.021733
47	48	0.056780	0.013868
48	49	0.052768	0.012888
49	50	0.186367	0.045517
49	54	0.329146	0.080388
50	51	0.331930	0.081068
51	52	0.163795	0.040004
52	53	0.121448	0.029662
54	55	0.146656	0.035818
55	56	0.299997	0.073269
55	57	0.168958	0.041265
58	59	0.145856	0.035623
58	64	0.143383	0.079451
59	60	0.104698	0.025571
59	61	0.788652	0.192615
61	62	0.414964	0.101348
62	63	0.421961	0.103057
64	65	0.148689	0.082392
64	66	0.055072	0.030516
66	67	0.084370	0.046751
67	68	0.134519	0.074540
67	69	0.015238	0.008444
69	70	0.120316	0.066670
70	71	0.033359	0.018485
71	72	0.049829	0.027612
72	73	0.123054	0.068187
73	74	0.030706	0.017015
73	75	0.054882	0.030411

Bus Plant Machine Load Fixed

Figure 9.14 (Continued)

Table 9.4 P_{Loss} reduction results at PSS/E with optimal DG – Mallawapitiya

Node	*Power from Grid (MW)*	*Power from DG (MW)*	*Load (MW)*	*Power Loss (kW)*	*Power Loss at Basecase (kW)*	*Power Loss Reduction %*
5	1.241	3.33	4.361	210.27	744.03	71.74
10	1.268	3.33	4.361	237.26	744.03	68.11
15	1.262	3.33	4.361	230.63	744.03	69.00
20	1.143	3.33	4.361	112.01	744.03	84.95
25	1.147	3.33	4.361	116.36	744.03	84.36
30	1.098	3.33	4.361	66.92	744.03	91.01
35	1.217	3.33	4.361	185.74	744.03	75.04
40	1.072	3.33	4.361	41.48	744.03	94.43
41	1.085	3.33	4.361	54.26	744.03	92.71
42	1.158	3.33	4.361	126.65	744.03	82.98
43	1.111	3.33	4.361	80.28	744.03	89.21
44	1.069	3.33	4.361	37.72	744.03	94.93
45	1.057	3.33	4.361	25.97	744.03	96.51
46	1.068	3.33	4.361	36.87	744.03	95.04
47	1.074	3.33	4.361	43.42	744.03	94.16
48	1.080	3.33	4.361	49.07	744.03	93.40
49	1.085	3.33	4.361	54.50	744.03	92.67
50	1.106	3.33	4.361	75.53	744.03	89.85
55	1.142	3.33	4.361	110.92	744.03	85.09
60	1.088	3.33	4.361	56.81	744.03	92.36
65	1.085	3.33	4.361	54.49	744.03	92.68
70	1.084	3.33	4.361	53.18	744.03	92.85
75	1.106	3.33	4.361	74.63	744.03	89.97

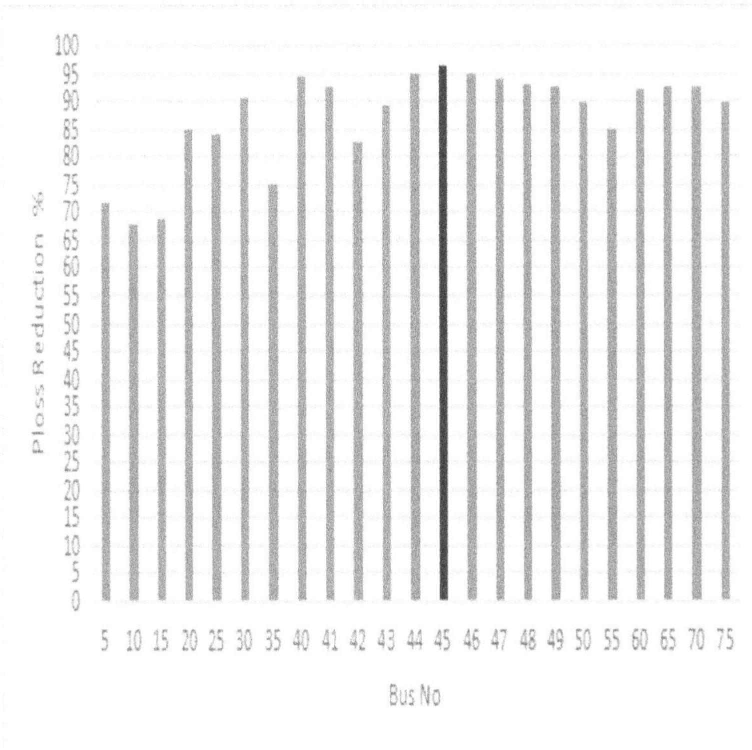

Figure 9.15 P_{Loss} reduction results on PSS/E with the optimal DG – Mallawapitiya

Table 9.5 Q_{Loss} reduction results at PSS/E with optimal DG – Mallawapitiya

Node	*Power from Grid (MVar)*	*Power from DG (MVar)*	*Load (MVar)*	*Power Loss (kVar)*	*Power Loss at Basecase (kVar)*	*Power Loss Reduction %*
5	1.222	1.596	2.702	115.13	411.95	72.05
10	1.214	1.596	2.702	107.60	411.95	73.88
45	1.207	1.596	2.702	100.71	411.95	75.55
20	1.167	1.596	2.702	60.74	411.95	85.26
25	1.169	1.596	2.702	63.15	411.95	84.67
30	1.142	1.596	2.702	35.79	411.95	91.31
35	1.179	1.596	2.702	72.17	411.95	82.48
40	1.145	1.596	2.702	21.71	411.95	90.73
41	1.135	1.596	2.702	28.79	411.95	93.01
42	1.153	1.596	2.702	46.58	411.95	88.69
43	1.129	1.596	2.702	22.62	411.95	94.51
44	1.126	1.596	2.702	19.65	411.95	95.23
45	1.120	1.596	2.702	13.84	411.95	96.64
46	1.127	1.596	2.702	20.52	411.95	95.02
47	1.129	1.596	2.702	21.50	411.95	94.48
48	1.129	1.596	2.702	22.90	411.95	94.44
49	1.131	1.596	2.702	24.22	411.95	94.12
50	1.132	1.596	2.702	25.28	411.95	93.86
55	1.140	1.596	2.702	33.98	411.95	91.75
60	1.132	1.596	2.702	21.06	411.95	93.89
65	1.135	1.596	2.702	28.95	411.95	92.97
70	1.132	1.596	2.702	28.22	411.95	93.15
75	1.147	1.596	2.702	40.11	411.95	90.26

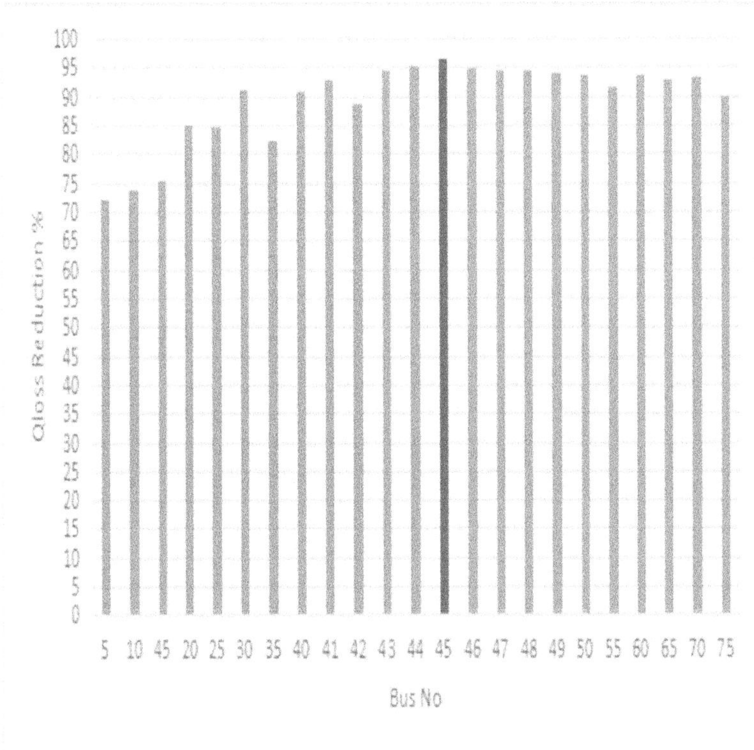

Figure 9.16 Q_{Loss} reduction results on PSS/E with the optimal DG – Mallawapitiya

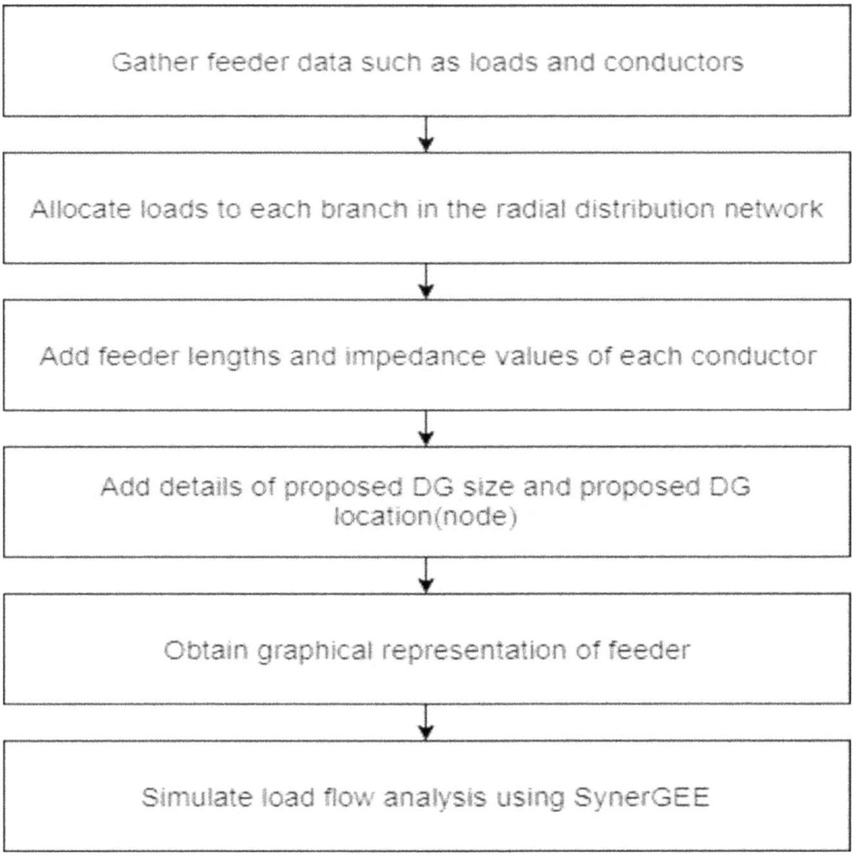

Figure 9.17 Procedure of DG planning using SynerGEE Electric software

Table 9.6 Comparison of proposed model and SynerGEE electric software modeling

Proposed method	*Distribution modeling software (SynerGEE)*
Can consider multi-objective optimization with ±6% voltage criteria Minimize power loss Minimize DG cost	Mainly consider voltage criteria (Check node voltage with acceptable range ±6%)
Can select the optimal size of DG using the given DG limits	Can use predefined DG sizes for analysis
Can select the optimal node (DG location) for DG installation	Can use predefined nodes (DG location) for analysis
Can obtain the optimal solution simply	Requires computational time for the optimal solution, but slow convergence

APPENDIX C

C.1: LOCATIONS AND TRANSFORMER DETAILS OF MALLAWAPITIYA FEEDER 8

Node	T/F Name	T/F (kVA)	Place	Node	T/F Name	T/F (kVA)	Place
1				**39**	4781083	100	Arankele
2				**40**			
3	4813013	400	Clothing (pvt) Ltd	**41**	4811084	100	Kurundugolla
4	4813415	100	Thimbirigasyaya Land Sale	**42**	4811085	100	Boraluwa Hewanegedara
5	4813084	100	Kanduluwa Junction	**43**	4781086	100	Clay Product Centre
6				**44**			
7	4813540	100	Karambayaya Watta	**45**	4811087	250	Kumbukgete
8	4813417	100	Godagama	**46**	4811088	100	Kumbukgete Town
9				**47**	4811162	400	Eco Transfer (Pvt) Ltd
10	4812014	160	Aragama	**48**	4811127	100	Koombuwa Kanda
11	4812419	100	Puddukkulama	**49**			
12	4812015	160	Kiribamuna	**50**	4781107	100	Ehetugaswewa
13				**51**	4781092	100	Bannaggama
14	4812558	100	Nelawa Deeyawa	**52**	4781093	100	Isak R/M
15				**53**	4781160	100	Athuruguruthenna
16	4812016	160	Nelawa Hekirilla Village	**54**	4781089	100	Dadapalagamuwa
17	4813474	100	Nelawa	**55**			
18	4812503	100	Wayamba Metal Works	**56**	4185091	100	Waduwawa
19				**57**	4185090	100	Ihalawaduwawa
20				**58**	4215094	100	Kasagala temple
21	4812456	100	Deeyawa	**59**			
22	4812425	100	Wewalagama	**60**	4215095	100	Hettigama Galwewa
23	4812066	100	Manapaha	**61**	4715096	160	Welgala
24				**62**	4215097	100	Nelliya 1
25	4812074	100	Rankirimada	**63**	4225096	100	Nelliya 2

(*continued*)

Node	*T/F Name*	*T/F (kVA)*	*Place*	*Node*	*T/F Name*	*T/F (kVA)*	*Place*
26	4812075	250	Kumbukwewa	**64**			
27	4812149	100	Hipawwa	**65**	4215099	100	Palugahakanda
28	4812170	160	Janapriya Rice Mall	**66**	4215164	100	Mandapola
29				**67**			
30	4811076	100	Kolongolla	**68**			
31	4811078	100	Nikagolla 1	**69**	4215150	100	Ranjith Watta
32				**70**	4215100	160	Doraweruwa
33	4811078	100	Nikagolla 2	**71**	4215145	160	Govijanasewa (near)
34	4811079	100	Kimbulwana Oya (unit 3)	**72**			
35	4821080	100	4 Canel Exit	**73**			
36				**74**	4184126	100	Kadubodagama
37	4811081	100	Thiththarawella	**75**	4184141	1000	Delmo Breeder Ltd
38	4811082	100	Japagama Medalanda				

C.2: LINE DATA OF MALLAWAPITIYA FEEDER 8

From	*To*	*Length (km)*	*R (per km)*	*R*	*R pu*	*X (per km)*	*X*	*X pu*
1	2	5.141	0.5192	2.66926	0.245111	0.2877	1.47909	0.135821
2	3	0.431	0.5192	0.22394	0.020564	0.2877	0.12409	0.011395
2	4	1.005	0.5192	0.52203	0.047937	0.2877	0.28927	0.026563
2	5	2.165	0.5192	1.12419	0.103232	0.2877	0.62294	0.057203
5	6	0.963	0.5192	0.49992	0.045906	0.2877	0.27702	0.025438
6	7	0.931	1.2963	1.20643	0.110783	0.3166	0.29465	0.027057
7	8	2.814	1.2963	3.64843	0.335026	0.3166	0.89107	0.081824
6	9	3.231	0.5192	1.67769	0.154057	0.2877	0.92964	0.085367
9	10	5.038	1.2963	6.53014	0.599645	0.3166	1.59488	0.146454
10	11	5.401	1.2963	7.00167	0.642944	0.3166	1.71004	0.157029
9	12	0.62	0.5192	0.32200	0.029569	0.2877	0.17843	0.016385
12	13	1.116	0.5192	0.57965	0.053228	0.2877	0.32120	0.029495
13	14	4.75	1.2963	6.15787	0.565461	0.3166	1.50396	0.138105
14	15	1.19	1.2963	1.54292	0.141683	0.3166	0.37683	0.034604
15	16	0.681	1.2963	0.88316	0.081098	0.3166	0.21570	0.019807
15	17	1.986	1.2963	2.57382	0.236347	0.3166	0.62861	0.057724
13	18	0.63	0.5192	0.32707	0.030034	0.2877	0.18124	0.016642
18	19	3.872	0.5192	2.01036	0.184606	0.2877	1.11398	0.102294
19	20	0.732	0.5192	0.37987	0.034882	0.2877	0.21049	0.019329
20	21	1.428	0.5192	0.74160	0.068099	0.2877	0.41094	0.037735
20	22	2.59	0.5192	1.34475	0.123485	0.2877	0.74516	0.068426
19	23	0.604	0.5192	0.31383	0.028818	0.2877	0.17390	0.015969
23	24	0.397	0.5192	0.20592	0.018910	0.2877	0.11411	0.010478
24	25	2.624	0.5192	1.36263	0.125127	0.2877	0.75506	0.069336
24	26	3.783	0.5192	1.96421	0.180368	0.2877	1.08841	0.099946
26	27	1.742	0.5192	0.90447	0.083055	0.2877	0.50119	0.046022
26	28	0.481	0.5192	0.24948	0.022909	0.2877	0.13824	0.012695
28	29	2.711	0.5192	1.40774	0.129269	0.2877	0.78006	0.071631
29	30	2.445	0.5192	1.26941	0.116566	0.2877	0.70341	0.064592
30	31	3.522	0.5192	1.82853	0.167909	0.2877	1.01323	0.093042
31	32	0.734	0.5192	0.38128	0.035012	0.2877	0.21127	0.019401
32	33	1.763	1.2963	2.28559	0.209880	0.3166	0.55822	0.051260
32	34	2.38	1.2963	3.08555	0.283338	0.3166	0.75360	0.069201
34	35	4.084	1.2963	5.29460	0.486189	0.3166	1.29312	0.118744
29	36	2.32	0.5192	1.20471	0.110625	0.2877	0.66756	0.061300
36	37	1.24	1.2963	1.60764	0.147625	0.3166	0.39264	0.036055
37	38	2.874	1.2963	3.72621	0.342168	0.3166	0.91007	0.083569
38	39	4.138	1.2963	5.36383	0.492546	0.3166	1.31003	0.120296
36	40	0.519	0.5192	0.26965	0.024761	0.2877	0.14942	0.013721
40	41	2.088	0.5192	1.08418	0.099557	0.2877	0.60077	0.055167
41	42	4.784	1.2963	6.20091	0.569413	0.3166	1.51447	0.139070
40	43	1.935	0.5192	1.00472	0.092261	0.2877	0.55674	0.051124
43	44	1.945	0.5192	1.00991	0.092738	0.2877	0.55962	0.051388
44	45	1.65	0.5192	0.85665	0.078664	0.2877	0.47469	0.043589
45	46	1.303	1.2963	1.68916	0.155111	0.3166	0.41255	0.037883
46	47	0.748	1.2963	0.96905	0.088985	0.3166	0.23667	0.021733

(continued)

From	To	Length (km)	R (per km)	R	R pu	X (per km)	X	X pu
47	48	0.477	1.2963	0.61833	0.056780	0.3166	0.15102	0.013868
48	49	0.443	1.2963	0.57464	0.052768	0.3166	0.14035	0.012888
49	50	1.566	1.2963	2.02953	0.186367	0.3166	0.49568	0.045517
50	51	2.788	1.2963	3.61472	0.331930	0.3166	0.88284	0.081068
51	52	1.376	1.2963	1.78372	0.163795	0.3166	0.43565	0.040004
52	53	1.02	1.2963	1.32257	0.121448	0.3166	0.32302	0.029662
49	54	2.765	1.2963	3.58440	0.329146	0.3166	0.87543	0.080388
54	55	1.232	1.2963	1.59709	0.146656	0.3166	0.39006	0.035818
55	56	2.52	1.2963	3.26697	0.299997	0.3166	0.79790	0.073269
55	57	1.419	1.2963	1.83995	0.168958	0.3166	0.44938	0.041265
45	58	0.989	0.5192	0.51351	0.047154	0.2877	0.28455	0.026129
58	59	1.225	1.2963	1.58837	0.145856	0.3166	0.38793	0.035623
59	60	0.88	1.2963	1.14016	0.104698	0.3166	0.27847	0.025571
59	61	6.625	1.2963	8.58842	0.788652	0.3166	2.09758	0.192615
61	62	3.486	1.2963	4.51896	0.414964	0.3166	1.10368	0.101348
62	63	3.545	1.2963	4.59516	0.421961	0.3166	1.12229	0.103057
58	64	3.007	0.5192	1.56144	0.143383	0.2877	0.86523	0.079451
64	65	3.119	0.5192	1.61923	0.148689	0.2877	0.89725	0.082392
64	66	1.155	0.5192	0.59973	0.055072	0.2877	0.33232	0.030516
66	67	1.77	0.5192	0.91879	0.084370	0.2877	0.50912	0.046751
67	68	2.821	0.5192	1.46491	0.134519	0.2877	0.81174	0.074540
67	69	0.32	0.5192	0.16594	0.015238	0.2877	0.09195	0.008444
69	70	2.524	0.5192	1.31025	0.120316	0.2877	0.72604	0.066670
70	71	0.7	0.5192	0.36328	0.033359	0.2877	0.20130	0.018485
71	72	1.045	0.5192	0.54264	0.049829	0.2877	0.30069	0.027612
72	73	2.581	0.5192	1.34006	0.123054	0.2877	0.74256	0.068187
73	74	0.644	0.5192	0.33439	0.030706	0.2877	0.18529	0.017015
73	75	1.151	0.5192	0.59766	0.054882	0.2877	0.33118	0.030411

C.3: LOAD DATA OF MALLAWAPITIYA FEEDER 8

	12.00 a.m. (mdngt)–3.00 a.m		*3.00 a.m.–6.00 a.m.*		*6.00 a.m.–9.00 a.m.*		*9.00 a.m.–12.00 p.m.*	
Noder	*P(akW)*	*Qr (akVAR)*	*P(akW)*	*Qr (kVAR)*	*P(akW)*	*Qr (kVAR)*	*P(akW)*	*Qr (akVAR)*
1	0.000	0.000	0.000	0.000	0.000	0.000	0.000	0.000
2	0.000	0.000	0.000	0.000	0.000	0.000	0.000	0.000
3	259.150	160.595	264.170	163.706	310.250	192.262	312.480	193.644
4	68.120	42.214	69.470	43.051	71.480	44.296	65.480	40.578
5	58.710	36.383	59.740	37.021	61.540	38.136	55.870	34.623
6	0.000	0.000	0.000	0.000	0.000	0.000	0.000	0.000
7	55.470	34.375	57.480	35.620	68.470	42.431	59.710	37.002
8	45.870	28.426	48.120	29.820	58.140	36.029	52.170	32.330
9	0.000	0.000	0.000	0.000	0.000	0.000	0.000	0.000
10	90.410	56.027	91.450	56.672	101.420	62.850	95.450	59.150
11	43.150	26.740	48.140	29.832	50.140	31.072	46.140	28.593
12	80.560	49.923	81.790	50.685	91.710	56.833	85.710	53.114
13	0.000	0.000	0.000	0.000	0.000	0.000	0.000	0.000
14	39.400	24.416	40.150	24.881	45.180	27.998	40.120	24.862
15	0.000	0.000	0.000	0.000	0.000	0.000	0.000	0.000
16	83.150	51.528	85.410	52.929	89.150	55.246	86.120	53.369
17	51.270	31.772	52.120	32.299	65.180	40.392	59.150	36.655
18	47.150	29.219	49.850	30.892	54.630	33.854	50.140	31.072
19	0.000	0.000	0.000	0.000	0.000	0.000	0.000	0.000
20	0.000	0.000	0.000	0.000	0.000	0.000	0.000	0.000
21	52.150	32.317	54.150	33.557	60.180	37.294	59.140	36.649
22	38.450	23.827	37.450	23.208	43.850	27.174	40.150	24.881
23	42.150	26.120	44.520	27.589	50.780	31.468	49.780	30.849
24	0.000	0.000	0.000	0.000	0.000	0.000	0.000	0.000
25	63.210	39.171	65.150	40.373	66.100	40.962	64.230	39.803
26	185.510	114.961	186.470	115.555	192.480	119.280	189.140	117.210
27	48.180	29.857	49.170	30.471	52.150	32.317	51.480	31.902
28	89.120	55.228	91.250	56.548	94.510	58.568	90.450	56.052
29	0.000	0.000	0.000	0.000	0.000	0.000	0.000	0.000
30	63.510	39.357	64.150	39.754	69.150	42.852	66.580	41.260
31	54.210	33.594	53.410	33.098	58.410	36.197	57.100	35.385
32	0.000	0.000	0.000	0.000	0.000	0.000	0.000	0.000
33	49.100	30.427	49.980	30.973	53.140	32.931	50.140	31.072
34	60.140	37.269	61.480	38.099	65.540	40.615	62.040	38.446
35	40.580	25.147	42.020	26.040	45.280	28.060	41.040	25.432
36	0.000	0.000	0.000	0.000	0.000	0.000	0.000	0.000
37	53.710	33.284	54.280	33.637	58.240	36.091	53.150	32.937
38	63.710	39.481	64.870	40.200	68.740	42.598	63.140	39.128
39	42.050	26.058	43.780	27.130	46.100	28.568	45.780	28.370
40	0.000	0.000	0.000	0.000	0.000	0.000	0.000	0.000
41	56.420	34.963	57.410	35.577	59.710	37.002	57.410	35.577
42	66.420	41.160	67.840	42.040	69.470	43.051	67.400	41.768
43	45.980	28.494	48.700	30.179	50.140	31.072	51.010	31.611

(continued)

	12.00 a.m. (mdngt)–3.00 a.m		*3.00 a.m.–6.00 a.m.*		*6.00 a.m.–9.00 a.m.*		*9.00 a.m.–12.00 p.m.*	
Noder	*P(akW)*	*Qr (akVAR)*	*P(akW)*	*Qr (kVAR)*	*P(akW)*	*Qr (kVAR)*	*P(akW)*	*Qr (akVAR)*
44	0.000	0.000	0.000	0.000	0.000	0.000	0.000	0.000
45	181.040	112.190	181.980	112.773	183.040	113.430	182.010	112.792
46	39.450	24.447	40.450	25.067	43.950	27.236	40.950	25.377
47	290.040	179.738	291.010	180.339	301.400	186.778	312.010	193.353
48	43.450	26.926	45.570	28.240	49.180	30.477	47.880	29.671
49	0.000	0.000	0.000	0.000	0.000	0.000	0.000	0.000
50	62.220	38.558	63.100	39.103	65.040	40.305	63.140	39.128
51	52.470	32.516	53.780	33.327	55.980	34.691	53.780	33.327
52	57.210	35.453	58.110	36.011	60.810	37.684	58.810	36.445
53	42.870	26.567	43.170	26.752	45.780	28.370	43.240	26.796
54	38.710	23.989	39.740	24.627	42.040	26.052	40.890	25.340
55	0.000	0.000	0.000	0.000	0.000	0.000	0.000	0.000
56	41.010	25.414	42.010	26.034	44.380	27.502	42.980	26.635
57	51.480	31.902	52.040	32.249	54.780	33.947	53.040	32.869
58	47.380	29.361	48.170	29.851	50.470	31.276	48.730	30.198
59	0.000	0.000	0.000	0.000	0.000	0.000	0.000	0.000
60	70.800	43.875	71.480	44.296	73.150	45.331	72.150	44.711
61	88.720	54.980	89.020	55.166	93.480	57.930	92.010	57.019
62	62.140	38.508	64.930	40.237	66.480	41.198	63.140	39.128
63	53.450	33.123	55.480	34.381	57.150	35.416	54.420	33.724
64	0.000	0.000	0.000	0.000	0.000	0.000	0.000	0.000
65	48.150	29.839	49.120	30.440	53.040	32.869	50.140	31.072
66	57.150	35.416	58.140	36.029	60.140	37.269	59.140	36.649
67	0.000	0.000	0.000	0.000	0.000	0.000	0.000	0.000
68	0.000	0.000	0.000	0.000	0.000	0.000	0.000	0.000
69	60.140	37.269	61.540	38.136	64.780	40.144	62.140	38.508
70	94.140	58.339	95.470	59.163	97.140	60.198	94.280	58.425
71	83.480	51.733	84.750	52.520	86.490	53.598	83.470	51.726
72	0.000	0.000	0.000	0.000	0.000	0.000	0.000	0.000
73	0.000	0.000	0.000	0.000	0.000	0.000	0.000	0.000
74	48.480	30.043	49.780	30.849	52.140	32.311	49.450	30.644
75	500.400	310.098	503.150	311.802	524.150	324.816	525.470	325.634

	12.00 p.m.–3.00 p.m.		*3.00 p.m.–6.00 p.m.*		*6.00 p.m.–9.00 p.m.*		*9.00 p.m–12.00 a.m. (mdngt)*	
Node	*P(akW)*	*Q(akVAR)*	*P(akW)*	*Q (akVAR)*	*P (akW)*	*Q(akVAR)*	*P (akW)*	*Q(akVAR)*
1	0.000	0.000	0.000	0.000	0.000	0.000	0.000	0.000
2	0.000	0.000	0.000	0.000	0.000	0.000	0.000	0.000
3	315.840	195.726	309.450	191.766	298.450	184.949	260.410	161.376
4	64.780	40.144	68.710	42.580	73.480	45.536	70.410	43.633
5	54.780	33.947	58.740	36.401	63.470	39.332	60.490	37.486
6	0.000	0.000	0.000	0.000	0.000	0.000	0.000	0.000
7	58.970	36.544	60.710	37.622	69.480	43.057	58.470	36.234
8	51.170	31.710	53.140	32.931	58.450	36.221	49.120	30.440
9	0.000	0.000	0.000	0.000	0.000	0.000	0.000	0.000
10	93.450	57.911	95.450	59.150	111.780	69.270	100.810	62.472
11	47.120	29.200	49.170	30.471	56.170	34.809	48.170	29.851
12	83.470	51.726	85.410	52.929	91.580	56.752	82.140	50.902
13	0.000	0.000	0.000	0.000	0.000	0.000	0.000	0.000
14	40.210	24.918	43.120	26.721	49.840	30.886	40.170	24.893
15	0.000	0.000	0.000	0.000	0.000	0.000	0.000	0.000
16	84.270	52.222	87.150	54.007	90.170	55.878	87.610	54.292
17	60.120	37.256	60.140	37.269	68.150	42.233	58.150	36.036
18	51.420	31.865	52.730	32.677	59.170	36.668	50.120	31.059
19	0.000	0.000	0.000	0.000	0.000	0.000	0.000	0.000
20	0.000	0.000	0.000	0.000	0.000	0.000	0.000	0.000
21	58.140	36.029	59.670	36.977	65.170	40.386	60.450	37.461
22	41.250	25.563	42.580	26.387	45.270	28.054	41.500	25.718
23	48.950	30.334	49.650	30.768	45.180	27.998	42.180	26.139
24	0.000	0.000	0.000	0.000	0.000	0.000	0.000	0.000
25	62.180	38.533	61.250	37.957	68.450	42.418	65.250	40.435
26	186.180	115.376	187.840	116.404	195.480	121.139	189.120	117.198
27	49.710	30.805	50.420	31.245	55.780	34.567	49.560	30.712
28	89.150	55.246	90.150	55.866	97.180	60.222	90.780	56.256
29	0.000	0.000	0.000	0.000	0.000	0.000	0.000	0.000
30	67.150	41.613	68.150	42.233	72.150	44.711	68.400	42.387
31	56.150	34.796	57.140	35.410	60.010	37.188	55.540	34.418
32	0.000	0.000	0.000	0.000	0.000	0.000	0.000	0.000
33	51.260	31.766	52.150	32.317	54.980	34.071	50.210	31.115
34	63.240	39.190	64.010	39.667	67.450	41.799	61.010	37.808
35	42.010	26.034	43.780	27.130	47.350	29.343	42.890	26.579
36	0.000	0.000	0.000	0.000	0.000	0.000	0.000	0.000
37	54.010	33.470	55.010	34.090	60.780	37.665	56.150	34.796
38	64.060	39.698	65.470	40.572	70.140	43.466	66.720	41.346
39	44.050	27.298	45.980	28.494	47.840	29.646	43.780	27.130
40	0.000	0.000	0.000	0.000	0.000	0.000	0.000	0.000
41	57.480	35.620	58.420	36.203	61.870	38.341	57.150	35.416
42	68.010	42.146	68.890	42.691	71.450	44.278	67.480	41.817
43	51.890	32.156	51.070	31.648	47.010	29.132	46.010	28.512
44	0.000	0.000	0.000	0.000	0.000	0.000	0.000	0.000
45	181.010	112.172	182.890	113.337	185.780	115.128	182.980	113.393
46	41.850	25.934	42.800	26.523	46.010	28.512	41.150	25.501
47	312.470	193.638	318.140	197.151	295.040	182.836	293.580	181.932

(continued)

	12.00 p.m.–3.00 p.m.		3.00 p.m.–6.00 p.m.		6.00 p.m.–9.00 p.m.		9.00 p.m–12.00 a.m. (mdngt)	
Node	P(akW)	Q(akVAR)	P(akW)	Q (akVAR)	P (akW)	Q(akVAR)	P (akW)	Q(akVAR)
48	48.540	30.080	48.990	30.359	50.280	31.159	44.040	27.292
49	0.000	0.000	0.000	0.000	0.000	0.000	0.000	0.000
50	62.870	38.961	63.780	39.524	67.070	41.563	63.140	39.128
51	52.740	32.683	53.980	33.451	57.470	35.614	53.720	33.290
52	58.740	36.401	59.040	36.587	63.140	39.128	59.780	37.046
53	42.040	26.052	43.890	27.199	47.850	29.653	43.010	26.653
54	40.720	25.234	41.010	25.414	43.010	26.653	40.420	25.048
55	0.000	0.000	0.000	0.000	0.000	0.000	0.000	0.000
56	42.010	26.034	43.410	26.901	45.040	27.911	43.180	26.759
57	52.140	32.311	53.470	33.135	55.140	34.170	53.140	32.931
58	48.790	30.235	48.990	30.359	52.010	32.231	50.340	31.196
59	0.000	0.000	0.000	0.000	0.000	0.000	0.000	0.000
60	72.480	44.916	71.480	44.296	74.830	46.372	71.480	44.296
61	91.010	56.399	92.890	57.564	95.780	59.355	93.400	57.880
62	63.870	39.580	64.550	40.002	67.480	41.817	63.150	39.134
63	54.250	33.619	55.450	34.362	58.140	36.029	54.180	33.575
64	0.000	0.000	0.000	0.000	0.000	0.000	0.000	0.000
65	50.780	31.468	51.120	31.679	55.020	34.096	53.010	32.850
66	58.470	36.234	59.140	36.649	62.010	38.428	59.140	36.649
67	0.000	0.000	0.000	0.000	0.000	0.000	0.000	0.000
68	0.000	0.000	0.000	0.000	0.000	0.000	0.000	0.000
69	62.870	38.961	63.100	39.103	66.980	41.508	62.890	38.973
70	93.480	57.930	94.820	58.760	98.470	61.022	96.480	59.789
71	82.460	51.100	83.470	51.726	87.640	54.311	85.640	53.071
72	0.000	0.000	0.000	0.000	0.000	0.000	0.000	0.000
73	0.000	0.000	0.000	0.000	0.000	0.000	0.000	0.000
74	50.480	31.282	51.800	32.100	54.150	33.557	51.140	31.691
75	527.150	326.675	522.140	323.570	512.140	317.373	502.140	311.176

Chapter 10

State of Charge Estimation of an Indian Real-Time Driving Cycle for an Electric Vehicle Battery Energy Storage System

S. Arun Mozhi, S. Charles Raja

10.1 INTRODUCTION

Over the past few decades, crude oil has served as the predominant energy source for vehicles, constituting a non-renewable resource with significant adverse effects on the environment. Economic fluctuations during periods of conflict or pandemics often result in heightened fuel prices. The advent of batteries has revolutionized the electronic sectors and transportation sectors, yet battery technology remains a critical bottleneck for the progression of electric vehicles (EVs). Electric propulsion diminishes both carbon emissions and the formation of smog, especially when renewable resources are a source of fuel, thereby offering a potential solution to vehicular pollution.

Despite the numerous economic and financial advantages associated with EVs, challenges persist. Notably, there is a shortage of charging infrastructure compared to traditional fuel stations, and batteries necessitate extended charging durations. Consequently, in the routine mobility of everyday life, the implementation of an effective and precise battery management system (BMS) has become indispensable. However, there are so many issues established in association with battery application and management that require resolution, with the most extensively researched and pivotal being state of charge (SoC) estimation. Consequently, the accurate range prediction based on the SoC has assumed paramount importance in development efforts.

SoC acts as a critical metric within battery management systems (BMSs), overseeing charging and discharging parameters that contribute to battery longevity and performance. While SoC lacks a perfect definition, it can be precisely described as the ratio of a battery's total available capacity to its maximum achievable charge storage capacity. Various implementation methods exist, including the integration of controller units and artificial intelligence (AI). AI, which emulates and executes actions akin to human intelligence, is widely applied in BMSs. Specifically, in order to forecast SoC for the future, neural network models (NNMs) are trained. One of the primary challenges confronting EVs relative to internal combustion engine

DOI: 10.1201/9781003537878-10

(ICE) vehicles is their limited driving range. Additionally, the availability of charging stations for EVs falls short in comparison to the abundance of fuel stations for non-EVs. Knowledge of the total traveled distance since a battery was fully charged allows for the estimation of the remaining charge.

In the contemporary world, automobiles have become widespread for transportation, reliant on either renewable sources or non-renewable sources, such as petrol and diesel. While the escalation in the adoption of renewable source-based vehicles is notable, particularly those using electric power, the numbers remain modest due to the high cost of EV batteries. In this chapter, we propose a potential solution for SoC estimation utilizing the real driving cycle (RDC). This approach aims to provide EV travelers with the ability to anticipate their journey's SoC, thereby addressing a crucial aspect of EV usability.

10.1.1 Driving Cycle

A driving cycle is a description of the speed-time pattern, delineating its movement along a predetermined route in a city or region, while conforming to specific conditions. Drive cycles serve various purposes:

- Performance assessment: They are employed to evaluate a vehicle's performance, including fuel economy and emission tests.
- Battery and energy analysis in EVs: Drive cycles play a crucial role in analyzing battery states and energy utilization in EVs.
- Simulation for the development of a vehicle model: Before the actual development phase begins, drive cycles are used to simulate the vehicle model.

For standard and real-time applications, there are two types of driving cycles available: modal drive cycle (MDC) and transient drive cycle (TDC). The cycle involves a constant speed and is typically utilized for specific actions, such as emission tests. An example is the New European Driving Cycle (NEDC). The type of cycle mirrors actual driving patterns, encompassing acceleration, deceleration, and velocity changes. The best example of TDC is the Federal Test Procedure, which was cultivated by the United States Environmental Protection Agency.

By distinguishing between these driving cycles, it becomes possible to address different testing and application needs, ensuring a comprehensive assessment of vehicle performance.

10.1.2 Indian Drive Cycle (IDC)

The IDC is a brief cycle, spanning only 108 seconds and encompassing six driving cycle modes. However, it does not comprehensively represent all the diverse conditions encountered on the road. It is used in testing two- and three-wheelers in India, and a modified version known as the Modified Indian Driving Cycle (MIDC) is employed to assess emissions from cars and light commercial vehicles (LCVs).

In contrast to the IDC, the MIDC is more aligned with the NEDC. The MIDC consists of an Extra-Urban Driving Cycle (EUDC2) and four ECE-15 Urban Driving Cycles (UDC1), offering a more thorough evaluation of vehicle performance in various driving scenarios. This adaptation ensures a more accurate and representative assessment of emissions and fuel efficiency for cars and LCVs on Indian roads.

10.1.3 Related Work

Based on various studies, estimating batteries' SoC can be approached through different methodologies, with many focusing on offline techniques, meaning when the battery is not actively in use. Recently, AI has been integrated with machine learning for BMSs. Python, an object-oriented programming language, is widely utilized across industries for automation tasks and handling large datasets. Estimation methods for SoC can be categorized into the direct approach, indirect approach, and book-compiling approach. The open-circuit method stands out among direct measurement methods, and involves the computation of a battery's open-circuit voltage at specific ambient temperatures [1], [2]. However, while accurate, this method requires significant resting time and relies on historical charging and discharging data due to the hysteresis of the open circuit voltage [3]. The most prominent SoC estimation method is the Kalman filter (KF) method [4]. This method comprises prediction and error correction stages, leveraging an equivalent circuit model to describe the behavior of the battery [5]. Although the KF method linearizes the estimation process, it overlooks nonlinear properties, leading to inevitable errors [6]. To deal with this, the extended KF (EKF) method was introduced, which is capable of estimating nonlinear properties [7], [8]. The importance of SoC in the BMSs of EVs cannot be overstated, as it significantly influences vehicle performance and efficiency.

10.2 PROBLEM DESCRIPTION

In today's world, vehicles are essential for transportation, with fuel serving as their lifeblood. Fuel sources can be broadly categorized into renewable and non-renewable. While there has been significant escalation in the adoption of renewable source-based vehicles, particularly electric ones, they still lag behind conventional vehicles powered by crude oil.

The primary deterrents to the wider adoption of EVs are the high cost of batteries and lengthy charging times. Unlike conventional vehicles that can be refueled in 10–20 minutes, it takes 3 hours to full charge an EV. This disparity in charging times often leads consumers to perceive conventional vehicles as more convenient, despite the environmental impact.

To address this challenge, we propose a solution aimed at improving battery efficiency and providing users with real-time visibility into their battery's SoC. By displaying the remaining battery charge on a mobile device in real time, users can plan their journeys more effectively and know when to recharge their EVs. This approach aims to enhance the usability and convenience of EVs, encouraging their wider adoption and contributing to environmental sustainability. The core objectives of the proposed research are as follows:

- to measure the real-time driving pattern of EVs with the help of an On-Board Diagnostic (OBD) II dongle
- to evaluate the real-time SoC of a portable EV, taking into account variables such as the actual driving cycle
- as effective battery management is crucial, the goal is to provide alerts for over-charging and discharging, ensuring optimal battery health.

10.3 PROPOSED METHODOLOGY

A series of trips are taken along predefined routes with the OBD dongle interlinked with the vehicle. This dongle features a built-in hotspot that connects to a mobile phone, allowing the monitoring of driving data via a diagnostic app like Torque Pro, which records information. Subsequently, this information is transferred to the cloud for Message Queuing Telemetry Transport (MQTT) for storage and analysis. The information stored in the cloud is then accessed by an ESP32, which processes it and sends pulse width modulation (PWM) signals, with the help of a controller, to the motor. The motor's operation is observed and, along with battery consumption, it is estimated with the voltage sensor connected to an ESP8266 module. The voltage and SoC are then computed, and the data are periodically monitored with the Arduino Integrated Development Environment (IDE). Figure 10.1 shows the block diagram of the hardware setup.

10.4 EXPERIMENTAL SETUP

10.4.1 Data Collection

Information is gathered by undertaking various trips along diverse routes within Madurai City, as shown in Figure 10.2. Utilizing an OBD-II dongle connected to a BS6 vehicle, the data are accessible and monitored through

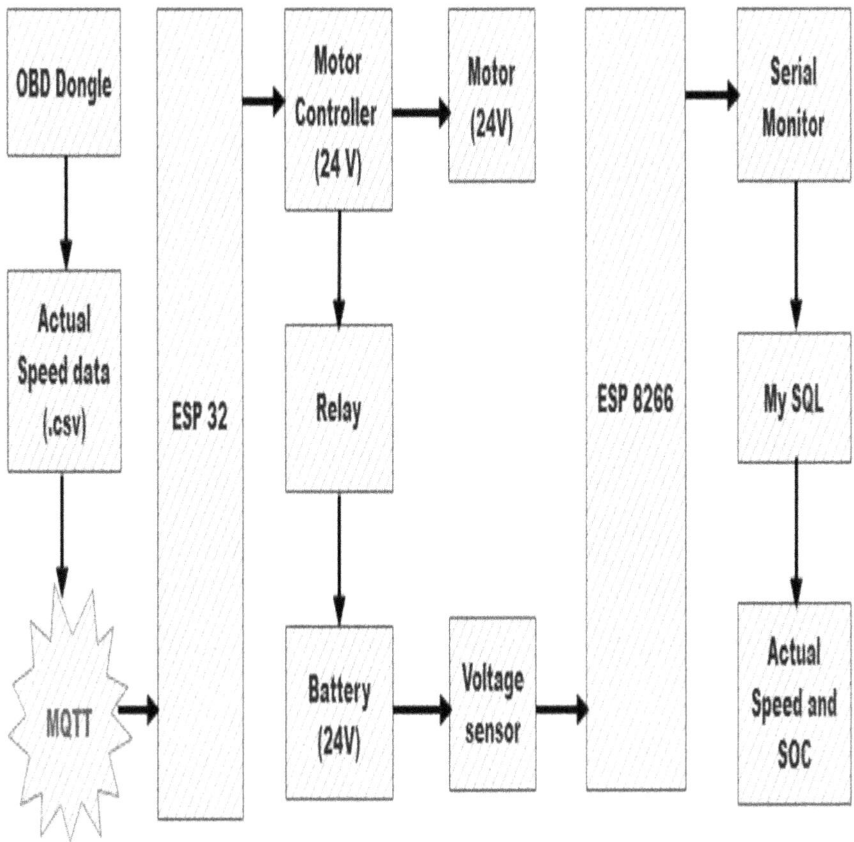

Figure 10.1 Architecture of the hardware setup

the Torque Pro app. Subsequently, the collected data are downloaded as a CSV file for further analysis. The routes for these trips include:

- Koripalayam to Mattuthavani
- Koodal Nagar to Iyer Bungalow
- Iyer Bungalow to Koodal Nager
- Periyar to Kaalavasal
- Thiruparankundram to Kaalavasal
- Iyer Bungalow to Thiruparankundram.

The real-time driving cycle of the predefined routes is formulated as a dataset for training the drive cycle pattern, which were created by conducting

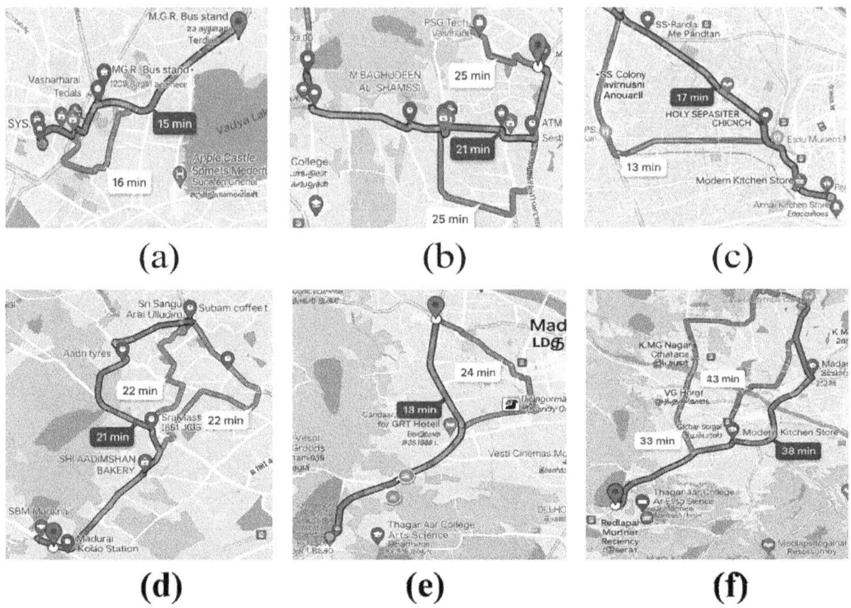

Figure 10.2 Trips taken: (a) Koripalayam to Mattuthavani, (b) Koodal Nagar to Iyer Bungalow, (c) Periyar to Kaalavasal, (d) Iyer Bungalow to Koodal Nagar, (e) TCE to Kaalavasal, and (f) Iyer Bungalow to TCE

multiple trips along various routes within Madurai city. These trips encompass different paths, and the routes taken during each trip are meticulously plotted and recorded for analysis.

10.4.2 Hardware Used

10.4.2.1 On-Board Diagnostic Dongle

An OBD is an external tool used to monitor the vehicle's subsystem. In ICE vehicles, it is used to regulate fuel emissions and engine oil levels. An OBD has a control unit called an electronic control unit (ECU), which controls the actuators with the help of various sensors in the vehicle subsystem, as input. As vehicles have become more and more computerized, OBD-II is now used to monitor the overall vehicle's scheme. OBD-II is an onboard computer with a 16 pin port known as a diagnostic link connector (DLC), located under the side dashboard. It provides electrical signal protocols and vehicle parameters. Diagnostic Trouble Codes (DTC) are stored in the vehicle's computer system and vary from one manufacturer to another. The OBD-II scan tool can connect to the vehicle through a port and read the DTC from the system's computer, as shown in Figure 10.3. This is mostly used by

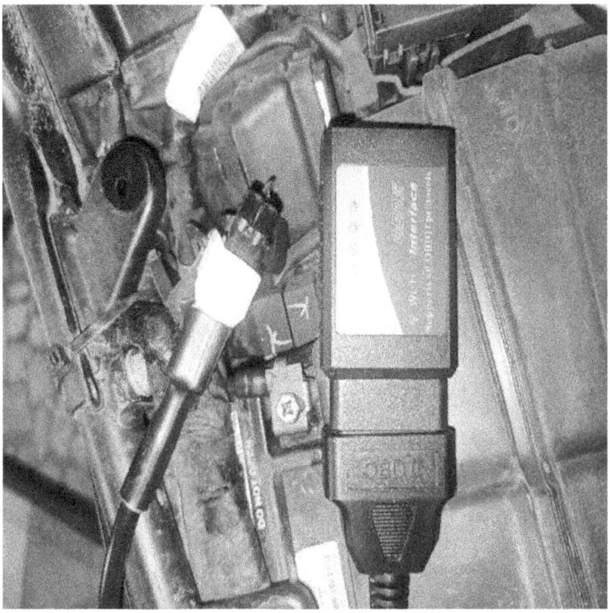

Figure 10.3 OBD-II connected to the vehicle

mechanics to find a fault in a vehicle. With the help of a diagnostic scanner tool, the parameters can be monitored on the user's personal devices.

10.4.2.2 Torque Pro

Torque Pro is a diagnostic scanner tool developed for Android devices to read the vehicle's control unit, as shown in Figure 10.4. It allows users to directly monitor the car's subsystem and retrieve various information from the sensors. It uses the user's phone's internal Global Positioning System (GPS) and accelerometer to graph information such as acceleration and speed. Torque also provides the capability to upload real-time data to the Torque Viewer website, enabling users to conveniently view their recorded OBD data online.

10.4.2.3 ESP32

ESP32 is an Arduino board that can perform as a standalone device or a slave device to the host microcontroller unit. ESP32 is in-built with TCP/IP networking software, which helps to interface with other systems through a Wi-Fi or Bluetooth connection.

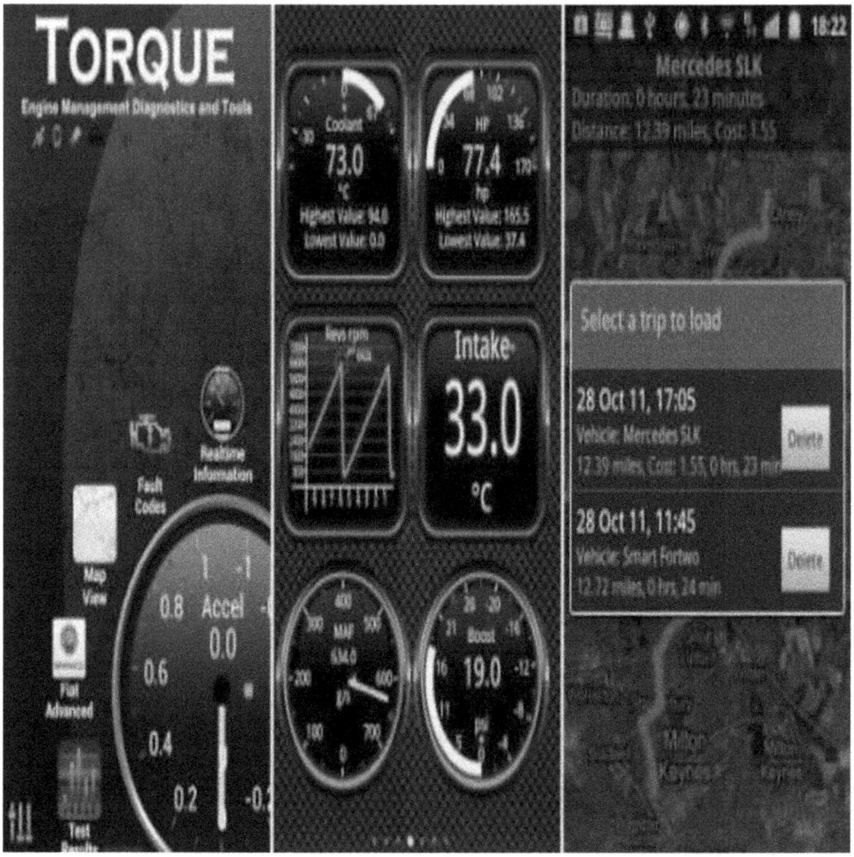

Figure 10.4 Torque Pro app

Specification

- 32-bit microprocessor with 240 MHz clock frequency, dual-core
- 34 programmable general-purpose input/outputs (GPIOs)
- supports Wi-Fi connectivity with speeds up to 150 Mbps
- also supports Bluetooth v4.2
- serial connectivity, which includes 3×UART.

10.4.2.4 ESP 8266

ESP8266 is a complete solution for self-containing Wi-Fi networking applications, which allows it to either communicate with other Wi-Fi networking processors or the host applications. ESP8266's storage capabilities allow it to be integrated with various sensors through its GPIO pins. The output of the sensors can be monitored in the Arduino IDE's serial monitor.

Specification

- VCC: +3.3V; can handle up to 3.6 V
- can be interfaced with microcontroller boards via serial ports
- 32-bit processor with low power consumption
- 17 GPIO pins
- 10-bit analog-to-digital converter (ADC)
- integrated TCP/IP protocol stack.

10.4.3 Software Used

10.4.3.1 MQTT.fx

MQTT represents a lightweight messaging protocol that operates on a publish and subscribe system. In this framework, a device or client has the capability to either subscribe to the published data or transmit data to the MQTT cloud service. Its functionality requires an internet protocol (IP) network, utilizing TCP/IP for communication between subscribers or publishers and a broker. MQTT is particularly advantageous for establishing connections with remote devices characterized by a small code footprint and operating in environments with limited bandwidth. This real-time communication protocol is well-suited for efficiently sharing data. Clients can subscribe to specific topics, ensuring that they receive the pertinent information they are seeking. The MQTT methodology is shown in Figure 10.5.

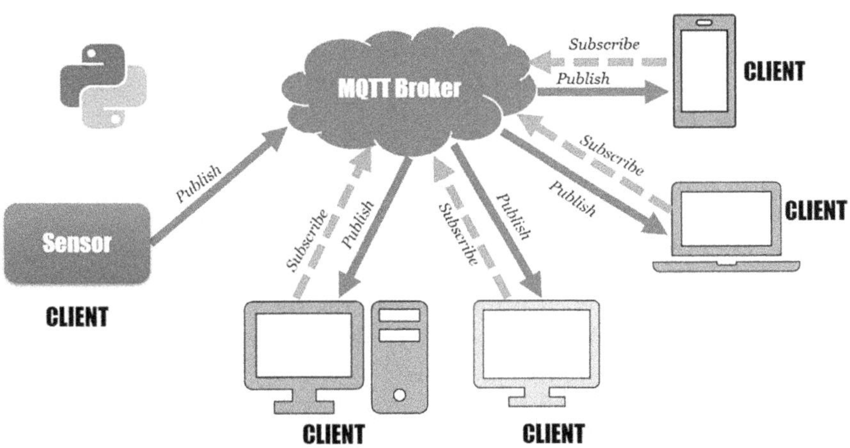

Figure 10.5 MQTT methodology

10.4.3.2 MySQL

A relational open-source database management system employing the SQL programming language belongs to MySQL. The collected data are transferred to Grafana, which serves as a MySQL database connection. For time series data, queries are required to include a time column that returns either an SQL date time or any numeric data type. Following this, the result sets of time series queries need to be organized in chronological order to facilitate proper visualization in panels. In the data frame query result, any column, excluding the time column or those of the string type, transforms into value fields. Meanwhile, string columns in the result become field labels in the data frame query output.

10.4.3.3 Arduino IDE

Arduino IDE is an open-source Arduino platform that connects to any Arduino hardware and communicates with it through programs or code. Arduino is capable of communicating with Bluetooth, Radio Frequency Identification (RFID), GPS units, cameras, motors, speakers, light-emitting diodes (LEDs), and the internet. The exchange of data with the connected boards can be viewed through a serial monitor.

To install Arduino IDE, users can check the Arduino official website and click the download link in [9], as is shown in Figure 10.6. A recent version of Arduino will be downloaded. The setup of all the necessary installations is to be done for the activation of the software.

Installing Libraries

Libraries are sketches that provide extra functionality when working with hardware. Once the Arduino IDE is installed, it is necessary to check for the libraries required for the respective hardware. In this chapter, ESP8266 and ESP32 are used for interfacing, so the respective libraries are to be installed in the board manager.

10.5 RESULTS AND DISCUSSION

10.5.1 MQTT Result

The speed data gathered is transmitted to the motor through the MQTT cloud using the ESP32. Consequently, the motor operates at a speed corresponding to the collected speed data. The result obtained from MQTT is shown in Figure 10.7.

Calculations are made using the battery's measured voltage and corresponding SoC. Using the Arduino IDE, the computed SoC is

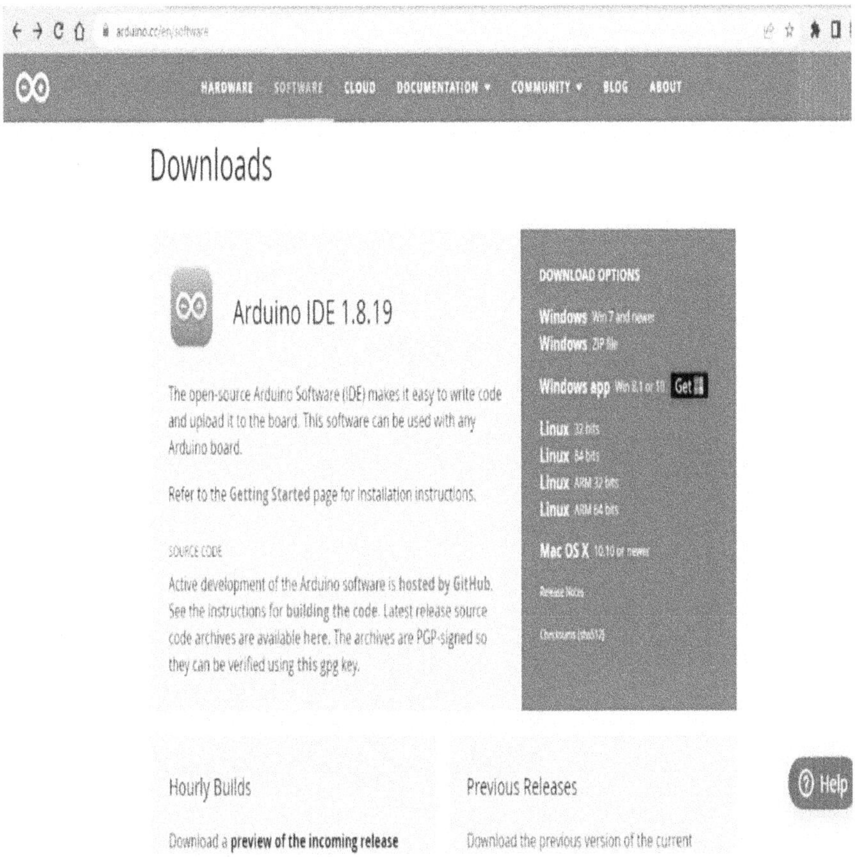

Figure 10.6 Arduino download

downloaded as a CSV file and observed in a serial monitor. The voltage and SoC are shown in Figure 10.8.

With the assistance of the Torque Pro app, the data transferred through MQTT are viewed. The real-time SoC and voltage of the EV battery are also monitored in this app. Hence, users come to know about the status of their vehicles and continue their journeys based on the data shown in the app, as shown in Figure 10.9.

10.6 CONCLUSION

Data are acquired through the implementation of an RDC, adhering to the specified parameters. These data are meticulously recorded and stored in a CSV file. The successful integration of the EV setup with other hardware

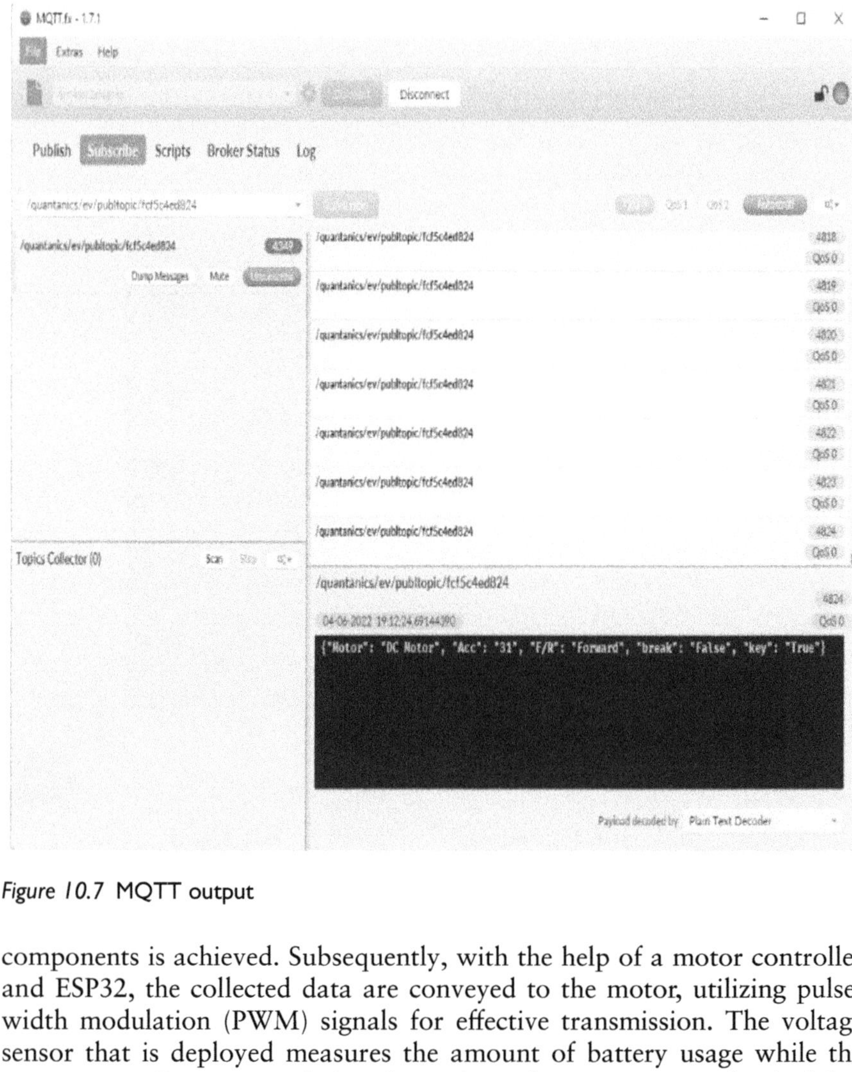

Figure 10.7 MQTT output

components is achieved. Subsequently, with the help of a motor controller and ESP32, the collected data are conveyed to the motor, utilizing pulse-width modulation (PWM) signals for effective transmission. The voltage sensor that is deployed measures the amount of battery usage while the motor runs. Data transmission from the voltage sensor to the Arduino IDE is made easier by the ESP8266 and voltage sensor together. A CSV file containing this sent data is then downloaded. Together with the voltage data, the SoC, which corresponds to the observed voltage, is computed and entered into a separate file for real data.

10.7 FUTURE SCOPE

We have implemented this for the urban part of Madurai city; however, it can be extended to the rural part and highway paths. It can be extended by

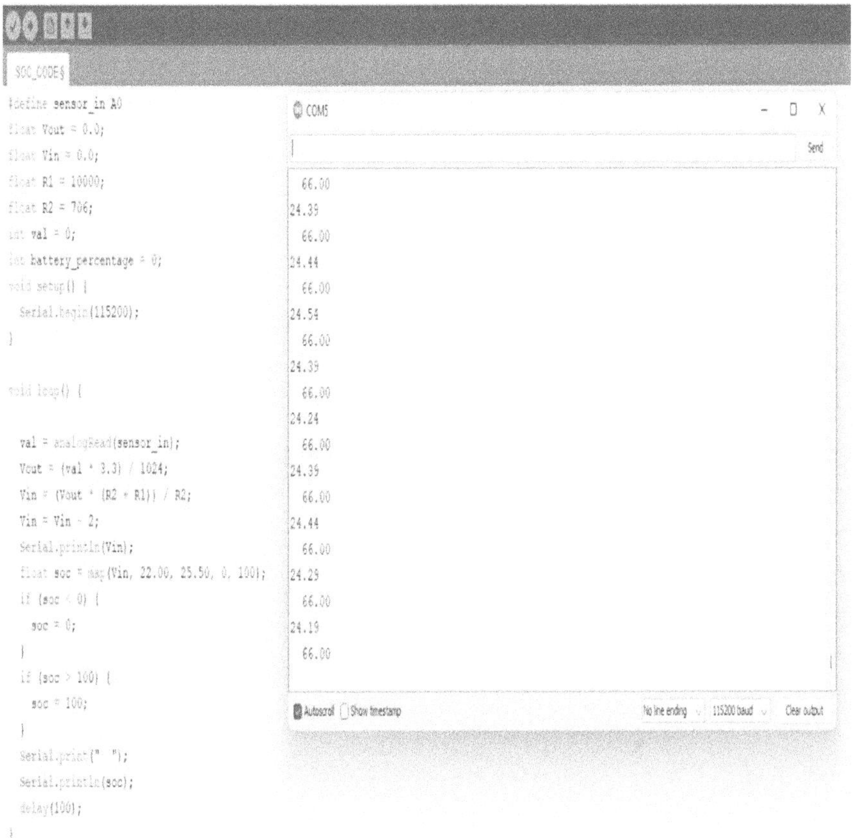

Figure 10.8 Voltage and SoC

predicting the SoC of the battery. The project's objective can be extended by displaying the suggested charging station.

ACKNOWLEDGMENT

The authors express their gratitude toward the management of Thiagarajar College of Engineering (TCE) for aiding them in conducting this research. They also thank the All India Council for Technical Education (AICTE), Government of India, for the AICTE Doctoral Fellowship (ADF) scheme (ADF scheme/U202119/I202120/S2023460) throughout this research.

A	B	C
Time	Relative Throttle Position(%)	Speed (OBD)(km/h)
10.9	13.33333302	0
11	13.72549057	0
11.1	13.72549057	0
11.2	13.72549057	0
11.3	13.72549057	14
11.4	13.72549057	14
11.5	13.72549057	14
11.6	16.47058868	14
11.7	16.47058868	14
11.8	16.47058868	14
11.9	16.47058868	14
12	16.47058868	20
12.1	16.47058868	20
12.2	16.47058868	20
12.3	16.47058868	20
12.4	16.47058868	20
12.5	16.47058868	20
12.6	0.39215687	20

Figure 10.9 Data collected: RDC

REFERENCES

[1] Attanayaka, A.M.S.M.H.S., Karunadasa, J.P. and Hemapala, K.T.M.U., 2019. Estimation of state of charge for lithium-ion batteriesA Review. *Aims Energy*, 7(2), pp.186–210.

[2] Rzepka, B., Bischof, S. and Blank, T., 2021. Implementing an extended Kalman filter for SoC estimation of a Li-ion battery with hysteresis: A step-by-step guide. *Energies*, 14(13), p.3733.

[3] Lee, S., Kim, J., Lee, J. and Cho, B.H., 2008. State-of-charge and capacity estimation of lithium-ion battery using a new open-circuit voltage versus state-of-charge. *Journal of Power Sources*, 185(2), pp.1367–1373.

[4] Barai, A., Widanage, W.D., Marco, J., McGordon, A. and Jennings, P., 2015. A study of the open circuit voltage characterization technique and hysteresis assessment of lithium-ion cells. *Journal of Power Sources*, 295, pp.99–107.

[5] Chen, Z., Fu, Y. and Mi, C.C., 2012. State of charge estimation of lithium-ion batteries in electric drive vehicles using extended Kalman filtering. *IEEE Transactions on Vehicular Technology*, 62(3), pp.1020–1030.
[6] Xie, Y., Wang, S., Zhang, G., Fan, Y., Fernandez, C. and Blaabjerg, F., 2023. Optimized multi-hidden layer long short-term memory modeling and sub-optimal fading extended Kalman filtering strategies for the synthetic state of charge estimation of lithium-ion batteries. *Applied Energy*, 336, p.120866.
[7] Wang, D., Yang, Y. and Gu, T., 2023. A hierarchical adaptive extended Kalman filter algorithm for lithium-ion battery state of charge estimation. *Journal of Energy Storage*, 62, p.106831.
[8] Wang, J., Yu, W., Cheng, G. and Chen, L., 2023. Online SOC estimation of Lithium-ion battery based on improved adaptive H infinity extended Kalman filter. arXiv preprint arXiv:2304.07748.
[9] Arduino (n.d.) *Arduino IDE software.* Available at: https://www.arduino.cc/en/software (Accessed: 23 January 2026).

Chapter 11

Network Optimization-Based Load Balancing Technique for an LV Electricity Distribution Network

K. T. M. U. Hemapala, Gayan Hemal Colamba Hewage, O.V. Gnana Swathika

11.1 INTRODUCTION

Electrification is still popular all over the world. Most countries are benchmarking their electrification target to the maximum. Sri Lanka is conducting a 100% electrification program and aims to improve accessibility to the distribution network. The Ceylon Electricity Board (CEB) and Lanka Electricity Company (LECO) are in the process of successfully reaching these targets [1], [2]. The usage of sensitive electrical devices demands an improvement to the quality of electricity distribution.

Electric vehicle charging and distributed generation entity addition into a low-voltage (LV) electricity distribution network shift the electrical behavior of the LV feeder to a complex pattern. The Sri Lankan government encourages domestic consumers to integrate solar systems to increase renewable energy usage [3]. The loads attached to the network are the key decision makers. Different types of consumers are rapidly connecting to the distribution network, and their consumption patterns vary considerably. This poses a key challenge in determining the accurate load prediction. This in turn creates unpredictable electricity behavior within the LV feeder. Sometimes electrical parameters such as voltage and current deviate drastically from pre-defined limits, which influences the healthy operation of the power distribution system. The unequal current distribution among phases is a key reason for the increase in the electrical energy loss. Sometimes this current goes beyond the current carrying capacity of the cable where large loads are connected. These unbalanced currents are returned through a neutral conductor path to the source and again create energy loss within the system. Additionally, a voltage drop is created through a neutral conductor and this affects the consumer end voltage. Single-phase consumers are usually the contributors for a significant neutral current.

Maintaining the network at the optimum condition turns out to be a huge challenge for the future for Sri Lankan utilities. The Public Utilities Commission of Sri Lanka (PUCSL) is the regulatory body for electricity

 DOI: 10.1201/9781003537878-11

utilities. It defines stipulated margins for the electricity distribution network [4] (e.g., voltage level at +/- 6% of the defined value and transformer loading at rated values).To maintain network performance at a considerable level, utilities are monitoring feeder end voltages, feeder currents, the per phase customer count, and instantaneous current measurement. These monitored parameters aid in load balancing and conducting energy loss reduction programs.

Three-phase four-wire neutral grounded overhead electricity distribution systems are used for electricity distribution in Sri Lanka. Different methods are used in the world to model such networks [5–10]. The suitable load flow analysis method aids in obtaining accurate results [10–13]. Most networks are in the transition stage from a traditional electricity distribution network to smart grids equipped with automatic phase changing. This transition may take a considerable time for developing countries such as Sri Lanka. However, it is necessary for utilities to maintain proper network optimization methods to cater to the present requirement and as a preliminary arrangement for an automated network. Some inherent properties exist for each electricity distribution system according to the standards maintained by certain companies/countries as per their objectives. As an example, three-phase four-wire overhead networks are used by LECO, which consist of a neutral conductor grounded at the feeder start (transformer) and feeder end using earth rods. CEB is using a three-phase four-wire overhead network with neutral grounded only at the transformer. Hence, the incorporation of such inherent properties during the calculation of the line impedance matrix and power flow analysis is useful for attaining realistic results. The precision of the line impedance matrix depends on these properties and on the assumptions made during modeling.

John Carson developed equations to find the impedances of the conductors, including the ground effect of such networks, which is used to model the system [14]. The neutral-to-earth voltage exists in different grounding techniques, except the solidly grounded condition. Such systems are modeled based on different assumptions and approximations. In a simulation, voltages and currents vary based on the way the distribution line and grounded systems are modeled [15].

Load flow analysis provides network voltages, phase angle, branch current, and other key parameters under a steady-state condition. Conventionally, nonlinear equations are used for performing load flow analysis. Many iterative techniques are used for load flow analysis [16],[17].

The forward/backward sweep method is commonly used for radial distribution networks. The association of the ground return path to the distribution system creates a requirement to modify the traditional forward/backward sweep method. If the ground-to-neutral voltages are used during calculations, they may pose a convergence problem. There are different proposals to separate ground impedance from Carson's equation to incorporate the current division calculation. These methods are based on some assumptions.

Network statuses vary based on the loads that are connected and disconnected from the network. The aggregated behavior of each consumer throughout the entire load profile needs to be evaluated for better analysis for the network condition. The ground path consideration is important for the completeness of the feeder energy loss calculation.

The objective of this chapter is to develop a network optimization-based load balancing technique for an LV electricity distribution network to provide a solution for the issues addressed in this section. A validated simplified power flow algorithm is used as the platform for evaluating the proposed algorithm. Section 11.2 discusses the proposed methodology in detail to develop a network optimization-based load balancing technique for an LV electricity distribution network. Section 3 showcases the application of the proposed load balancing technique in detail for an LV feeder.

11.2 METHODOLOGY

Following steps are followed to reach the objective of this research:

1. Propose an impedance model for an electricity distribution network with ground return.
2. Select the appropriate power flow algorithm.
3. Validate the applicability of the selected power flow algorithm for a three-phase four-wire neutral grounded overhead radial distribution feeder.
4. Develop an algorithm for network optimization based on load balancing.
5. Apply the proposed load balancing technique to an LV feeder.
6. Evaluate the performance of the proposed method.

11.2.1 Development of the Power Flow Algorithm for the Considered Network Model

The power flow algorithm is developed for the network model as indicated in Figure 11.1.

11.2.2 Validation of the Proposed Power Flow Algorithm

The applicability of the proposed power flow algorithm for the considered feeder model is dependent on the closeness of the results of steady-state analysis to the corresponding field measurements. The proposed power flow algorithm is implemented using MATLAB R2008a software. The source voltage, constant real and reactive power load consumed at each node,

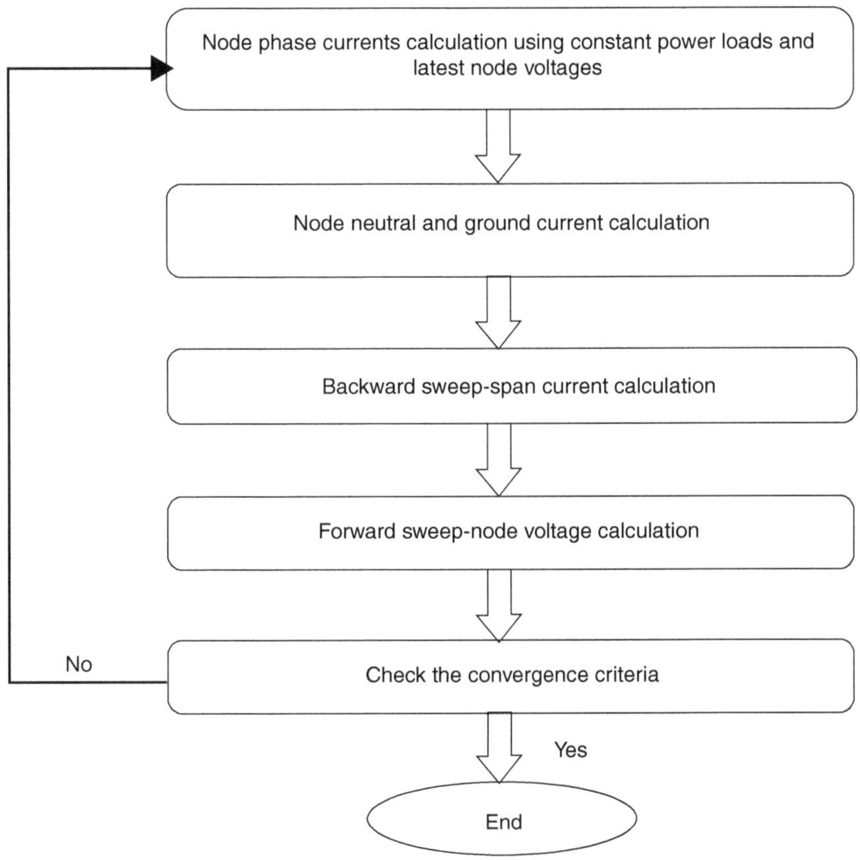

Figure 11.1 Power flow algorithm

primitive impedance matrix, and span lengths are inputs to the program. The node voltages, node currents, calculated power, and span currents are outputs obtained at the completion of the iteration process. The radial distribution feeder of LECO is selected for analysis. This feeder is attached to distribution transformer number AZ-0043 located at the Kelaniya branch in the Dalugama customer service center area.

11.2.2.1 Primitive Impedance Matrix for the Considered Feeder

LECO employs an aerial bundle conductor (ABC) for the LV electricity distribution system. The primitive impedance matrix is derived as follows:

$$[Z_{abcn}] = \begin{bmatrix} 0.00065934+0.00079237i & 4.9345e-005+0.00067703i & 4.9345e-005+0.00067703i & 4.9345e-005+0.00071154i \\ 4.9345e-005+0.00067703i & 0.00065934+0.00079237i & 4.9345e-005+0.00067703i & 4.9345e-005+0.00071154i \\ 4.9345e-005+0.00067703i & 4.9345e-005+0.00067703i & 0.00065934+0.00079237i & 4.9345e-005+0.00071154i \\ 4.9345e-005+0.00071154i & 4.9345e-005+0.00071154i & 4.9345e-005+0.00071154i & 0.0010293+0.00080054i \end{bmatrix} \quad (11.1)$$

Consumer Arrangement Details

Nine consumers are connected to the considered feeder as indicated in Table 11.1. The overview of how the meters are arranged in the field is given in Figure 11.2. The actual placement of the energy meters are represented in Figures 11.3–11.6. Data recordable meters are installed in the considered feeder and all the measurements are programmed to record with a 15-minute average time interval. The installed meters are 2 W three-element meters and are compatible with IEC 62053-11. Reactive power is calculated using the measured parameters. The same symbolic representation is used for both the meter and consumer connection (C_i),

Table 11.1 Consumer connection arrangement

Phase	*Node-01*	*Node-02*	*Node-03*	*Node-04*	*Node-05*	*Node-06*	*Node-07*	*Node-08*
R		C1,C2			C7,C8		C9	
Y			C3,C4	C5,C6				
B								

where C_i is the consumer connection of the ith domestic consumer, i = 1, 2, 3,..., 9.

11.2.2.2 Field Measurements Taken for the Validation Process

The field measurements are indicated in Table 11.2.

11.2.2.3 Validation Method

The calculated and measured values are compared to check the validity of the proposed power flow algorithm. The above procedure is repeated for

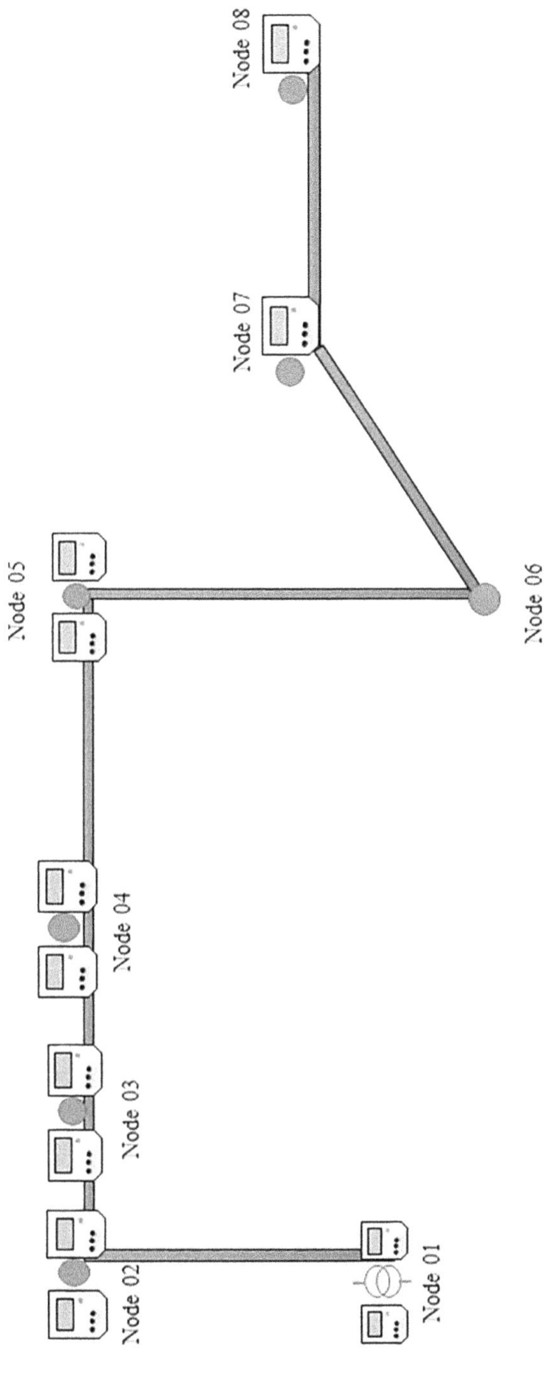

Figure 11.2 Meter arrangement in the field

Table 11.2 Details of data recordable meters and their application

Location	*Application*	*Meter type*	*Quantity of meter installed*	*Measured parameters*
Node-02 Node-03 Node-04 Node-05 Node-07	Single-phase domestic consumer information collection	Direct connected	9	• Neutral to phase voltages (V) • Phase current (A) • Active power (kW) • Energy consumption (kWh) • Demand (kVA) • Power factor
Node-08	LV feeder information collection (at end)	Direct connected	1	• Neutral to phase voltages (V)
Node-01	LV feeder information collection (at start – source)	Direct connected	1	• Neutral to phase voltages (V) • Phase current (A) • Active power (kW) • Energy consumption (kWh) • Demand (kVA) • Power factor
Node-01	Neutral conductor information collection (at start – source)	External CT connected	1	• Neutral current(A)

the dataset taken on December 18, 2016, as shown in Figure 11.7. The measured and calculated parameters are slightly deviated and that could happen due to many reasons. The voltage measurements are taken using two energy meters installed for two customers in the same phase closely at the same node and it indicates different values for a particular time slot. This is illustrated in Figure 11.8. However, this is not an acceptable result as it indicates a time synchronization issue for the energy meters. Initial time checking is conducted for all the meters and corrected. However, slight time shifting is observed after the installation. The calculated impedance matrix values may vary from the actual value available in the conductor.

Voltage errors are defined as the difference between the calculated and measured values. Figures 11.9–11.11 indicate how the voltage errors vary throughout the day. Figure 11.12 showcases how the current errors vary throughout the day. The current errors are defined as the difference between the calculated and measured values.

According to the observations from the voltage error variation graphs, most values fall between -0.5V and +0.5V. Table 11.3 is a summary of the number of errors distributed in the observed range. According to the observations from the current error variation graphs, most values fall

Figure 11.3 Energy meter installation near the transformer

Figure 11.4 Energy meters installed at the transformer for source detail collection

Figure 11.5 Energy meter installation for a domestic consumer

Figure 11.6 Energy meter installation at the feeder end

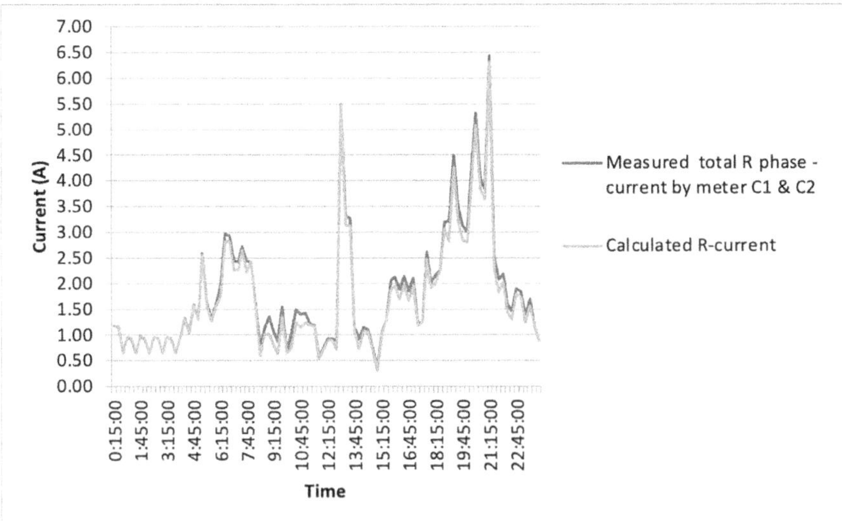

Figure 11.7 Node-02 current profiles taken on December 18, 2016

between -0.2V and +0.2V Table 11.4 is a summary of the number of errors distributed in the observed range. Hence, based on the aforementioned analysis, we may draw the conclusion that the proposed network model and power flow algorithm provide an acceptable accuracy for using the network status evaluation in subsequent selections.

11.2.3 Development of the Algorithm for the Network Optimization-Based Load Balancing Technique

11.2.3.1 Introduction

Voltages, currents, and network losses within the feeder vary throughout the day based on the consumer arrangement among phases at the nodes and their consumption. Network statuses differ according to the consumer connected pattern, although the consumption pattern follows the same style. There are a number of possible patterns that may be assigned to a certain set of consumers among phases along the LV feeder. The consumer connected pattern is a combination of the consumer arrangement at each node. However, consumers cannot be interchanged between nodes according to practical limitations during combination generation. The network status depends on the aggregated result of individual consumers' electricity consumption for a specific pattern. The results generated from each consumer's connected pattern are evaluated individually and the most appropriate pattern is identified based on the optimum condition for the

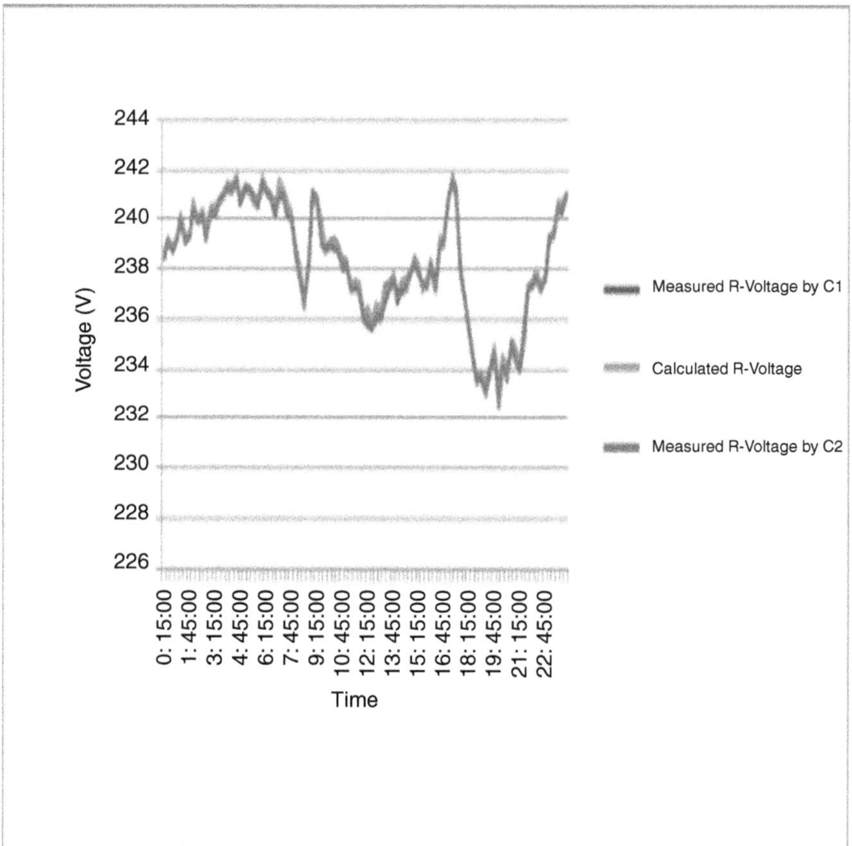

Figure 11.8 Node-02 voltage profiles taken December 18, 2016

network. The algorithm for the proposed network optimization-based load balancing technique is illustrated in Figure 11.13.

11.2.3.2 Generation of Consumer Connection Arrangements among Phases at a Node

Figure 11.14 describes the details of possible consumer connection arrangements at Node-*i* using the proposed algorithm for four consumer connections.

Consumer Connection Arrangements

A set of consumer connections that belong to a certain node are arranged among Phases "R," "Y," or "B" in different ways. Each consumer

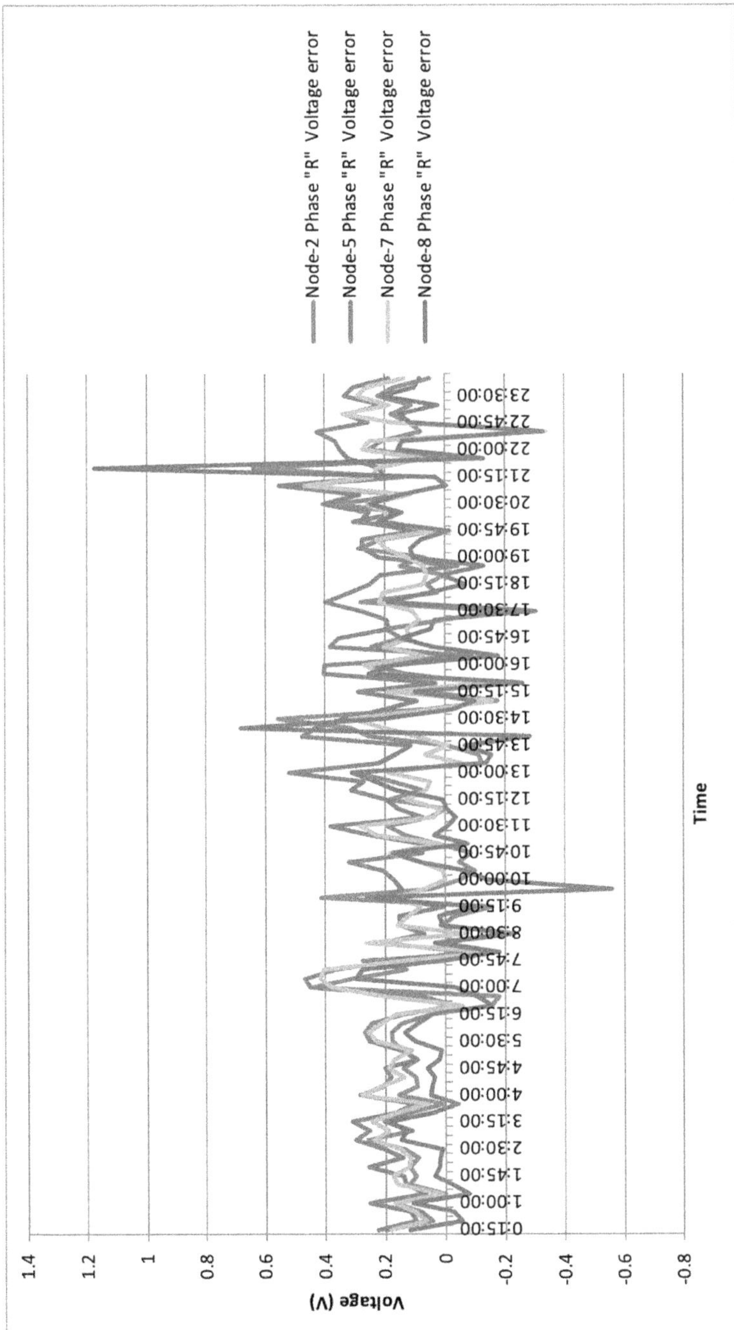

Figure 11.9 Phase “R” voltage error variation

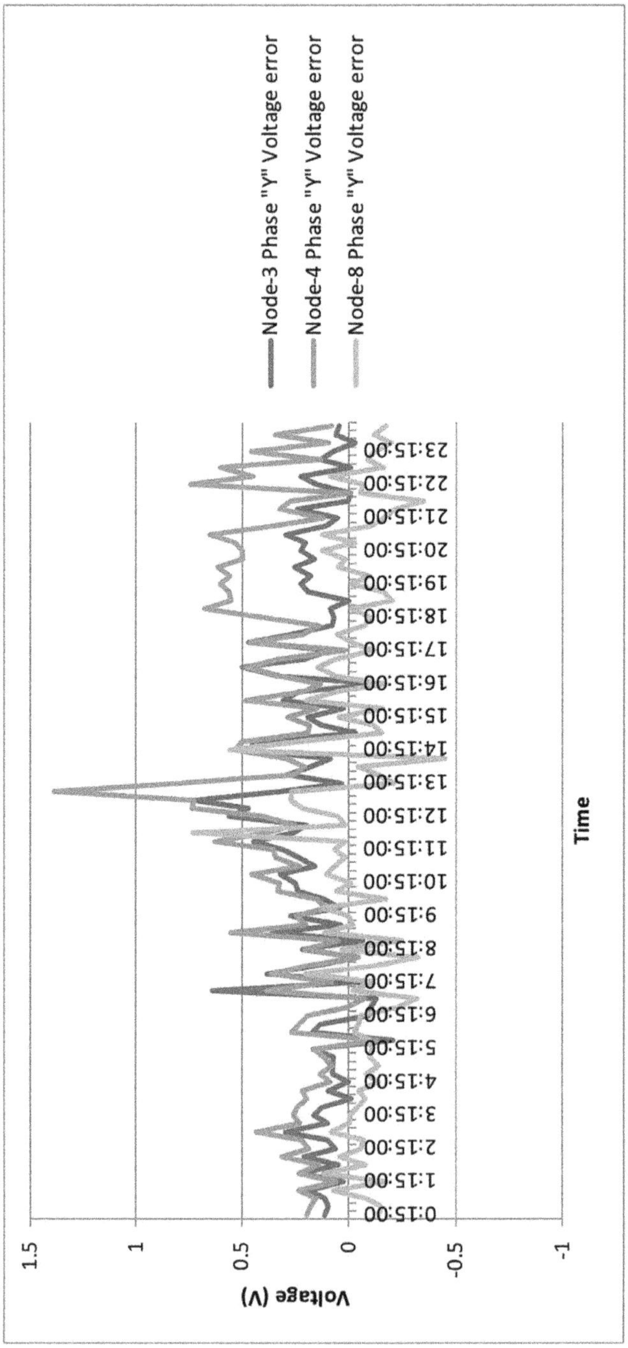

Figure 11.10 Phase "Y" voltage error variation

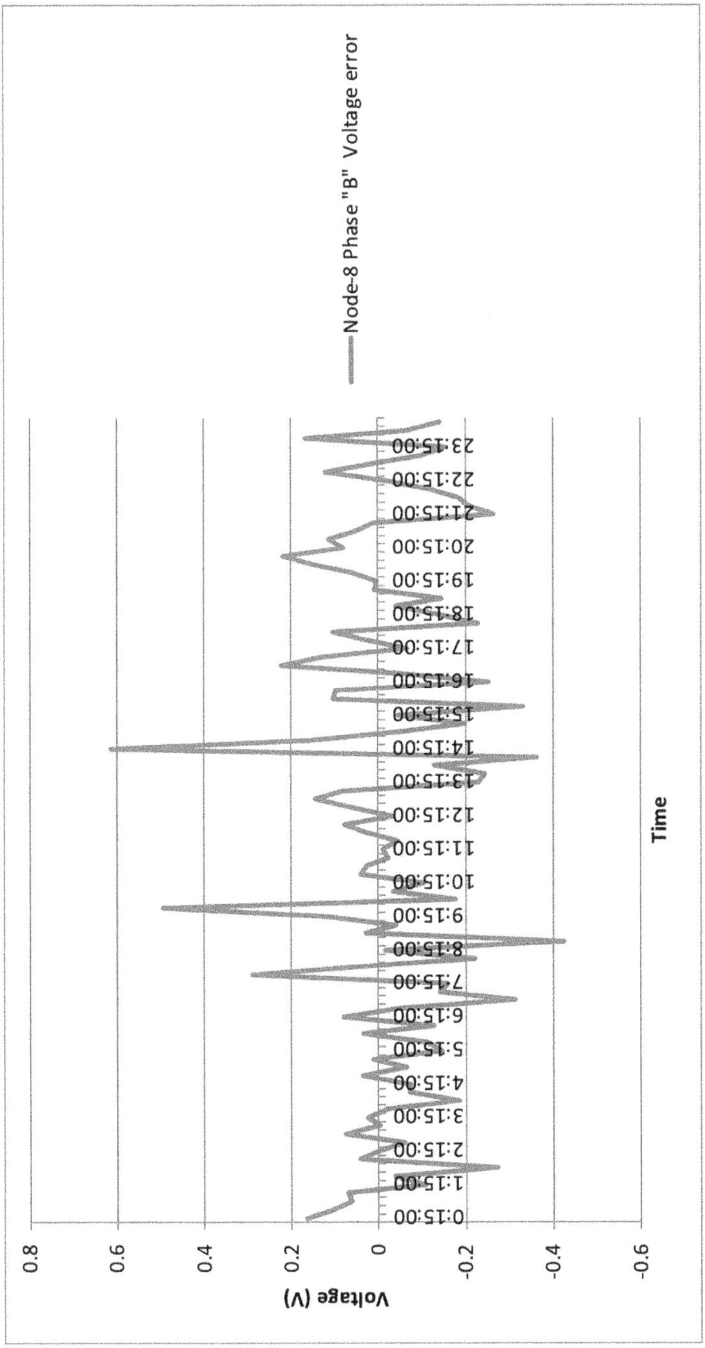

Figure 11.11 Phase "B" voltage error variation

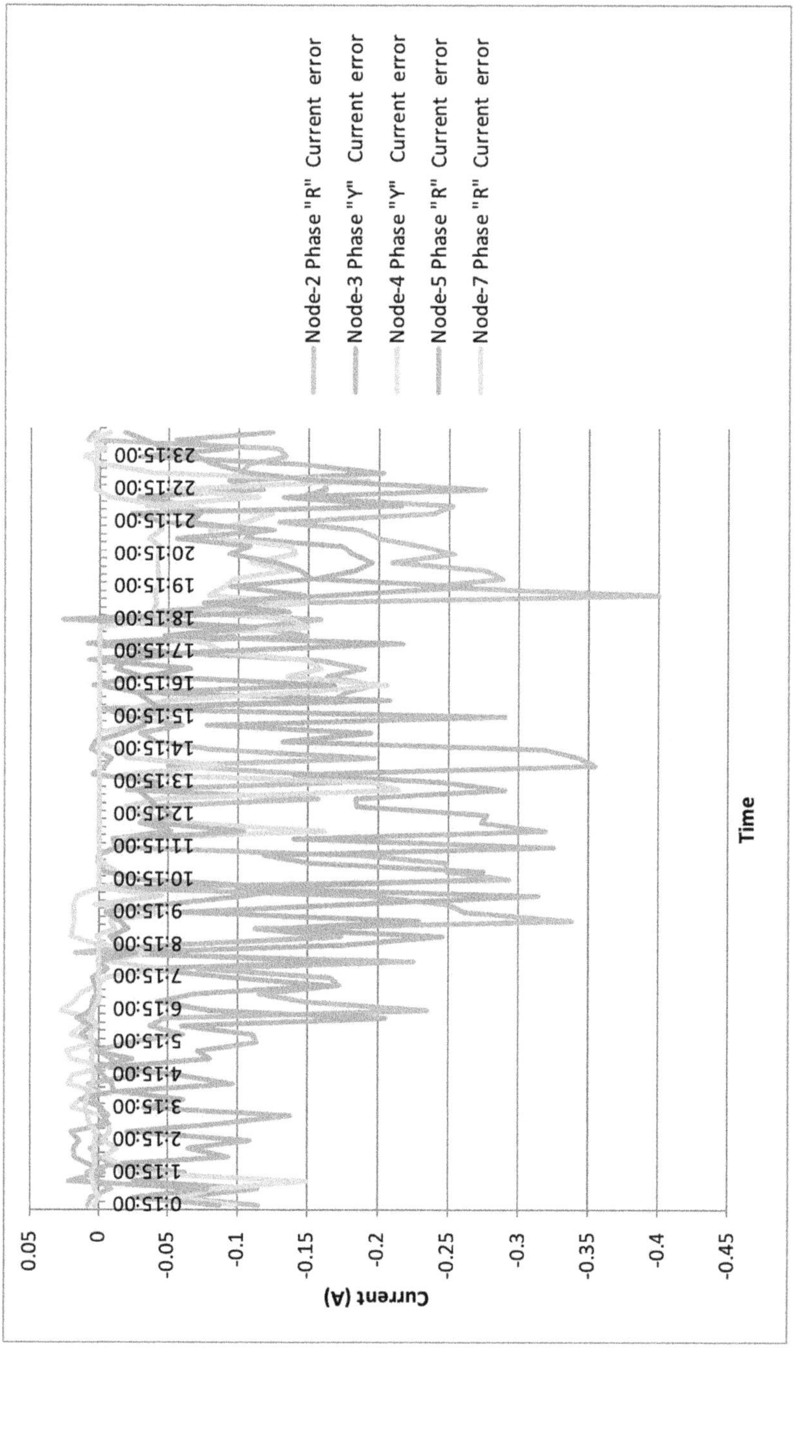

Figure 11.12 Current error variation

Table 11.3 Summary of number of voltage error distribution

Number of measurements taken	*Percentage of the number of error values less than -0.5 V*	*Percentage of the number of error values between -0.5 V and +0.5 V*	*Percentage of the number of error values less than +0.5V*
768	0.1302	95.8333	4.0365

Table 11.4 Summary of number of current error distributions

Number of measurements taken	*Percentage of number of error values less than -0.2A*	*Percentage of number of error values -0.2A between +0.2A*	*Percentage of number of error values less than +0.2A*
480	8.75	91.25	0

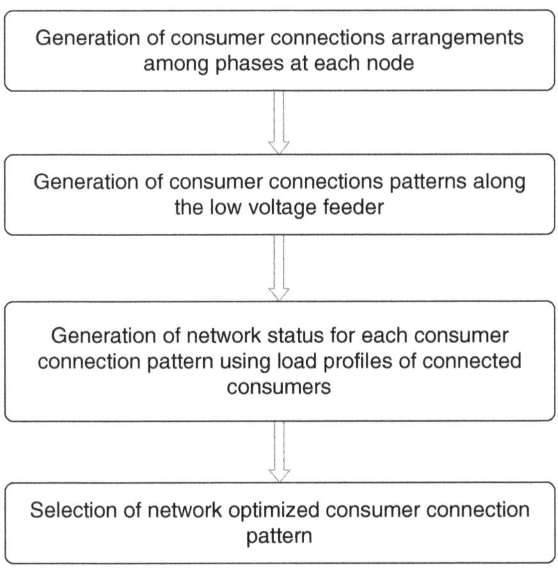

Figure 11.13 Algorithm for the network optimization-based load balancing technique

connection that belongs to a certain node is given an opportunity to be placed at all the phases during possible consumer arrangement generation within the node. The consumer connections are stored in a row matrix. The number of consumer connections is a triple value that generates all possible arrangements among Phase "R," "Y," and "B". Additionally, zero elements are added to complete the number of connections to generate a triple value

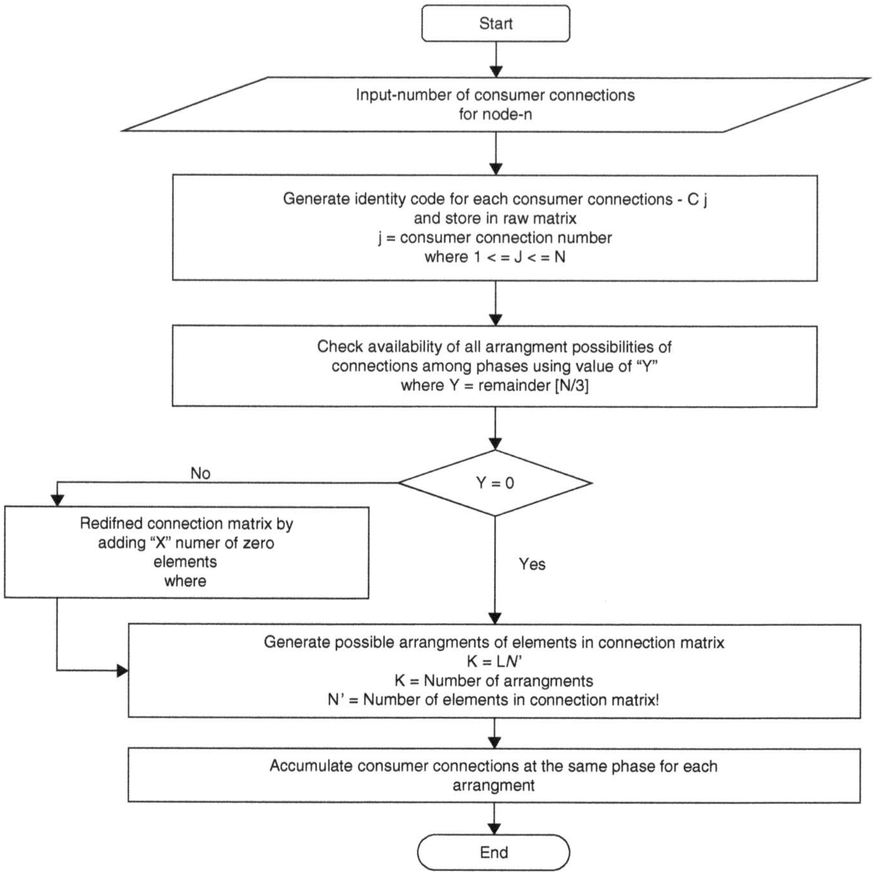

Figure 11.14 **Algorithm for the generation of possible consumer arrangements among phases at a node**

if the criteria are not matched. The elements of such a redefined dataset are ordered in different ways and mathematically known as permutations. For "n" number of items per dataset, $(n) \times (n-1) \times (n-2) \times (n-3) \ldots \times (1) = \lfloor n$ permutations are generated (factorial of "n"). Each zero element is identified as a unique element during data ordering and due to that, some of the arrangements are duplicated. The consumer connections at the same phase for a certain arrangement are accumulated for the corresponding phase before the completion of the generation of possible consumer arrangements at each node.

Table 11.5 Symbolic representations of four consumer connections at Node-*i*

Node-i			
C_1	C_2	C_3	C_4

Table 11.6 Redefined set of consumer connections

R	*Y*	*B*	*R*	*Y*	*B*
C_1	C_2	C_3	C_4	0	0

Table 11.7 Sample of generated connection arrangements

R	*Y*	*B*	*R*	*Y*	*B*
C_1	C_2	C_3	C_4	0	0
C_1	C_2	C_3	0	C_4	0

Step 1:
Table 11.5 represents the symbolic representation of the four consumer connections at Node-*i*.

Step 2:
Check the value of Y = Remainder (N/3), where $N = 4$ for a given case. $Y = 1$ and $X = 3 - Y$; so, $X = 2$. Add two zero elements to the connections matrix.

Step 3:
The consumer connections matrix is redefined by adding the number of zero elements equal to X as shown in Table 11.6.

Step 4:
All possible arrangements for the consumer connections are generated by implementing the above algorithm using MATLAB R2008a. Possible arrangement = $\lfloor 6 = 720$. A sample of a generated connection arrangement is shown in Table 11.7.

Step 5:
Accumulate the consumer connections at the same phase for each arrangement as shown in Table 11.8.

Table 11.8 Accumulated connection arrangements

R	Y	B
$C_1 + C_4$	C_2	C_3
C_1	$C_2 + C_4$	C_3

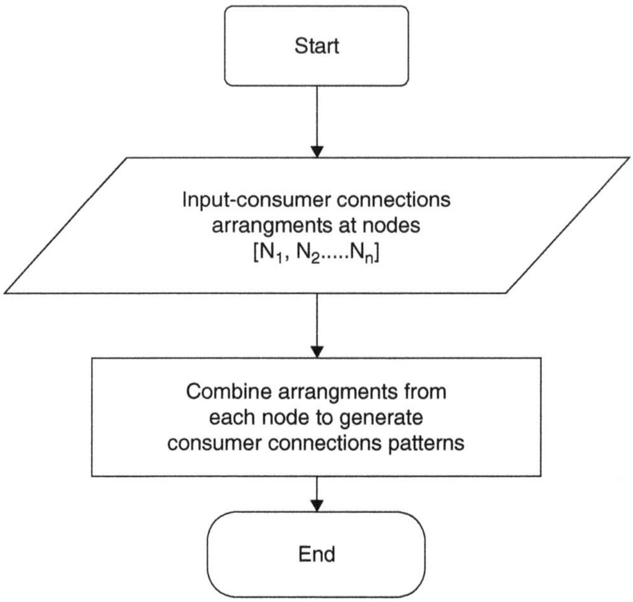

Figure 11.15 Algorithm for consumer connection pattern generation

11.2.3.3 Generation of Consumer Connection Patterns along the LV Feeder

Possible consumer connection patterns are generated by combining one consumer arrangement from each node at a time, as indicated in Figure 11.15. This process continues until all the combinations are generated. The connections are not allowed to swap between poles, which are originally reserved. The total number of generated combinations is the multiplication of the number of phase arrangements at each node:

$$\text{number of patterns} = (N_1) \times (N_2) \times \ldots \times (N_n),$$

where
$N_1, N_2, \ldots, N_n$ number of possible arrangements at each node
$1, 2, \ldots, n$ node number.

As an example, three nodes of the network consist of one consumer each and generate $\lfloor 3 \times \lfloor 3 \times \lfloor 3 = 216$ of consumer connection patterns. The possible patterns for consumer connections are generated by implementing the algorithm using MATLAB R2008a. Table 11.9 indicates a sample of patterns generated by the algorithm.

11.2.3.4 Generation of the Network Status for Each Consumer Connection Pattern Using the Load Profiles of a Connected Consumer

Load profiles are fed into a load flow program to generate the network status for each consumer connection pattern as per Figure 11.16. The load profile is an indication of the electricity consumption of a particular consumer. The time-synchronized load profiles of each consumer belonging to a certain feeder are used to generate the network status for a period when the load profile is defined. The load profiles that show similar or approximately similar behaviors of a consumer are grouped and they are represented by a certain characteristic curve. However, the characteristic curves used for network status generation to represent each consumer are derived from a similar time frame.

11.2.3.5 Selection of the Network Optimized Consumer Assignment Pattern

The selection of the optimum consumer connection pattern depends on the evaluation of the network status generated for each consumer connection pattern for a particular time frame. The objective of distribution network optimization is to minimize the total feeder loss generated through conductors by satisfying the constraints of maintaining node voltages to fulfill the statutory requirements and the span currents below the current carrying capacities of the conductors. The load profiles must belong to the same time frame that is used to select the consumer connection pattern that creates the optimum condition for the network. Cross evaluation is conducted for different time frames to find the best customer assignment.

The optimum status generated pattern is the one that satisfies the objective function and its constraints:

$$\text{Minimize} \sum_{x=a,b,c} \left[\sum_{t=1}^{T} \left[\sum_{i=1}^{n} \left(I_{t,i,x} \right)^2 R_{i,x} \right] \right] \tag{11.2}$$

Subjected to $I_{t,i,x} \leq I_x'$

$$V_{min} \leq V_{t,k,y} \leq V_{max}$$

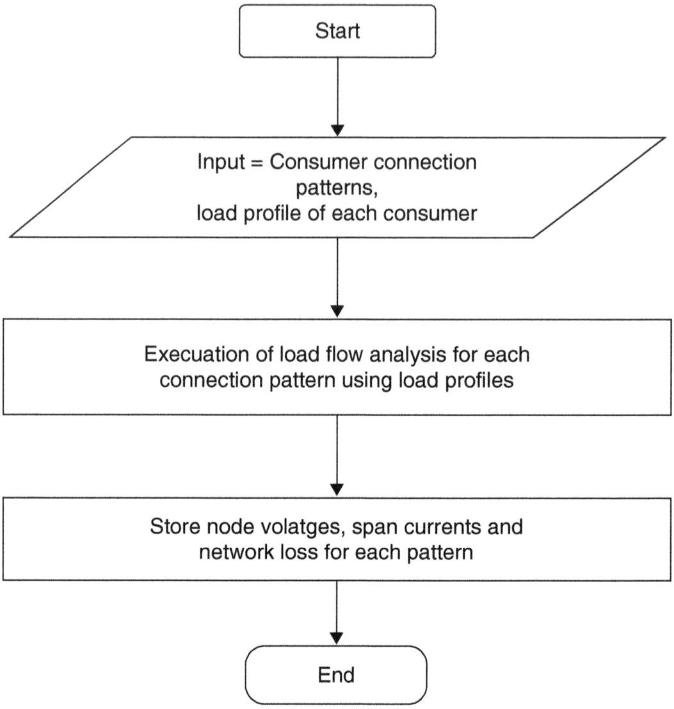

Figure 11.16 Algorithm for network status generation

Table 11.9 Sample of patterns generated by the algorithm

Node-1			*Node-2*			*Node-3*		
R	*Y*	*R*	*R*	*Y*	*B*	*R*	*Y*	*B*
C_1	0	0	C_2	0	0	C_3	0	0
0	C_1	0	0	0	C_2	0	C_3	0

where

I'_x = maximum allowed current at the conductor of Phase x
$I_{t,i,x}$ = current at the conductor of the ith span of Phase x at time t
$t = 1, 2, 3\ldots, Tx = a, b, c, n$
$R_{i,x}$ = resistance of the conductor of the i^{th}span of Phase x
i = span number, $i = 1, 2, 3\ldots, n$
$V_{t,k,y}$ = voltage at k^{th} node of Phase y at time t
$y = a, b, ck$ = node number

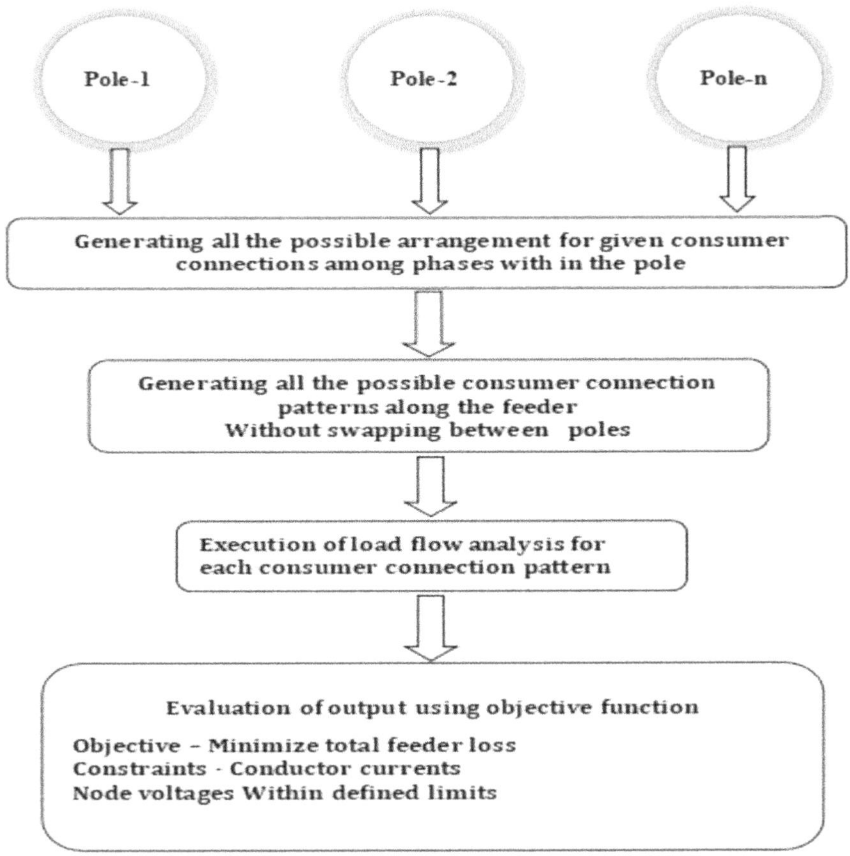

Figure 11.17 Summary of the load balancing algorithm

V_{max} = maximum allowed voltage
V_{min} = minimum allowed voltage.

A summary of the load balancing algorithm is given in Figure 11.17.

11.3 RESULTS AND DISCUSSION

11.3.1 Application of the Proposed Load Balancing Technique for an LV Feeder

11.3.1.1 Existing Status Analysis of a Sample Feeder

The LV feeder used for the power flow algorithm validation in Section 11.2 is selected to analyze the performance of the proposed load balancing algorithm. The voltages, currents, and power consumptions for the connected consumers and the status of the feeder are recorded for the period of December 18, 2016 to January 11, 2017. The electrical characteristic of the considered feeder is analyzed to understand the existing status before further improvement.

The variation of the phase current drawn from the source and neutral current return as shown in Figures 11.18 and 11.19, respectively, provide some indication of the load arrangement within the feeder. The considered feeder is newly constructed and consumers are attached only for Phases "R" and "Y." The current profiles taken on December 18, 2016 are shown in Figure 11.20. It is evident that a considerable amount of return current exists relative to the number of consumers connected according to Figure 11.21. An analysis of the energy loss in such a feeder is an important area to study and correlate with the current profiles of the conductors. The voltage variation is another important factor to be considered for such a network.

The Phase "R" and Phase "Y" voltages at the feeder end follow a decreasing pattern compared to the feeder start, but Phase "B" shows the feeder end voltage rise compared to the feeder start when the neutral current magnitude is considerable, as shown in Figure 11.22. These unbalanced neutral currents originate due to the consumer connection pattern along the feeder, unpredictable nature of the power consumption of connected

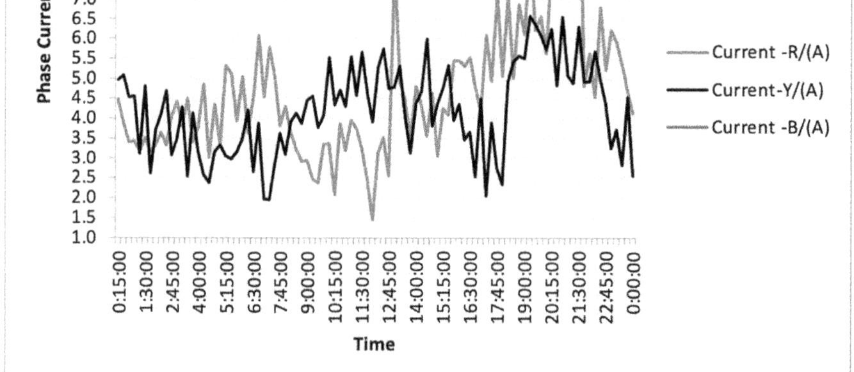

Figure 11.18 Phase current variation of the feeder start taken on December 18, 2016

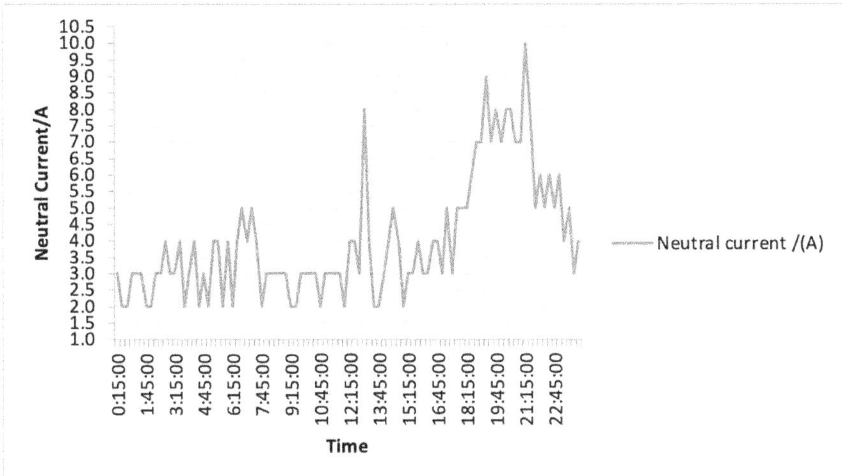

Figure 11.19 Neutral current variation of the feeder start taken on December 18, 2016

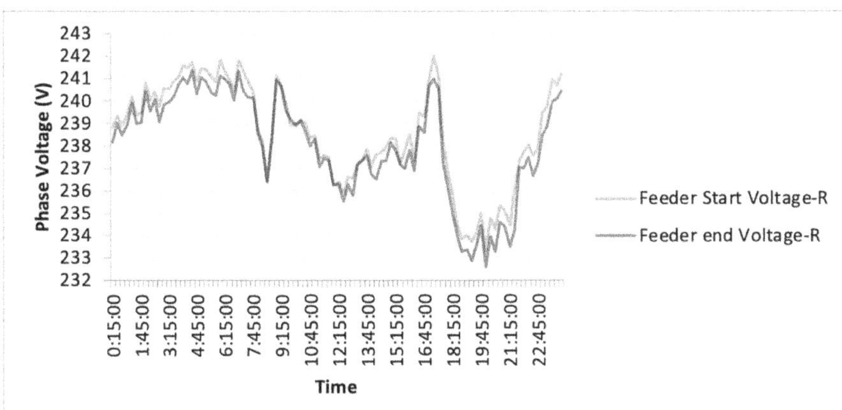

Figure 11.20 Voltage variation at the feeder start and end of phase R taken on December 18, 2016

individual consumers, and harmonic currents. However, less probability is available for harmonic current generation from considered domestic loads.

Hence, the consumer connection pattern and stochastic nature of power consumption contribute significantly to unbalancing situation creation for less harmonic polluted feeders. Figure 11.23 shows the current profiles taken on December 18, 2016 when nine consumers are connected. The stochastic nature of power consumption is clearly identified from these profiles

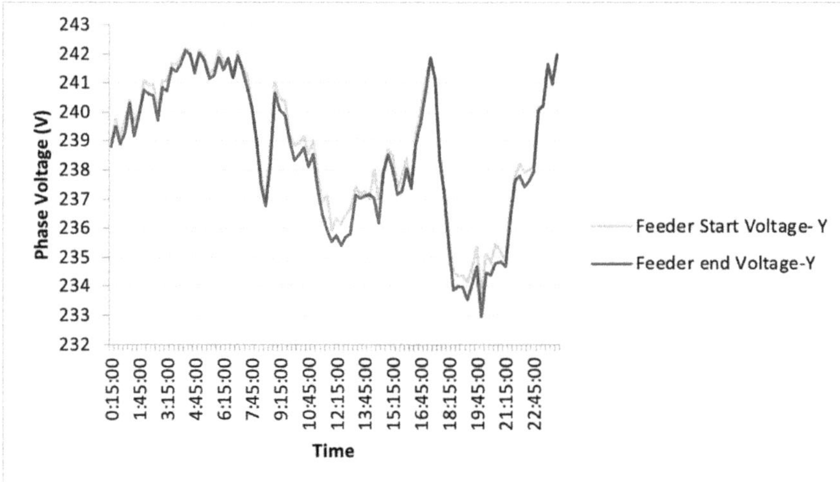

Figure 11.21 Voltage variation at the feeder start and end of phase Y taken on December 18, 2016

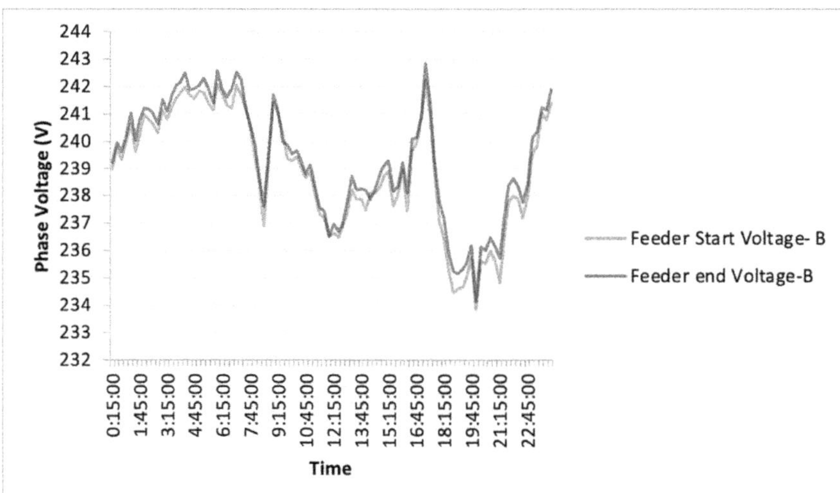

Figure 11.22 Voltage variation at the feeder start and end of phase B taken on December 18, 2016

and the characterized load profile generation is challenging in nature. The objective of this research is to analyze the status of feeders for different consumer connection patterns using the characterized electricity consumption profile of connected consumers. Generally, the arrangement of consumer connections is referred to as load balancing within this context. The voltages and currents vary along the feeder throughout the day. The frequency of the deviation of voltages and currents from the defined ranges and energy losses due to the resistance of the feeder for certain connection patterns are used for the consumer connection pattern evaluation process.

11.3.1.2 Application of the Proposed Algorithm

Consumer Assignment Patterns

In the considered network, there are nine single-phase consumers attached to the feeder. The proposed pattern generation algorithm is implemented using MATLAB R2008a. Tables 11.10–11.13 indicate the summarized output of the proposed pattern generation algorithm. This information consists of possible connection arrangements at each node and the number of consumer connection patterns along the LV feeder.

Sample of Consumer Arrangements at Nodes

11.3.1.3 Generation of the Network Status for Each Consumer Connection Pattern Using the Load Profiles of Connected Consumers

The accuracy of the characterized load profile is important for the generation of a more realistic network status for the evaluation. The load profiles of a certain consumer may vary due to the consumer's stochastic nature of electricity consumption. The most reasonable characteristic curves that are used to represent a certain consumer are yielded for a more realistic output after load flow analysis for each consumer assignment pattern. This leads to further study on load clustering and that is beyond the scope of this chapter. However, a simple method, as in Figure 11.24, is proposed for the generation of consumer characteristic curves to evaluate the performance of the proposed algorithm.

Sample of Consumer Connection Patterns

Collection of Actual Load Details

The active power and data for the reactive power calculation are recorded at the consumer premises. They are used as input to the characteristic curve generation process.

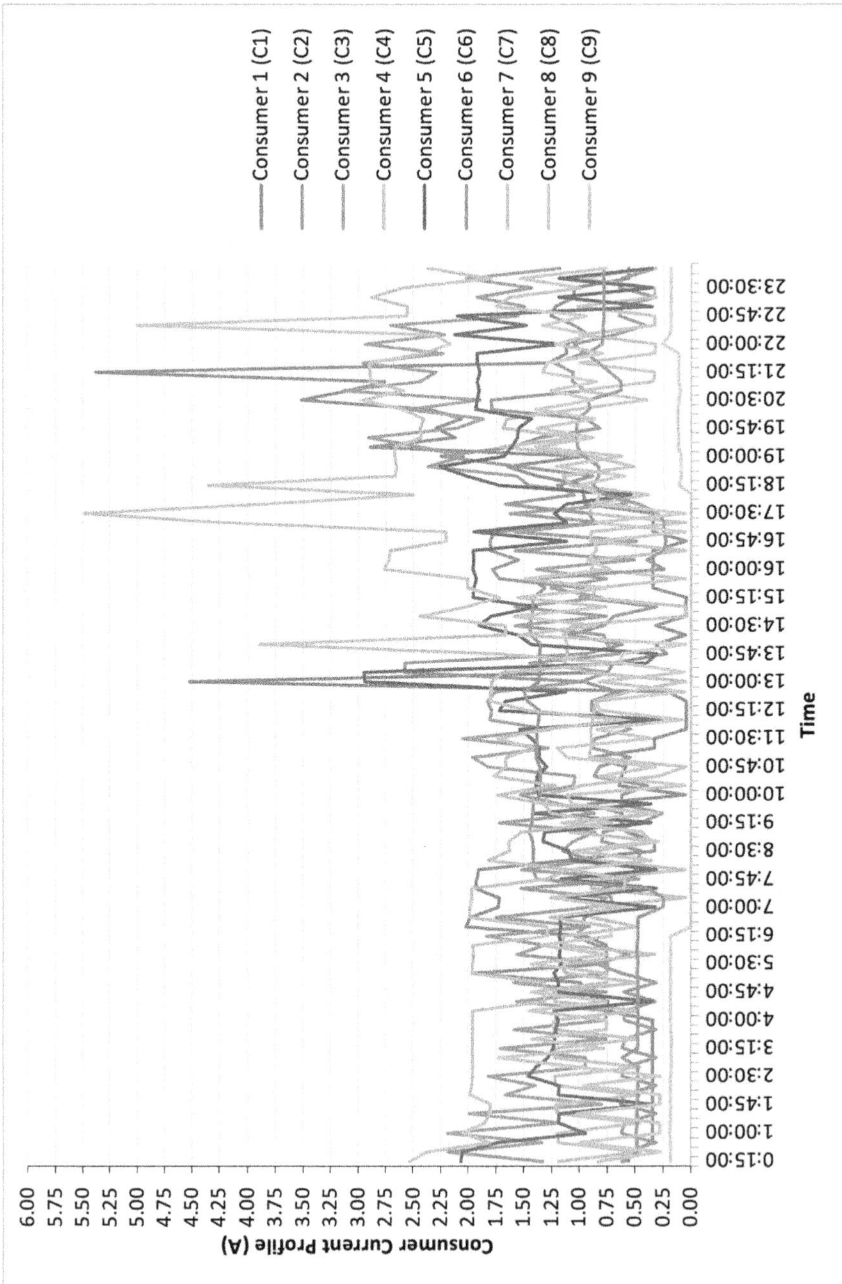

Figure 11.23 Consumer current profile taken on December 18, 2016

Table 11.10 Summarized output of the pattern generation algorithm

Node number	*Number of connections at the node*	*Number of connection arrangements at each node after redefining of dataset*	*Number of consumer connection patterns*
Node-01			6 × 6 × 6 × 6 × 6 = 7776
Node-02	2	Factorial(3) = 6	
Node-03	2	Factorial(3) = 6	
Node-04	2	Factorial(3) = 6	
Node-05	2	Factorial(3) = 6	
Node-06			
Node-07			
Node-08	1	Factorial(3) = 6	
Node-09			

Table 11.11 Sample of consumer arrangements at Node-5

	Node-5	
R	*Y*	*B*
0	C8	C7
0	C7	C8
C8	0	C7
C8	C7	0
C7	C8	0
C7	0	C8

Table 11.12 Sample of consumer arrangements at Node-7

	Node-7	
R	*Y*	*B*
0	0	C9
0	C9	0
0	0	C9
0	C9	0
C9	0	0
C9	0	0

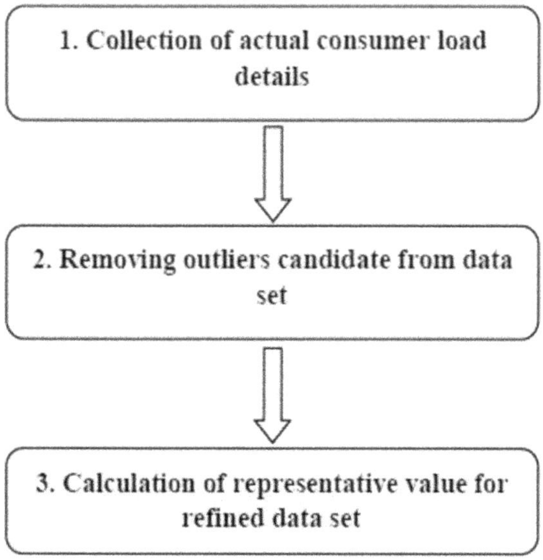

Figure 11.24 Characteristic curve generations for consumers

Removing Outliers from the Dataset

The difference between the maximum and median values of the sorted data is used for removing outliers from the dataset belonging to a particular time interval. Due to the stochastic electrical consumption behavior of consumers, values greater than 25 times the interquartile range of the sorted dataset are considered as outliers for that dataset. The defined range for outlier removal is selected to generate the characteristic curves that provide a monthly energy consumption that approximately matches the measured energy consumption for that particular month. The same procedure is followed for outliers removed that belong to the minimum values of the measured dataset. These values are removed from the original dataset.

Calculation of the Representative Value for a Refined Dataset

The centroid of each dataset is calculated to select the representative value for the considered dataset that belongs to a particular time interval. The above procedure is repeated for each time interval of the load profiles to generate the characteristic curve for the considered period. The active power and reactive power datasets are considered separately and combined to

Table 11.13 Sample of consumer connection patterns along the LV feeder

Node-2			*Node-3*			*Node-4*			*Node-5*			*Node-7*		
R	*Y*	*B*	*R*	*Y*	*B*	*R*	*Y*	*B*	*R*	*Y*	*B*	*R*	*Y*	*B*
C1	0	C2	C3	0	C4	C6	C5	0	C8	0	C7	0	C9	0
C1	0	C2	C3	0	C4	C6	C5	0	C8	0	C7	C9	0	0
C1	0	C2	C3	0	C4	C6	C5	0	C8	0	C7	C9	0	0
C1	0	C2	C3	0	C4	C6	C5	0	C8	C7	0	0	0	C9
C1	0	C2	C3	0	C4	C6	C5	0	C8	C7	0	0	C9	0
C1	0	C2	C3	0	C4	C6	C5	0	C8	C7	0	0	0	C9
C1	0	C2	C3	0	C4	C6	C5	0	C8	C7	0	0	C9	0
C1	0	C2	C3	0	C4	C6	C5	0	C8	C7	0	C9	0	0
C1	0	C2	C3	0	C4	C6	C5	0	C8	C7	0	C9	0	0
C1	0	C2	C3	0	C4	C6	C5	0	C7	C8	0	0	0	C9
C1	0	C2	C3	0	C4	C6	C5	0	C7	C8	0	0	C9	0
C1	0	C2	C3	0	C4	C6	C5	0	C7	C8	0	0	0	C9
C1	0	C2	C3	0	C4	C6	C5	0	C7	C8	0	0	C9	0
C1	0	C2	C3	0	C4	C6	C5	0	C7	C8	0	C9	0	0
C1	0	C2	C3	0	C4	C6	C5	0	C7	C8	0	C9	0	0
C1	0	C2	C3	0	C4	C6	C5	0	C7	0	C8	0	0	C9
C1	0	C2	C3	0	C4	C6	C5	0	C7	0	C8	0	C9	0
C1	0	C2	C3	0	C4	C6	C5	0	C7	0	C8	0	0	C9
C1	0	C2	C3	0	C4	C6	C5	0	C7	0	C8	0	C9	0
C1	0	C2	C3	0	C4	C6	C5	0	C7	0	C8	C9	0	0
C1	0	C2	C3	0	C4	C6	C5	0	C7	0	C8	C9	0	0
C1	0	C2	C3	0	C4	C5	C6	0	0	C8	C7	0	0	C9
C1	0	C2	C3	0	C4	C5	C6	0	0	C8	C7	0	C9	0
C1	0	C2	C3	0	C4	C5	C6	0	0	C8	C7	0	0	C9

develop an apparent power characteristic curve. The characteristic curves are generated for each consumer for the working day using this method. Figures 11.25 and 11.26 are the characteristic curves generated using the above proposed methods. The current profile at the feeder start for the existing connection pattern is shown in Figure 11.27.

It is important to consider the network status for the existing consumer connection pattern using the generated characterized load profiles in order to compare it with those for the other possible consumer connection patterns. The nominal voltages are used as source voltages for each phase. Table 11.14 shows a summary of the evaluation for the considered feeder using characteristic curves that represent the working days from each consumer. From the results, 0.097561 kWh is the loss for the day and the monthly value is 2.9268 kWh. A considerable number of consumers are connected to the LV feeder and the length is around 200 m to 400 m. The value of the loss for

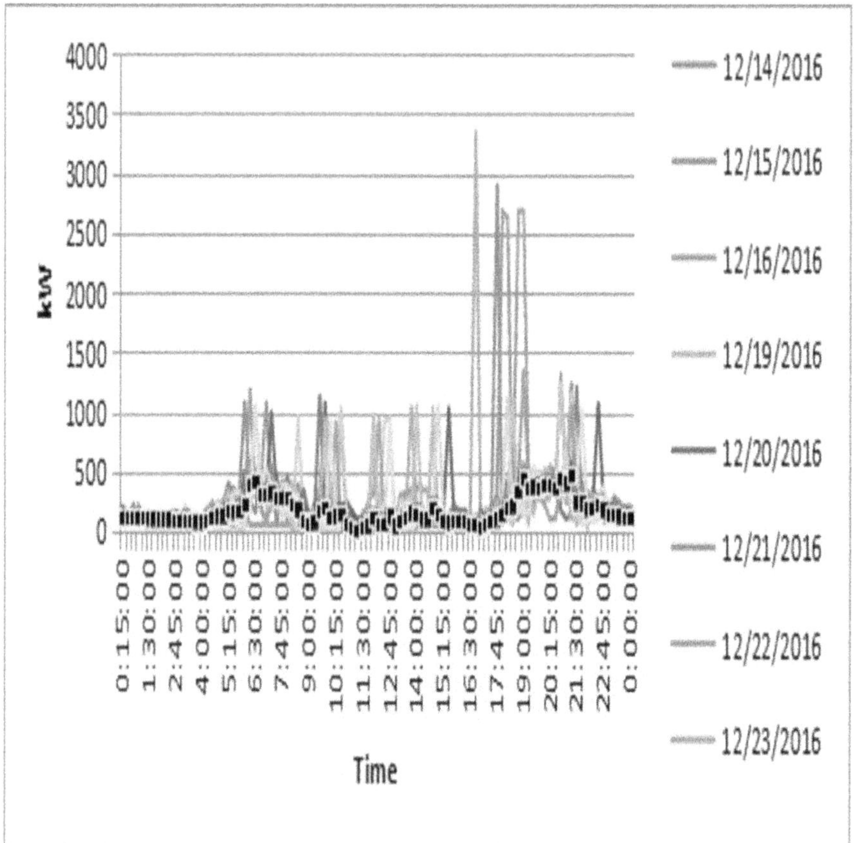

Figure 11.25 Measured and characteristic kW profile – consumer C1

such a feeder is considerable and the total impact for the entire distribution area is significant.

11.3.1.4 Generation of the Network Status for Each Consumer Assignment Pattern Using Representative Load Profiles

A total of 7776 consumer assignment patterns are generated for the selected LV feeder, which comprises nine consumers. The network status for these patterns for a certain time frame is evaluated for a comprehensive analysis of this feeder. The proposed algorithm is implemented using MATLAB 7.6.0 (R2008). The network status for a sample of 1000 patterns are generated and stored for evaluation.

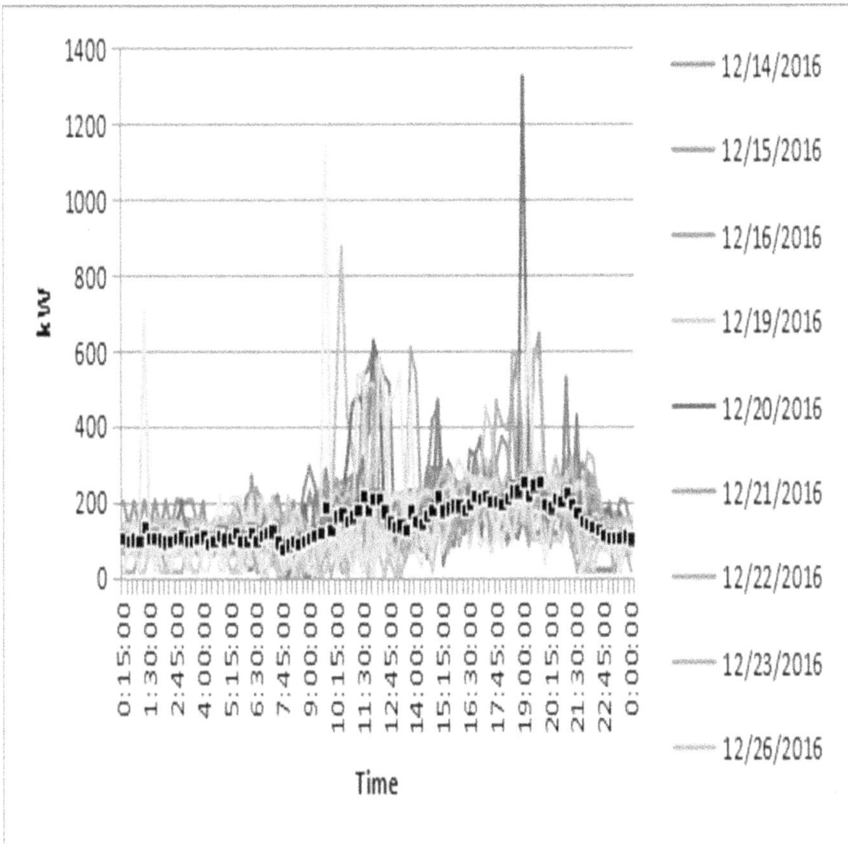

Figure 11.26 Measured and characteristic kW profile – consumer C2

11.3.2 Evaluation of Patterns

The stored dataset includes the feeder loss, frequency of violation of the maximum voltage threshold, frequency of violation of the minimum voltage threshold, and frequency of violation of the conductor current capacity.

11.3.2.1 Sample Results of the Proposed Algorithm

Table 11.15 includes the sample results from the 1000 consumer assignment pattern evaluation. The network losses are relevant to a characteristic load profile of nine consumers for a period of one day for the considered patterns that are generated by the proposed algorithm. The feeder loss

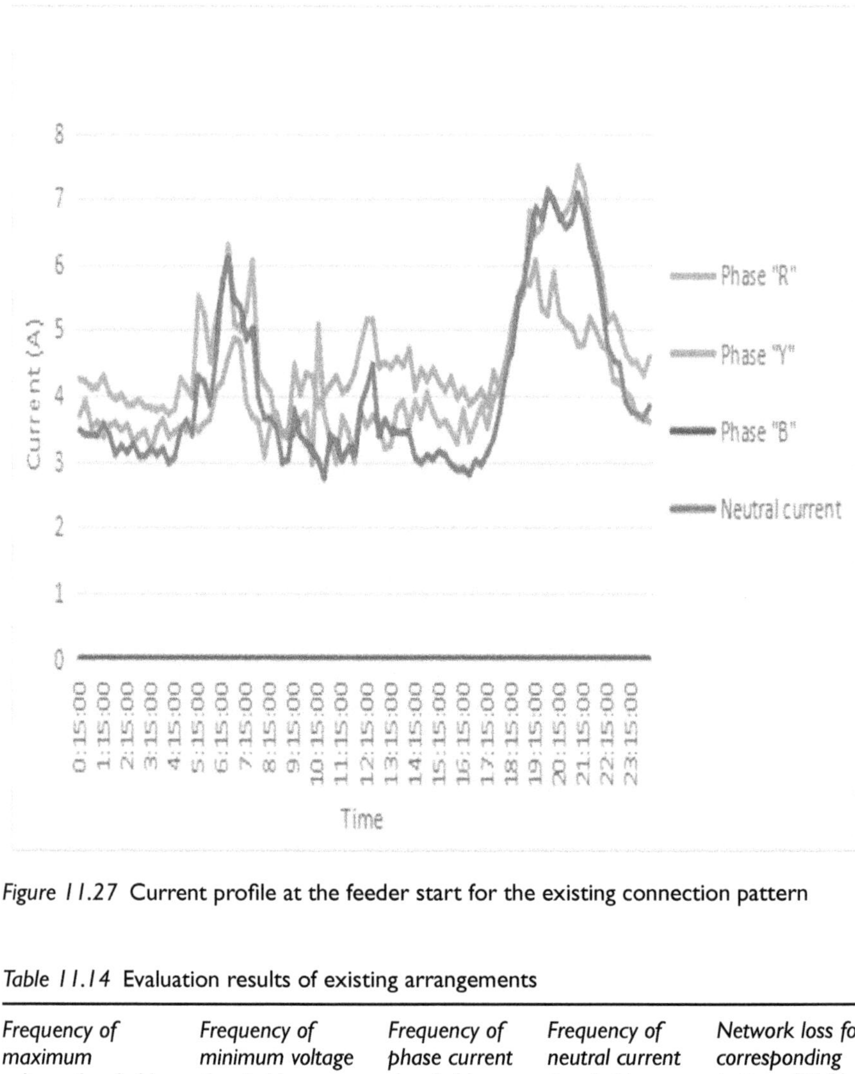

Figure 11.27 Current profile at the feeder start for the existing connection pattern

Table 11.14 Evaluation results of existing arrangements

Frequency of maximum voltage threshold violations	*Frequency of minimum voltage threshold violations*	*Frequency of phase current threshold violations*	*Frequency of neutral current threshold violations*	*Network loss for corresponding pattern (kWh) per day*
0	0	0	0	0.097561

variation relevant to a sample pattern is as shown in Figure 11.28. The minimum loss occurrence pattern as shown in Table 11.16 is selected as per the definition of the objective function and constraints. The minimum loss that occurred among the evaluated 1000 pattern is 0.041564 kWh

Table 11.15 Sample results of the proposed algorithm

Pattern No.	*Frequency of maximum voltage threshold violations*	*Frequency of minimum voltage threshold violations*	*Frequency of phase current threshold violations*	*Frequency of neutral current threshold violations*	*Network loss to corresponding pattern (kWh) per day*
1	0	0	0	0	0.099603
2	0	0	0	0	0.097118
3	0	0	0	0	0.099603
4	0	0	0	0	0.097118
5	0	0	0	0	0.095344
6	0	0	0	0	0.095344
7	0	0	0	0	0.089491
8	0	0	0	0	0.089172
9	0	0	0	0	0.089491
10	0	0	0	0	0.089172
11	0	0	0	0	0.085863
12	0	0	0	0	0.085863
13	0	0	0	0	0.087036
14	0	0	0	0	0.08347
15	0	0	0	0	0.087036
16	0	0	0	0	0.08347
17	0	0	0	0	0.084282
18	0	0	0	0	0.084282
19	0	0	0	0	0.061796
20	0	0	0	0	0.063001
21	0	0	0	0	0.061796
22	0	0	0	0	0.063001
23	0	0	0	0	0.060766
24	0	0	0	0	0.060766
25	0	0	0	0	0.044793
26	0	0	0	0	0.045377
27	0	0	0	0	0.044793
28	0	0	0	0	0.045377
29	0	0	0	0	0.045293
30	0	0	0	0	0.045293

per day. So, the evaluation of the entire possible pattern is an important requirement for decision-making about the consumer arrangements of the LV feeder to maintain the network at the optimum condition as indicated in Figure 11.29.

Hence, the currents that are shared between the phases and neutral current are also reduced compared to the existing arrangement. The feeder losses are related to conductor currents and these new current profiles help to improve the network condition.

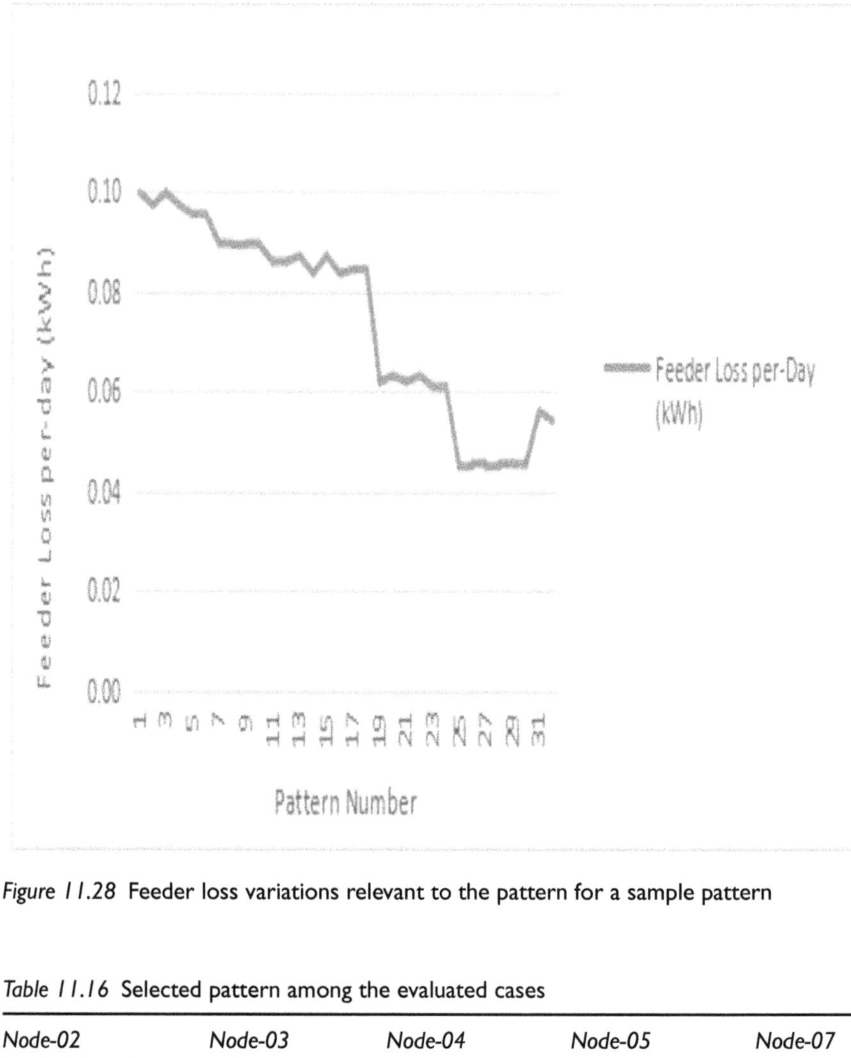

Figure 11.28 Feeder loss variations relevant to the pattern for a sample pattern

Table 11.16 Selected pattern among the evaluated cases

Node-02			*Node-03*			*Node-04*			*Node-05*			*Node-07*		
R	*Y*	*B*	*R*	*Y*	*B*	*R*	*Y*	*B*	*R*	*Y*	*B*	*R*	*Y*	*B*
0	C2	C1	C4	C3	0	0	C5	C6	C7	0	C8	0	C9	0

11.4 CONCLUSION AND RECOMMENDATIONS

Electricity distribution network modeling is one of the critical steps in the steady-state analysis of a distribution system. In this chapter, the system is modeled as five wires and the ground path is limited to a resistive

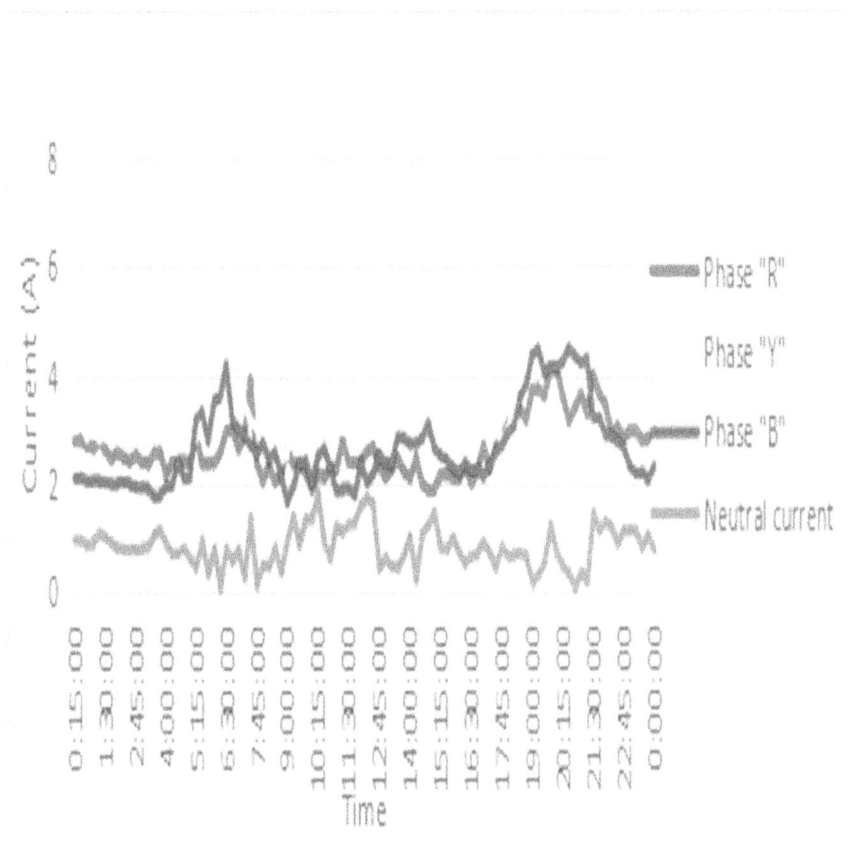

Figure 11.29 Current profile at the feeder start for the selected connection pattern

component. A customized load flow analysis method for the considered network model is developed. The current division between the neutral conductor and the ground path is calculated in the backward sweep at each iteration step. The self-impedance of the ground path is a required parameter for this calculation. Different assumptions are used by researchers in their scope and their validity is limited. The ground path is approximated to a resistive component. The validation of the proposed load flow program is done. The calculated and measured values are compared with each other. The distribution of error values of the voltages and currents are around 95% voltage errors that fall between -0.5V and +0.5V, and 91% current errors that fall between -0.2V and +0.2V. The accuracy of the impedance model and errors in the measurements significantly contribute to deviations. The measurements are averaged values for a 15-minute time interval. Time synchronization of all meters is required for a better analysis of a particular time interval during validation. However, the developed network model and load flow analysis method is sufficient for the evaluation of the proposed network optimization method. Load balancing is defined as the consumer connection arrangement within the electricity distribution context. Different consumer connection patterns create different statuses in the LV feeder. The network status varies throughout the day based on the electricity consumption of each consumer for a particular consumer attached pattern. The proposed algorithm generates the entire consumer connection pattern for a particular set of consumers. The voltages and currents deviate from the defined ranges of each pattern at different time intervals. The network statuses for each consumer connection pattern are generated using the characteristic curves of the electricity consumption of each consumer. The total feeder loss is recorded for each consumer pattern. These values are used to select the optimum consumer connection patterns for the network. This chapter also includes the application of the proposed method to a sample feeder. A detailed analysis of the relationship between consumer connection patterns and network status is an important area in the electricity distribution sector for providing better service to customers. In this chapter, sample analysis is done only for nine consumers connected to a small feeder and considerable feeder loss is witnessed. When considering the entire distribution network, the total energy loss is a significant value and the economic impact cannot be neglected. So, the proposed method may be conveniently utilized by distribution network operators to maintain their network at the optimum condition.

REFERENCES

[1] Ceylon Electricity Board, Statistical Digest, 2013.
[2] Ceylon Electricity Board, Annual report, 2012.
[3] Sri Lanka Energy Sector Development Plan for a Knowledge-based Economy 2015-2025.
[4] Public Utilities Commission of Sri Lanka. *Distribution code of Sri Lanka*, 2012.
[5] Kersting WH. "Distribution system modeling and analysis". In *Electric power generation, transmission, and distribution* (pp. 26–31). CRC press, 2018.
[6] Anderson PM. *Analysis of Faulted Power Systems* (pp.71–83). IEEE Press Power Systems Engineering Series, 1995, ISBN 0-7803-1145-0.
[7] Carson JR. Wave propagation in overhead wires with ground return. ;*The Bell System Technical Journal*, ;1926, *5*(4), 539–54.
[8] Short TA. *Electric power distribution handbook*. CRC Press, 2004. - ISBN 0-8493-1791-6.
[9] Horton R, Sunderman WG, Arritt RF, Dugan RC. Effect of line modeling methods on neutral-to-earth voltage analysis of multi-grounded distribution

feeders. In *Power Systems Conference and Exposition (PSCE), 2011 IEEE/PES* Mar 20 (pp. 1–6). IEEE, 2011.

[10] Kersting WH. "A three-phase unbalanced line model with grounded neutrals through a resistance". In *Power and Energy Society General Meeting-Conversion and Delivery of Electrical Energy in the 21st Century, 2008 IEEE* Jul 20 (pp. 1–2). IEEE, 2008.

[11] Urquhart AJ. *Accuracy of low voltage electricity distribution network modelling* (Doctoral dissertation, © Andrew Urquhart).

[12] Dashti R, Daisy M, Shaker HR, Tahavori M. Impedance-based fault location method for four-wire power distribution networks. *IEEE Access*, 2018, 6, 1342–9.

[13] Murari K, Padhy NP. An efficient graphical method for load-flow solution of distribution systems. *Arabian Journal for Science and Engineering*, 2019, *44*(3), 1791–1808.

[14] Bazrafshan M, Gatsis N. Convergence of the Z-bus method for three-phase distribution load-flow with ZIP loads. *IEEE Transactions on Power Systems*. 2018, 33(1), 153–65.

[15] Olivier F, Fonteneau R, Ernst D. Modelling of three-phase four-wire low-voltage cables taking into account the neutral connection to the earth. In *Proceedings CIRED Workshop,* 2018.

[16] Afolabi OA, Ali WH, Cofie P, Fuller J, Obiomon P, Kolawole ES. Analysis of the load flow problem in power system planning studies. *Energy and Power Engineering*, 2015, 7(10), 509.

[17] Rupa JM, Ganesh S. Power flow analysis for radial distribution system using backward/forward sweep method. *International Journal of Electrical, Computer, Electronics and Communication Engineering*, 2014, 8(10), 1540–4.

Chapter 12

IoT-Based Monitoring and Protection of the Transformer

Aadi Ashutosh Chauhan, Rohan Bhojwani, Aayush Karthikeyan, K. T. M. U. Hemapala, O.V. Gnana Swathika

12.1 INTRODUCTION

Transformers play a crucial role in energy systems, distribution systems, smart grids, and microgrids. Transformers are static electromechanical devices that are very costly and are an indispensable part of the transmission system. As transformer applications are increasing, it is necessary to protect transformers.

Due to daily population growth, the electrical load on the power grid is increasing rapidly. As load demand increases, the transformer becomes overloaded, causing various types of internal failures in the transformer. There is an increasing need to control transformer overload and fault current trips to avoid transformer damage. Transformers are protected with differential protection, actuators, mechanical switches, programmable logic controller (PLC) switches, and microprocessor-based protection systems. However, microprocessor-based protection systems are preferred during the development of hardware apparatus in this chapter. The objective is to make a system that detects a fault in the system and signals to the main server so that required actions are taken accordingly; hence, it helps in increasing the lifespan of the transformer. Moreover, it also helps to make the system a bit less complex and helps to reduce manpower so that the chances of the occurrence of a fault are reduced and solving the problem becomes easier.

12.2 LITERATURE REVIEW

Temperature sensors, potential sensors, and current sensors are used to monitor real-time factors, such as temperature, current, and potential, so that immediate steps may be taken to avert problems. The Global System for Mobile Communications (GSM)-based data transmission system is the most important technology used in this case, as it allows for a more reliable network for data transmission [1].

The Internet of Things (IoT)-based monitoring system, which checks parameters such as current, voltage, temperature, and load capabilities with

 DOI: 10.1201/9781003537878-12

the help of various sensors, are used for preserving transformer health. If any of these parameters exceed the threshold value, information is sent to an Internet Protocol (IP) address via the ESP8266 module using the HTTP protocol, which displays real-time data in chart form on any web-connected device. The system is also set up in such a way that it corrects some flaws [2]. IoT-based edge computing technology is used to check transformer health on a regular basis. The researchers use a windowing approach to split the buzzing sound into 2 s segments after collecting data from both normal and damaged transformers. Mel-frequency cepstral coefficients are used to extract information from these sounds and a dataset is created. The support vector machine (SVM) classifier is then used to train this dataset, allowing it to determine the values for which the transformer is healthy or not [3].

Some common tests must be performed on the transformer to ensure correct operation and increase its operational time. Regular diagnostic tests utilizing IoT sensors make remote monitoring and proper transformer operation possible [4]. When it comes to transformer health, defining suitable parameters is critical. Conditional anomaly detection, correlation, evidence combination, diagnostics, health index estimation, and prognostics modules have been used to gain a better understanding of transformer health monitoring analytics. These tools are applied to the sensor's data and aid in the early detection of malfunctions [5].

The development of a transformer health index using neural networks has been found to be quite effective. To determine health indicators, a feed forward artificial neural network (FFANN) is utilized with parameters such as acidity, water content, hydrogen content, breakdown voltage, and total solids in oil for each transformer [6].

IoT can be used to make transformer health monitoring easier using sensor monitor temperature, oil level, and vibration. This information is given to the control, which checks if any of the threshold values are exceeded, and if so, a message is delivered to the appropriate body [7].

Dissolved gas analysis (DGA), dielectric loss, breakdown voltage, acid value, micro-water content, furfural content, and other elements are subjected to correlation analysis. An index system and a health status checking model for the distribution transformer are built using a combination of qualitative analysis and the analytic hierarchy process approach to establish the weight of the index [8].

Voltage, current (overvoltage, undervoltage, and overcurrent), and other equipment parameters are all measured with sensors. These data are sent to the microcontroller, which validates the parameters before transmitting them to the IoT web server using Adafruit software, guaranteeing that the operator obtains the correct data [9].

Internal transformer faults can be identified with the use of IoT. Sensors are used to detect transformer malfunctions, which send a specific signal frequency to notify irregularities [10].

Temperature, humidity, and the rate of loading are all monitored, and a transformer autonomous relay switch mechanism has been proposed and executed. If the transformer is not serviced quickly, it separates the low-importance loads and keeps the high-importance loads; if the transformer is not able to provide the high-importance loads, it disconnects all loads and remains in no-load mode, in which the transformer autonomously monitors various conditions and returns the loads to the service if all parameters return to normal [11].

It is impossible to overestimate the importance of real-time monitoring for defect detection. As a result, the transformer's current and voltage, as well as the oil temperature and level, and whether there is any spark, flame, or smoke, are all monitored on a regular basis, and the necessary warning message is sent [12].

A lot of data are collected due to the usage of edge computing and an agent-based system design. Without relying on the cloud, autonomous choices are performed at the edge with locally accessible data, assisting in the management of transformer health [13]. IoT technology is used to collect data such as temperature, current, and oil level. The data are then sent to the internet through the TCP/IP protocol, and if any parameter exceeds a threshold value, an alert is produced, which may be displayed on an Android app [14].

To monitor real-time errors in the transformer, many transformer characteristics, such as voltage imbalances, current volume, transformer oil levels, temperature, and vibration, are recorded. The fail state and working state are predicted using the artificial neural network (ANN) network technique based on these factors [15]. The transformer is monitored by taking into account several characteristics, such as voltage, temperature, current, and oil level, using various sensors and a microprocessor. Operational data with a unique transformer status are received using the IoT server and these data are stored on a computer server with the help of a specific web address [16].

Two new methods for forecasting volume have been discovered. In both procedures, a parametric scale is utilized to describe the unique usage pattern, and a neural network and backward vector support are used to calculate the total amount of day use that is redistributed according to the day pattern and establish the final loading pattern [17]. The system notifies the engineer using the GSM module via the Short Message Service (SMS) for any restrictions. In addition, automation is introduced into the transformer cooling machine, where the cooler operating capacity is controlled by a microcontroller-supported response system, thus reducing human dependence to a great extent. The results from experiments and simulations are analyzed to predict the performance of automatic IoT-based transformers and are compared with the existing model, and real barriers to the same performance are discussed [18].

The real-time monitoring of the distribution transformer utilizing IoT has occurred, including measurements of parameters, such as oil level, load current, line voltage, and temperature, as well as the installation of the Message Queuing Telemetry Transport (MQTT) protocol on the IoT divide for reduced system latency [19].

12.3 PROPOSED METHODOLOGY

The aim is to practically build a prototype of a real-time transformer monitoring system that monitors the transformer remotely and also integrate machine learning to optimize the recognition of the threshold value. To achieve this aim, all real-life parameters are to be considered and all pragmatic constraints are to be taken into consideration to avoid further hindrances. Moreover, the aim is the robust testing of IoT protocols to establish fast and efficient wireless communication. Considering the vastness of execution, the study is divided into two segments: the real-time monitoring of transformers (consists of the construction of a hardware prototype, IoT layer, and data monitoring unit) and machine learning analysis.

The real-time transformer monitoring system is an approach for monitoring the transformer and simultaneously creating a dashboard that helps the user to remotely monitor the transformer and its various parameters, and also control the relays in the case of malfunction.

Machine learning analysis is an algorithmic way to analyze data in which an algorithm takes existing data, uses it to update itself with each iteration of data in the machine learning model, and uses this information gained to predict future outcomes. Various machine learning algorithms on real-time transformer data are trained and tested, and used to predict future transformer health based on the historical data used to train the machine learning model.

12.3.1 Design of a Real-Time Transformer Monitoring System

The principal idea is to create a hardware unit that is directly mounted on the transformer so that different parameters, such as oil lever, current, and smoke, are monitored. The design is such that these observations are wirelessly communicated to the remote device without any human interference. To do so, the microcontroller is connected to a Raspberry Pi, which acts as a server and helps in establishing the IoT protocol so that easier and faster communication is possible. The Raspberry Pi also plays an important role by creating a user interface (UI) for the monitoring of the data. The hardware unit consists of different sensors and relays that are connected to an ESP8266 NodeMCU, which acts as a microcontroller unit. This microcontroller is further wirelessly connected to the Raspberry Pi with the help of the MQTT protocol. The Raspberry Pi acts as a server and helps in

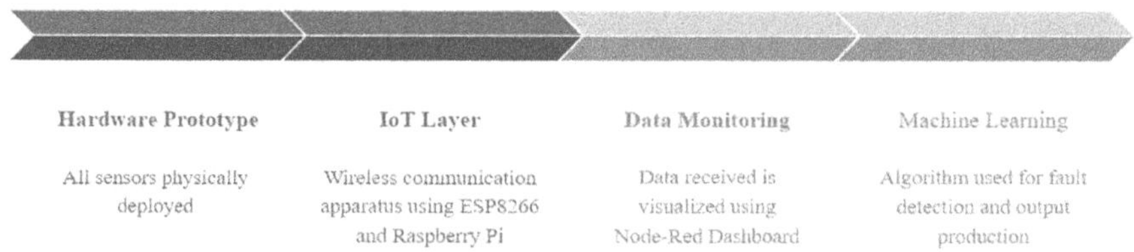

Figure 12.1 Block diagram of the workflow and functioning of the apparatus

establishing a publish-subscribe model between the dashboard and every sensor on the hardware unit using the Mosquitto broker. The Raspberry Pi is then connected to Node-RED, which helps in creating a dashboard so that the data sent by the hardware are easily visualized. This apparatus, shown in Figure 12.1, has two-way communication, that is, data are sent and received wirelessly at the same time. The system architecture is shown in Figure 12.2.

12.3.2 Machine Learning Analysis

Five machine learning algorithms are used for the analysis of transformer health monitoring. The dataset shown in Figure 12.3, made up of various transformer parameters, is used to train various machine learning algorithms. After the machine learning model is trained, it is used to predict transformer health. After the five machine learning models are created, every model is compared with the others and the optimum is found. To achieve this aim, the dataset is obtained from Kaggle, which is the world's largest data science community. These data are collected via IoT devices from June 25, 2019 to April 14, 2020 that are updated every 15 minutes.

The constraints for comparison in the machine learning algorithm are precision, recall, F1 score, and support. Precision is defined as the ratio of true positives to the sum of true positives and false positives. Recall is defined as the ratio of true positives to the sum of true positives and false negatives. Support is the count of the actual presence of a class in the dataset

The machine learning algorithms used and analyzed are as follows:

Gradient boosting: This is a type of machine learning. It relies on the intuition that the best possible next model in combination with the previous model minimizes the overall prediction error. The key idea is to set the target results for the next model to minimize errors.

Support vector machine: The SVM is a supervised machine learning algorithm used for both classification and regression. The goal of the SVM algorithm is to find a hyperplane in *n*-dimensional space that uniquely classifies data points.

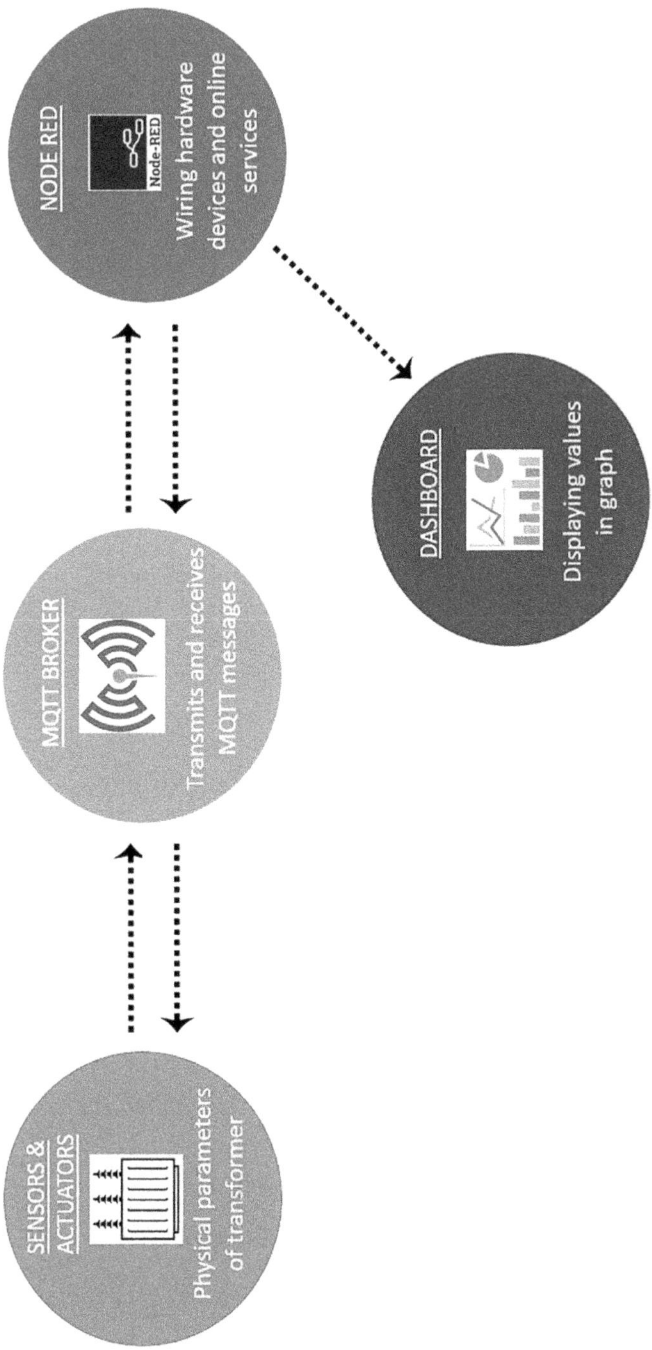

Figure 12.2 System architecture and interface

	Oil Temperature Indicator	Winding Temperature Indicator	Ambient Temperature Indicator	Oil Level Indicator	Oil Temperature Indicator Alarm
count	20316	20316	20316	20316	20316
unique	59	2	34	65	2
top	34.0	0.0	28.0	100.0	0.0
freq	1542	14821	1774	8175	20215

Figure 12.3 Dataset showing different parameters

Logistic regression: This is a method of modeling the chances of a discrete outcome given an input variable. The most common logistic regression models a binary outcome, such as true/false.

Random forest classifier: This is a classifier that holds a number of decision trees on different subsets of the given dataset and takes the average to improvise the predictive accuracy of that dataset.

12.4 DESIGN APPROACH AND SPECIFICATIONS

12.4.1 Design of a Real-Time Transformer Monitoring System

The hardware unit consists of a microcontroller(ESP8266 NodeMCU), which acts as the central unit and plays the important role of sending the data received from the sensors and receiving instructions to control the relays, as shown in Figure 12.4. This microcontroller also plays the important role of powering all the components connected to it. The sensors are connected to ESP8266.

The major task of establishing MQTT between the server and sensors is done by the Mosquitto broker, which is installed on the Raspberry Pi and acts as a broker in the MQTT protocol. Moreover, it sends signals received by the dashboard to operate relays. It also helps to send data to the dashboard using Node-RED.

For testing purposes, a step-down transformer with 240 V AC input and 12 V AC output is chosen. The output of this transformer is connected to a full bridge rectifier such that AC is converted to DC. The output of this bridge rectifier is given to a 12 V DC motor. Ammeter and ACS712 sensors are connected in series between the rectifier and load to measure current. A DHT 22 (temperature and humidity sensor) and SW 420 (vibration sensor) are mounted on the transformer.

The hardware specifications of the components used are as shown in Figures 12.5–12.11 and Table 12.1.

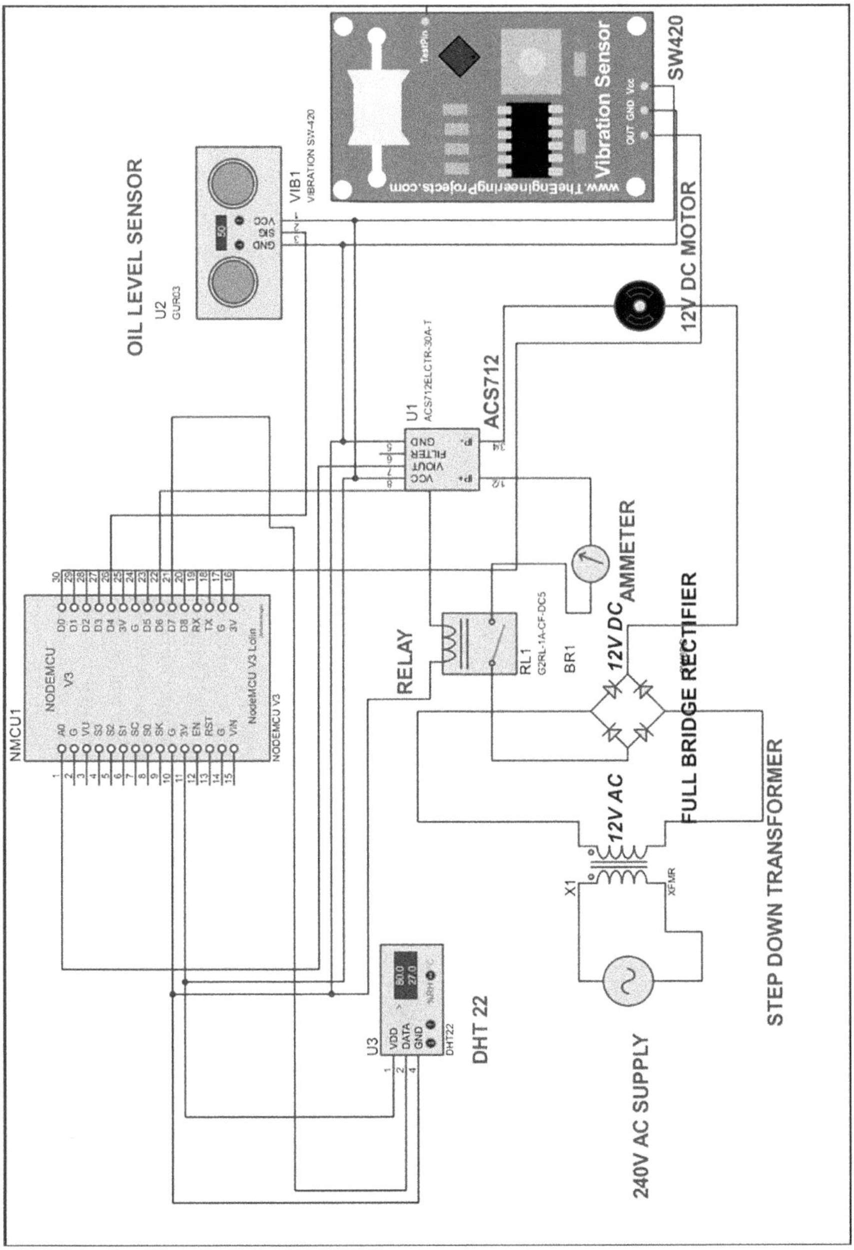

Figure 12.4 Circuit diagram of the hardware setup

Table 12.1 Hardware specification

S. No.	*Component name*	*Component pin layout*	*Specification and usage*
1	ESP8266 NodeMCU	Figure 12.5: ESP8266 pin layout	It is a low-power on-chip microcontroller with integrated Wi-Fi, therefore decreasing dependability on the external integration of the Wi-Fi module. This microcontroller has multiple digital pins, which are general-purpose input/output (GPIO) pins, and sufficient for the number of sensors and relays used in the hardware. It also has one analog pin, which is used to take input from the current sensor. This is further connected to the Raspberry Pi over Wi-Fi. The constraint with ESP8266 is that it works on 3.3 V digital logic and also has a single inbuilt analog-to-digital converter (pin A0).
2	Raspberry Pi 3B+	Figure 12.6: Raspberry Pi 3B+	It is an inexpensive minicomputer used to execute multiple IoT applications. It is a low-power consuming device, which is connected to a mouse, keyboard, and screen. The Raspberry Pi serves as an MQTT broker, which helps in transmitting and receiving data. The Mosquitto broker, Node-RED, and Node-RED dashboard are installed on the Raspberry Pi.
3	Temperature and humidity sensor: DHT 22	Figure 12.7: DHT 22 Pin layout	It is a digital temperature and humidity sensor. In this application, it is used to measure the oil temperature and humidity in the transformer. Oil temperature is a key identifier in life expectancy. If the oil temperature exceeds a threshold limit, this may cause the deterioration of the paper insulating material. Increased oil temperature also results in poor cooling from the heat generated by the losses. The moisture content in transformer oil also tends to reduce the life and performance. The sensor used has a range of -40°C to 80°C temperature readings with ±0.5°C accuracy. It has a digital output, which acts as input for ESP8266.
4	Current sensor: ACS712	Figure 12.8: ACS712 pin layout	It gives the real-time current value from the transformer. This hardware system uses only one sensor for ease of application, but in the actual execution of hardware coupled with a transformer, all three phases are monitored using the current sensor. Reading the current value from a transformer is also used as a safety measure in the system. When the current value is greater than the nominal value of current allowed to flow through the transformer, a relay is actuated that disconnects the load side from the

			transformer. The ACS712 current sensor module uses the Hall effect principle to measure the value of AC/DC current. It has a sensitivity of 100 mV/A and current measuring range of -30 A to 30 A. The output of ACS 712 is analog in nature and therefore is connected to pin A0 of ESP8266. NOTE: This sensor needs a 5 V operating voltage. Because ESP8266 does not have a 5 V output pin, external 5 V DC has to be given to this sensor.
5	Oil level sensor-ultrasonic sensor: HC SR04	Figure 12.9: HC SR 04 pin layout	Oil acts as the insulator and the cooling agent in a transformer. The older the transformer, the lesser is the oil content present in it. It is of utmost importance to maintain the optimum oil level in the conservator tank else it may cause overheating and electrical flashovers. The ultrasonic sensor measures the distance to the target by emitting ultrasonic sound waves. It has a measuring range of 2 cm to 150 cm, with a 3 mm accuracy. The output by this sensor is digital in nature.
6	Vibration sensor: SW-420	Figure 12.10: SW-420 pin layout	Transformer noise is caused by a phenomenon called magnetostriction. Winding distortion creates extra electromagnetic force between the windings and the core, and in turn gives rise to the change in the transformer vibration signature. The vibration signatures inside the oil tank of a transformer have a wideband frequency ranging from 20 Hz to 1000 Hz. The output given by this sensor is digital in nature.
7	Gas sensor: MQ2	Figure 12.11: MQ2 pin layout	The MQ2 sensor is the sensor that measures methane, propane, liquefied petroleum gas (LPG), and hydrogen in the surroundings. This sensor detects smoke on the basis of the amount of change in these gasses. Because during any faulty circumstances smoke may get released, this sensor is used to detect smoke. This sensor works when both analog and digital pins are connected to the microcontroller, but output is read by any one of the pins at a time. Due to the unavailability of multiple analog pins in ESP8266, the digital output of the MQ2 sensor is used to detect the sensor. Note: The operational voltage of this sensor is 5 V and because there is no 5 V output pin on ESP8266, an external supply of 5 V DC is given to this sensor.

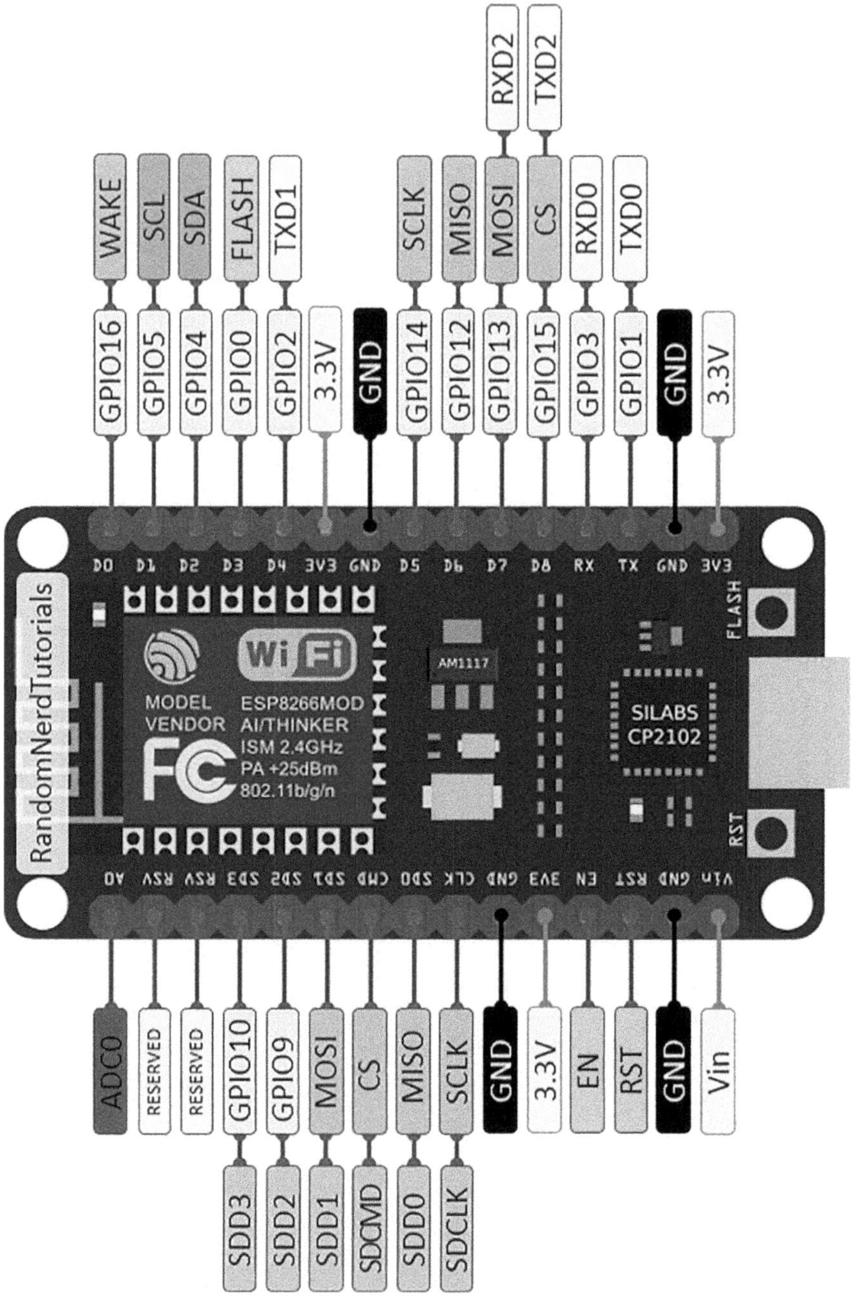

Figure 12.5 ESP8266 pin layout

Figure 12.6 Raspberry Pi 3B+

DHT22 pins	
1	VCC
2	DATA
3	NC
4	GND

Figure 12.7 DHT 22 pin layout

12.4.2 IoT Layer and Standards

A layer of the IoT is established between NodeMCU and the Raspberry Pi so that wireless communication shown in Figure 12.12 is established between the hardware deployed on the transformer and dashboard. To do so, the MQTT protocol is used.

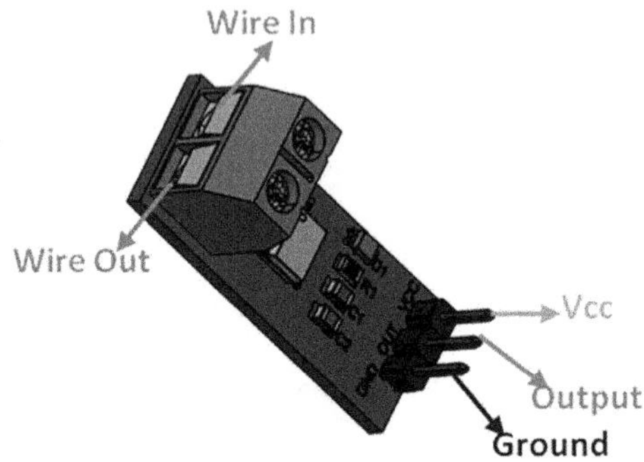

Figure 12.8 ACS712 pin layout

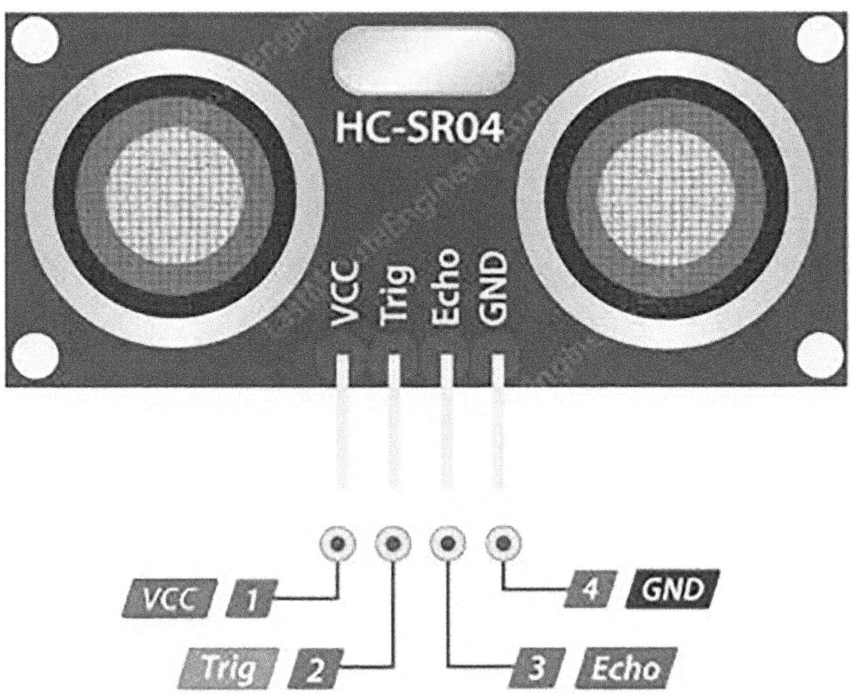

Figure 12.9 HC SR 04 pin layout

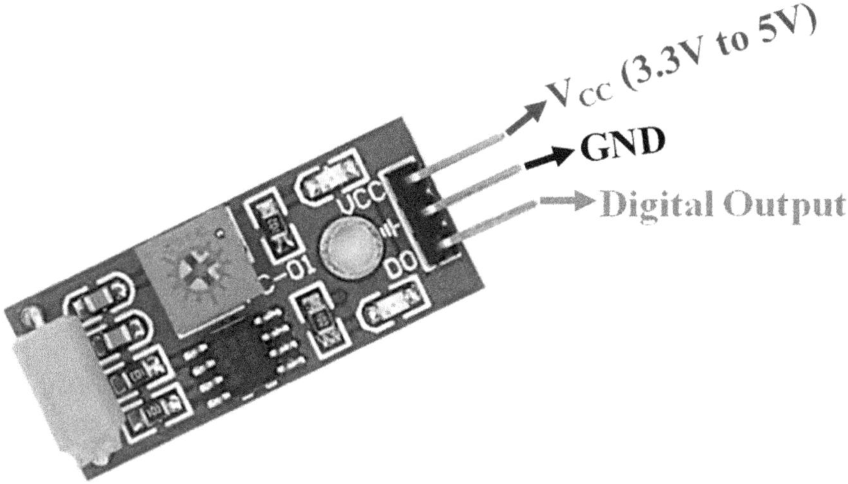

Figure 12.10 SW-420 pin layout

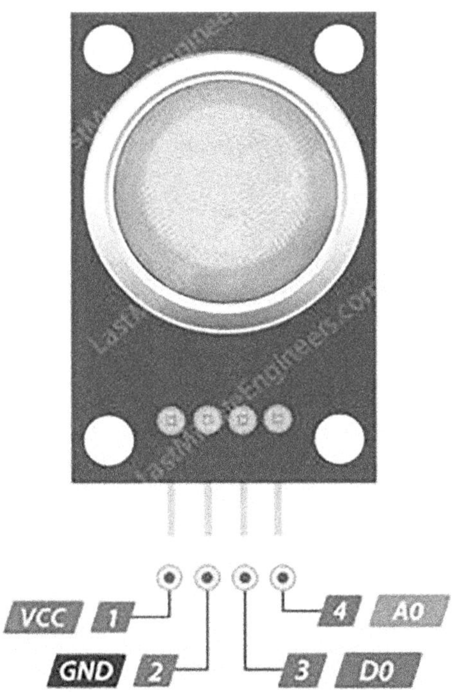

Figure 12.11 MQ2 pin layout

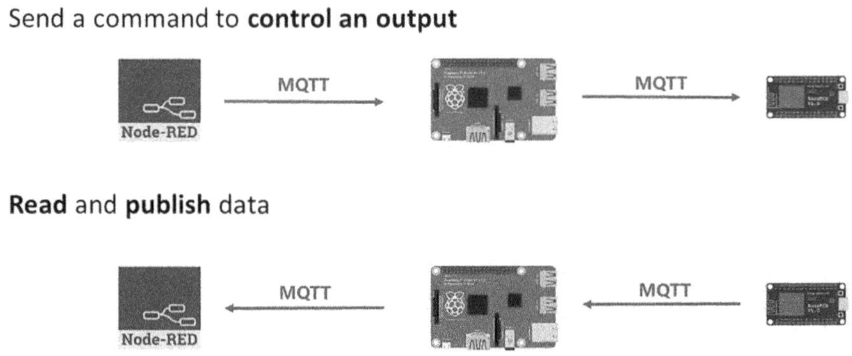

Figure 12.12 Bidirectional communication of the hardware setup and dashboard UI

MQTT is a lightweight messaging protocol that works on the publish-subscribe model. As a lightweight protocol, MQTT helps in the faster and efficient transfer of data between the sensors, relays, and dashboard.

A "broker" is an application that is used to receive and transmit MQTT messages. The messages are also filtered and delivered to the subscriber using this broker. In this study, a pre-installed MQTT broker known as Mosquitto is used on the Raspberry Pi. A "client" connects to the central broker to send and receive messages. The client is able to subscribe to various sensor values as topics in the system. The client is allowed to publish commands to operate the relays when required. In a publish-subscribe model, a device publishes a message on a topic or subscribes to a topic in order to receive messages.

12.4.3 Software

The software specification used is as follows:

- Arduino Integrated Development Environment (IDE)
  Arduino software (IDE) provides an open-source platform for writing code and uploading to the board. Code is written and debugged using Arduino IDE. Arduino IDE provides a simple and clear programming environment that allows the use of integrated libraries with external hardware support. To run the code for the ESP8266 microcontroller, additional libraries and board managers are installed.
- Node-RED
  Node-RED is built on Node.js, which is a visual programming tool, and is used for connecting hardware devices, application programming interfaces (APIs), and online services. Visualization becomes

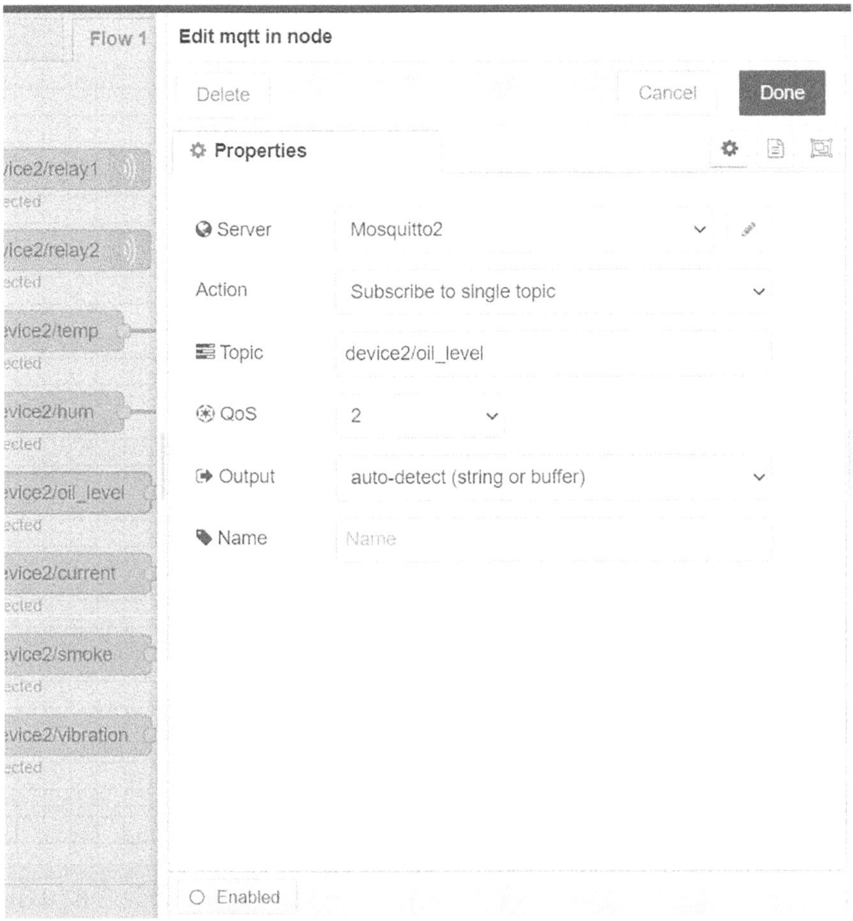

Figure 12.13 MQTT node setup

easy with Node-RED with the inbuilt dashboard plugin. A user-friendly platform is created to visualize the real-time data and monitor transformer health through a personal computer or mobile devices. Data from all the sensors of the network are sent to the dashboard every few seconds and displayed in gauges.

Once received by the Raspberry Pi from the sensors using MQTT shown in Figure 12.13, the data are forwarded to Node-RED shown in Figure 12.14, which is installed on the Raspberry Pi.

Node-RED is a JavaScript-based flow-based development tool. Node-RED has a node called the "MQTT node," which receives data by the name

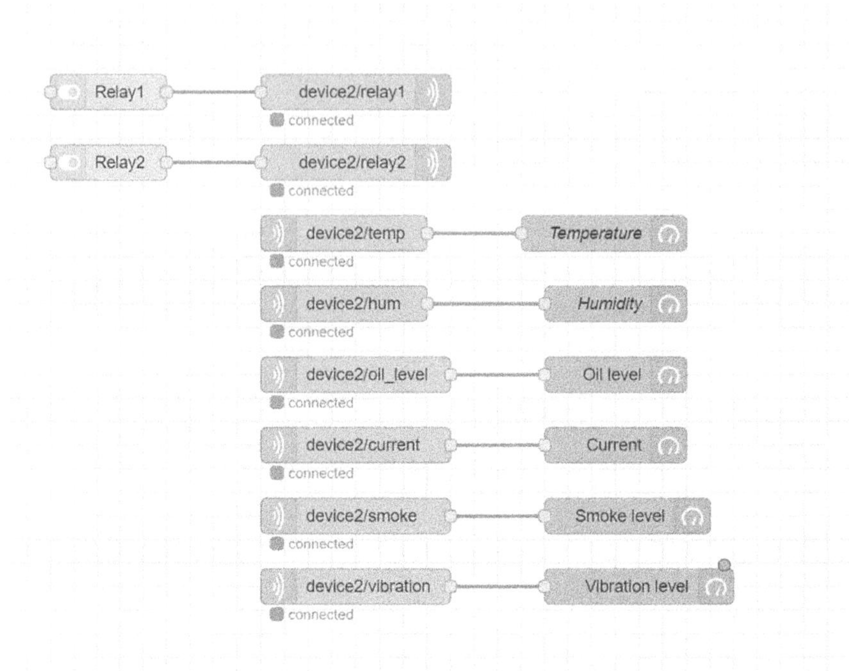

Figure 12.14 Node-RED flow

of the topic. For instance, as shown in Figure 12.14, the oil level sensor publishes data for the topic "device2/Oil Level." Hence, in Node-RED, the MQTT node is assigned the topic "device2/Oil Level."

The MQTT node is further connected to the dashboard node. The dashboard is a library palette that has to be specifically downloaded for the visualization of data received from the MQTT node. In the dashboard palette, the Gauge node is used to display the data received from the sensors. In the case of relay switching, as shown in Figure 12.15, the signal has to be published from the dashboard. The Switch node (part of the dashboard palette) is used to receive signal from the dashboard and forward it to the MQTT node with the designated topic.

12.4.4 Machine Learning Analysis

Various parameters of the transformer are used for training and testing the machine learning algorithms shown in Figure 12.15. The parameters are as follows:

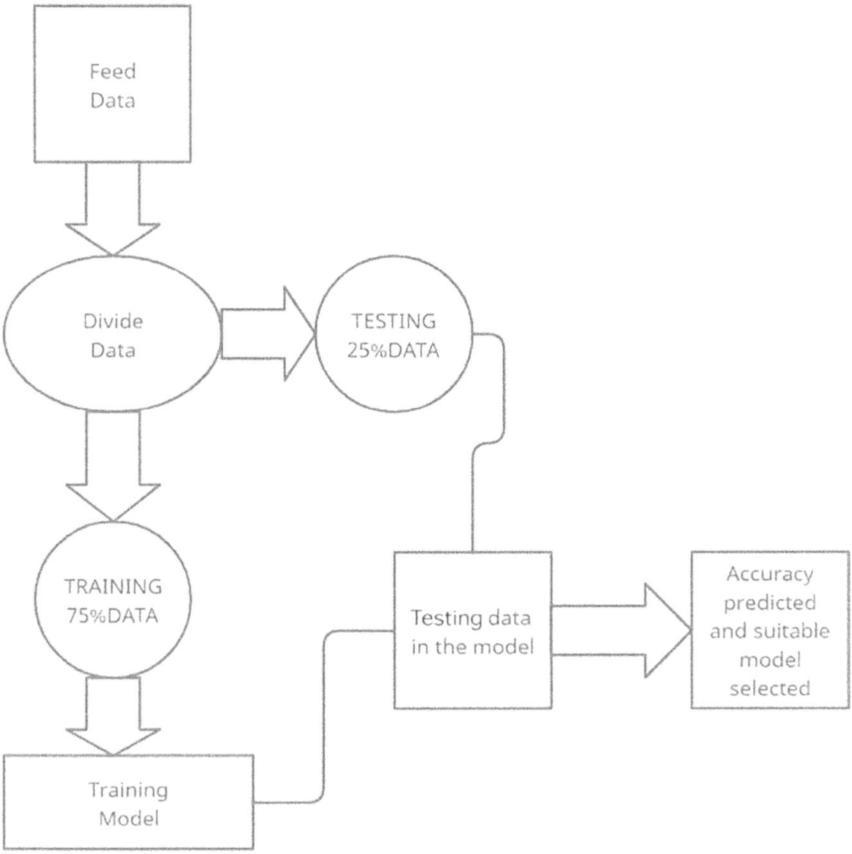

Figure 12.15 Functioning of an algorithm

Oil temperature indicator: This is the oil hotspot temperature indicator, which is a critical parameter to measure within a transformer

Winding temperature indicator: This indicates the winding temperature and operates the buzzer, relays/switches, and cooler control contacts. This instrument operates on the principle of thermal imaging.

Ambient temperature indicator: Ambient temperature is a measure of the temperature around a given asset or piece of equipment or other object.

Magnetic oil gauge indicator: A magnetic oil level gauge (MOG) is a device used to indicate the position of the transformer insulating oil level in the conservator of a transformer. This is a mechanical device.

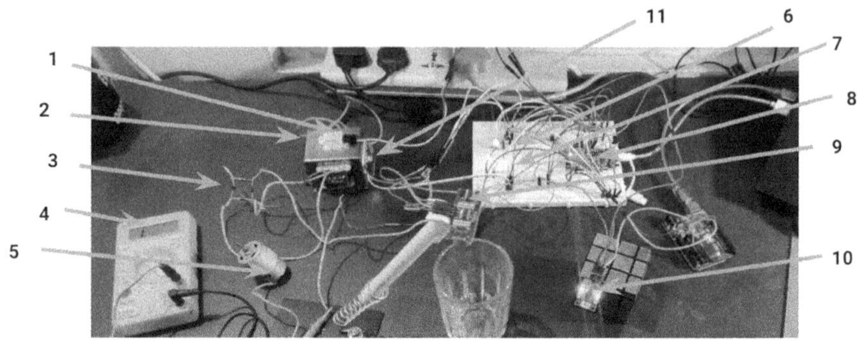

Figure 12.16 Hardware execution

Table 12.2 In reference to the markers in Figure 17

Marker number	*Name*
1	DHT 22
2	Step-down transformer
3	Full bridge rectifier
4	Multimeter used as an ammeter
5	12V DC motor
6	LED 1
7	LED 2
8	ESP8266 NodeMCU
9	Oil level sensor
10	Smoke sensor
11	Vibration sensor

The data are assessed for five machine learning algorithms: gradient boosting, SVM, *k*-neighbors classifier, logistic regression, and random forest classifier.

The oil level indicator, oil temperature indicator alarm, and magnetic oil gauge indicator are the parameters used as output for predictive analysis. The data are divided into two parts: 75% for training the model and 25% of data for testing.

12.5 OUTPUT AND RESULTS

12.5.1 Hardware Execution

For the execution of the monitoring system shown in Figure 12.16 on the actual transformer, a 240 V AC to 12 V AC transformer is used. The output of this transformer is 12 V AC and, due to unavailability of AC loads with an operational voltage of 12 V, a full bridge rectifier is used to convert the 12 V AC to 12 V DC. The output of the full bridge rectifier is then connected to the relay, multimeter, and ACS712 (current sensor) in series, which is then connected to the 12 V DC motor as the load. The DHT22 Sensor and SW420 sensor are directly mounted on the transformer to measure its parameters.

Due to the unavailability of the conservator tank, a cylindrical glass is used to measure the oil filled in it.

The smoke sensor is placed slightly at a distance to measure smoke in the surroundings.

Light-emitting diode (LED) 1 (representing relay 1, which is used to disconnect the load from the transformer in the case of malfunction) is placed on the breadboard. LED 2 (representing relay 2, which is used to disconnect the transformer primary from its input supply) is placed on the breadboard.

12.5.2 Node-RED Dashboard

The components of the dashboard shown in Figure 12.17 are:

- Relay 1: toggle switch to control LED 1
- Relay 2: toggle switch to control LED 2
- temperature gauge: displays the temperature of the transformer in degrees Celsius
- humidity gauge: displays humidity in percentage
- current gauge: displays current in amperes
- oil level: displays the amount of the conservator tank filled in centimeters
- vibration level: displays the amount of vibration on the scale of 0 to 1000, with 0 being the least
- smoke level: displays smoke in parts per million (ppm).

12.5.3 Sensor Testing and Results

The sensors are as follows:

- Temperature and humidity sensor
  The temperature and humidity gauge shows no value when the transformer is in the OFF state, hence no observation takes place. When the transformer is the ON state and the load is connected, the temperature and humidity gauge starts displaying the respective values on the gauges, as shown in Figure 12.18.

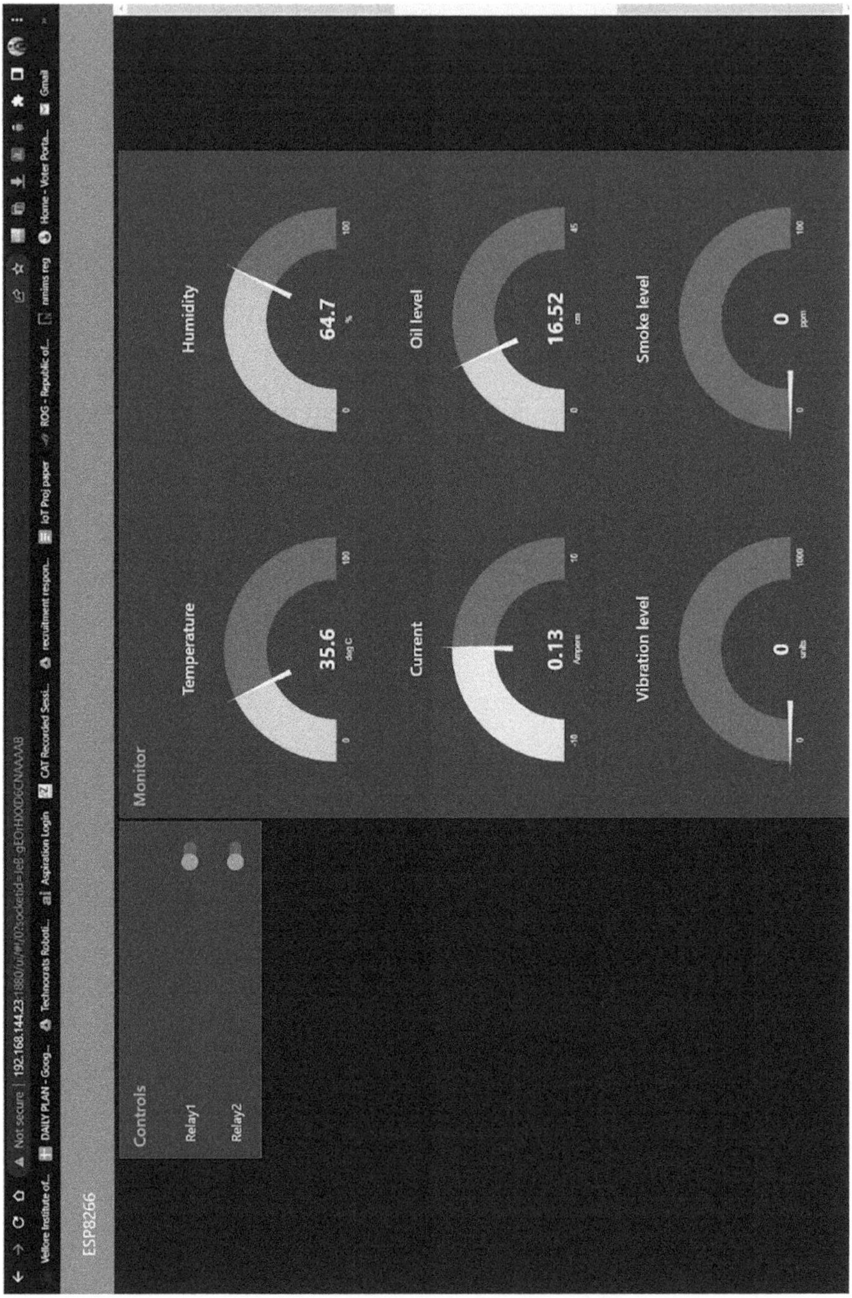

Figure 12.17 Dashboard UI

Figure 12.18 Multimeter displaying current passing through the 12 V DC motor

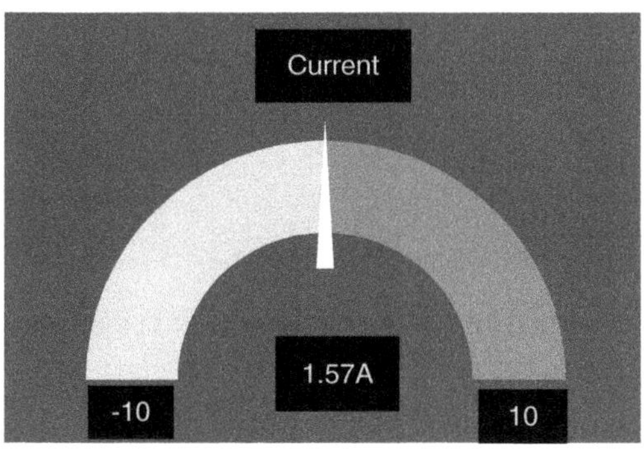

Figure 12.19 Current gauge displaying the load current

- Current sensor
  The current gauge displays no values when the transformer is in the OFF state. As and when the transformer is in the ON state and the load is connected, the multimeter and the current gauge display the value of the current passing through the load in amperes as a unit, as shown in Figures 12.18 and 12.19.

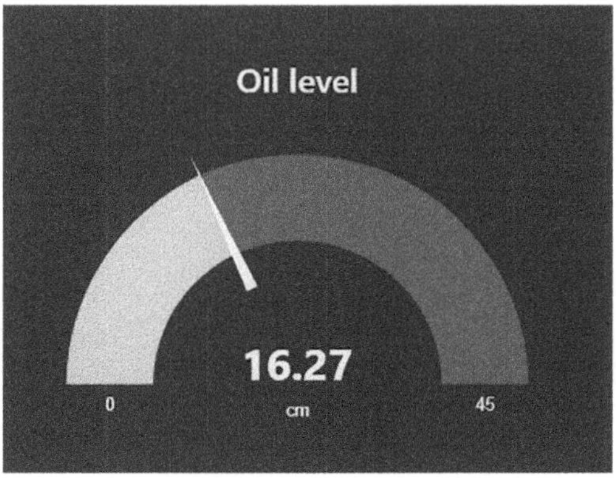

Figure 12.20 Oil level when the container is empty

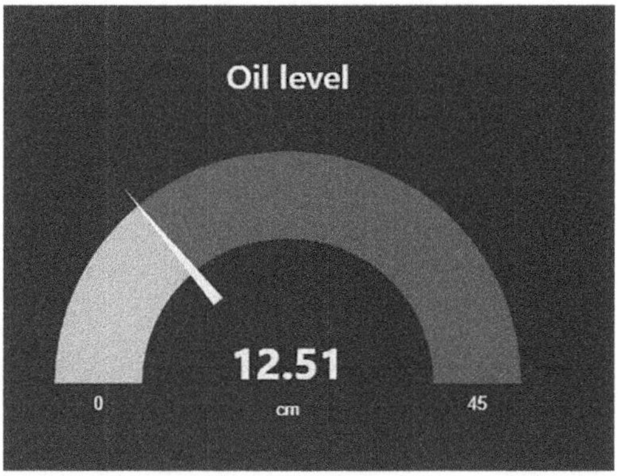

Figure 12.21 Oil level decreasing due to increasing liquid in the container

- Oil level sensor
  The oil level sensor measures the maximum distance between the sensor and the depth of the container when there is no liquid filling the container, as shown in Figures 12.20 and 12.21. As and when the container is filled with liquid and the length between the sensor and the surface starts decreasing, the value displayed by the oil level gauge starts decreasing. The unit of measurement is centimeters.
- Vibration sensor
  The vibration level gauge shows "0" when the transformer is in the OFF state and therefore no vibration is observed, as shown in Figure 12.22. As and when the transformer is in the ON state and vibration is measured, the sensor starts displaying a value between 1 and 1000, with 1 being the minimum and 1000 being the maximum.

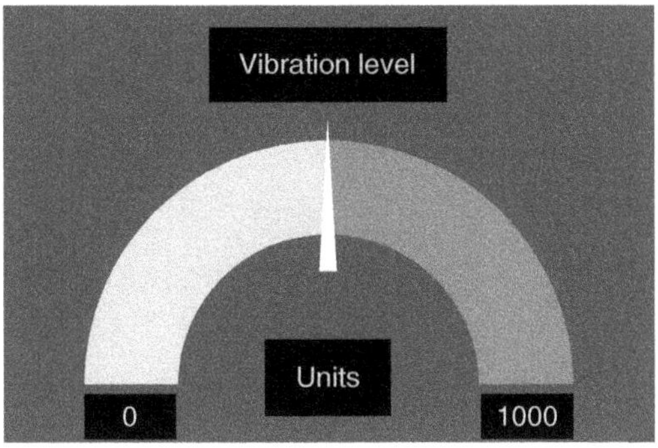

Figure 12.22 Vibration gauge on the dashboard

Figure 12.23 Smoke provided to the smoke sensor

- Smoke sensor
  The smoke sensor shown in Figure 12.23 detects no smoke at the time that the transformer is in the ON state and there is no smoke in the atmosphere. As and when a smoke source is placed near the smoke sensor, the smoke sensor detects the smoke and displays it on the smoke gauge shown in Figure 12.24. The unit of measurement is ppm.
- Relay control
  Figure 12.25 signifies when the dashboard control of Relay 1 and Relay 2 are in the OFF state and the respective LEDs are in the OFF state.

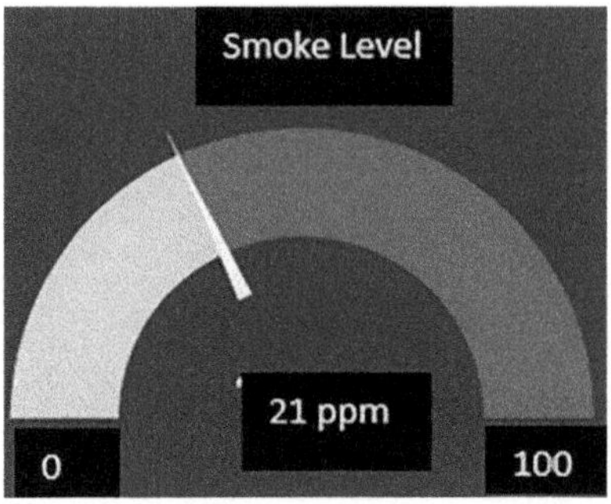

Figure 12.24 Smoke gauge when smoke is detected

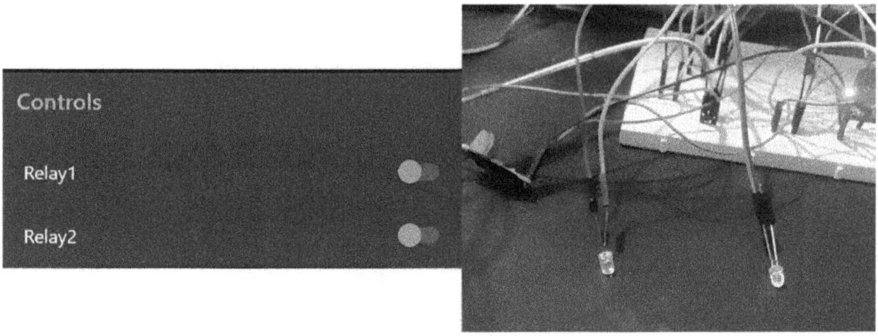

Figure 12.25 When Relay 1 and Relay 2 are in the OFF state

Figure 12.26 signifies when both switches on the dashboard are in the ON state and both LEDs are in the ON state.

12.5.4 Machine Learning Algorithm Results

After careful consideration and brief analysis shown in Figure 12.27, the inference is that the random forest classifier algorithm is the most efficient algorithm when it comes to this type of data. Therefore, the random forest classifier is used ahead.

12.6 CONCLUSION

This method of online transformer monitoring helps users to know the real-time condition of distribution transformers. The six gauges on the Node-RED dashboard that display various physical parameters helps to indicate the values. With the help of the MQTT protocol, the system also

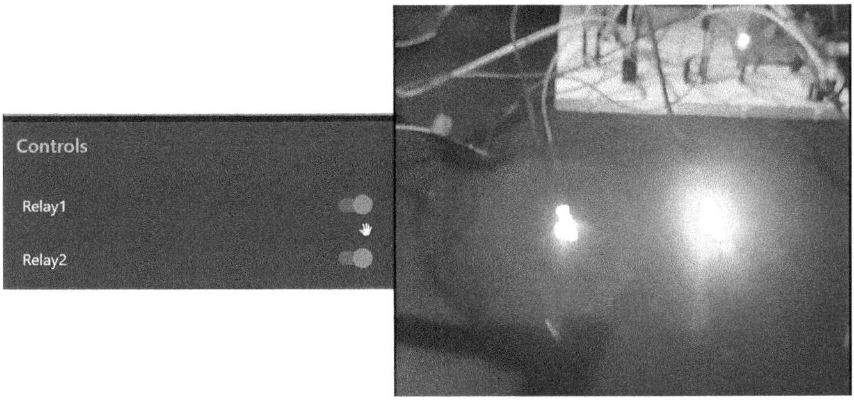

Figure 12.26 When Relay 1 and Relay 2 are in the ON state

	precision	recall	f1-score	support
0.0	1.00	1.00	1.00	5053
1.0	0.93	0.54	0.68	26
accuracy			1.00	5079
macro avg	0.97	0.77	0.84	5079
weighted avg	1.00	1.00	1.00	5079

Figure 12.27 Accuracy analysis using the random forest classifier

establishes a two-way interaction between the monitoring end and the distribution transformer. The actuators/relays connected on the distribution transformer end are controlled by the engineer at the monitoring end, therefore disconnecting the load from the transformer and transformer from the primary input.

With the help of machine learning analysis, the optimum algorithm, that is, the random forest classifier, was found for the use case of transformer data, which helped to further determine the threshold values of the parameters tested. The accuracy and precision of this algorithm was also tested, indicating the reliability of the algorithm.

REFERENCES

[1] Srivastava, D. and Tripathi, M.M., 2018, October. Transformer Health Monitoring System Using Internet of Things. In *2018 2nd IEEE*

International Conference on Power Electronics, Intelligent Control and Energy Systems (ICPEICES) (pp. 903–908). IEEE.

[2] Bethalsha, C., Jennifer, A., Karthik, M.S. and Sreejavijay, M., 2020. Real-Time Transformer Health Monitoring Using IOT. *International Journal for Research in Applied Science and Engineering Technology*, 8(9), pp.521–526.

[3] Ahmad, I., Singh, Y. and Ahamad, J., 2020, October. Machine Learning Based Transformer Health Monitoring Using IoT Edge Computing. In *2020 5th International Conference on Computing, Communication and Security (ICCCS)* (pp. 1–5). IEEE.

[4] Sharma, R.R., 2021. Design of Distribution Transformer Health Management System using IoT Sensors. *Journal of Soft Computing Paradigm*, 3(3), pp.192–204.

[5] Aizpurua, J.I., Catterson, V.M., Stewart, B.G., McArthur, S.D., Lambert, B., Ampofo, B., Pereira, G. and Cross, J.G., 2017, April. Determining appropriate data analytics for transformer health monitoring. In *10th International Topical Meeting on Nuclear Plant Instrumentation, Control, and Human-Machine Interface Technologies, NPIC and HMIT 2017* (pp. 1–11). American Nuclear Society.

[6] Abu-Elanien, A.E., Salama, M.M.A. and Ibrahim, M., 2011, June. Determination of Transformer Health Condition Using Artificial Neural Networks. In *2011 International Symposium on Innovations in Intelligent Systems and Applications* (pp. 1–5). IEEE.

[7] Suja, K. and Yuvaraj, T., 2021, February. Transformer Health Monitoring System Using Android Device. In *2021 7th International Conference on Electrical Energy Systems (ICEES)* (pp. 460–462). IEEE.

[8] En-Wen, L. and Bin, S., 2014, October. Transformer Health Status Evaluation Model Based on Multi-Feature Factors. In *2014 International Conference on Power System Technology* (pp. 1417–1422). IEEE.

[9] Khandait, A.P., Kadaskar, S. and Thakare, G., 2017. Real time monitoring of transformer using IOT. *International Journal of Engineering Research & Technology (IJERT)*, 6(03), pp.146–149.

[10] Kamlaesan, B., Kumar, K.A. and David, S.A., 2017, August. Analysis of Transformer Faults Using IoT. In *2017 IEEE International Conference on Smart Technologies and Management for Computing, Communication, Controls, Energy and Materials (ICSTM)* (pp. 239–241). IEEE.

[11] Hasan, W.K., Alraddad, A., Ashour, A., Ran, Y., Alkelsh, M.A. and Ajele, R.A., 2019, August. Design and Implementation Smart Transformer Based on IoT. In *2019 International Conference on Computing, Electronics & Communications Engineering (iCCECE)* (pp. 16–21). IEEE.

[12] Kumar, R.K., Thilagaraj, M., Vengatesh, P., Rajalakshmi, J. and Babul, M.M., 2021. Remote Transformer Faults Analyzing System Using IoT. *International Journal of Modern Agriculture*, 10(2), pp.2390–2402.

[13] Thangiah, L., Ramanathan, C. and Chodisetty, L.S., 2019, April. Distribution Transformer Condition Monitoring Based on Edge Intelligence for Industrial IoT. In *2019 IEEE 5th World Forum on Internet of Things (WF-IoT)* (pp. 733–736). IEEE.

[14] Aralikatti, S., Pradyumna, R., Reddy, R., Sanjay, B.R. and Reddy, S.N.K., 2021, September. IoT-Based Distribution Transformer Health Monitoring

System Using Node-MCU & Blynk. In *2021 Third International Conference on Inventive Research in Computing Applications (ICIRCA)* (pp. 1–4). IEEE.

[15] Khairnar, V., Kolhe, L., Bhagat, S., Sahu, R., Kumar, A. and Shaikh, S., 2020. Industrial Automation of Process for Transformer Monitoring System Using IoT Analytics. In *Inventive Communication and Computational Technologies* (pp. 1191–1200). Springer, Singapore.

[16] Yamuna, R., Geetha, R., Gowdhamkumar, S. and Jambulingam, S., 2019. Smart Distribution Transformer Monitoring and Controlling Using IoT. *International Research Journal of Multidisciplinary Technovation*, pp.111–115.

[17] Singh, M.J., Agarwal, P. and Padmanabh, K., 2016, December. Load Forecasting at Distribution Transformer Using IoT Based Smart Meter Data from 6000 Irish Homes. In *2016 2nd International Conference on Contemporary Computing and Informatics (IC3I)* (pp. 758–763). IEEE.

[18] Subramanian, P.V., Boddapati, V. and Daniel, S.A., 2022. Automated Real-Time Transformer Health Monitoring System Using the Internet of Things (IoT). In *Advancement in Materials, Manufacturing and Energy Engineering*, Vol. I (pp. 503–511). Springer, Singapore.

[19] Roy, T.K. and Roy, T.K., 2018, February. Implementation of IoT: Smart Maintenance for Distribution Transformer Using MQTT. In *2018 International Conference on Computer, Communication, Chemical, Material and Electronic Engineering (IC4ME2)* (pp. 1–4). IEEE.

Chapter 13

CI-Based SEPIC DC/DC Converter for Photovoltaic Applications

Kanimozhi G., Varun Mukesh Lal, R. Jeyaraman, Ravi Samikannu

13.1 INTRODUCTION

The emergence of novel applications such as renewable energy (RE) [1], electric motors, power factor adjustment, and light-emitting diode (LED) drivers has boosted the need for power supply at various voltage levels. As a result, there is a requirement for a broad variety of voltage conversions, with DC/DC converters playing a key part. There are three main types of DC/DC converters based on their operation: buck, boost, and buck-boost. The load voltage of these converters is maintained by controlling the conducting period of the controlled active semiconductor switches, such as a metal-oxide-semiconductor field-effect transistor (MOSFET) and insulated-gate bipolar transistor (IGBT). The buck-boost converter provides the flexibility to perform not only the buck but also the boost operation using a single converter topology [2]. One of the derived configurations of the buck-boost family of DC-to-DC converters is the single-ended primary inductor capacitor (SEPIC) converter. SEPIC/SEPIC derived converters are commonly used for applications such as RE [3], vehicle-to-grid [4], power factor correction [5], electric vehicles (EVs), plug-in EVs [2], LED lighting [6], and motor drives [7]. For applications such as RE (as shown in Figure 13.1), high voltage gain (VG) converters are necessary to enhance the DC bus voltage. Connecting the solar cells in series [8] is one option, but inefficient as the power generation decreases as the numeral of solar cells linked in series rises. Similarly, high voltage levels can be accomplished by operating the converter topologies at a maximum duty ratio with limitations. Apart from high gain, for a power converter to be best suitable for a particular application, generally it should have certain merits, such as a reduced amount of voltage stress across the switches, ripple in currents and voltages, and high efficiency.

For high-power applications [9], for the protection of the power devices, load, and source, it is desirable to have galvanic isolation between the load and source. Non-isolated converters are an ideal choice for applications that

 DOI: 10.1201/9781003537878-13

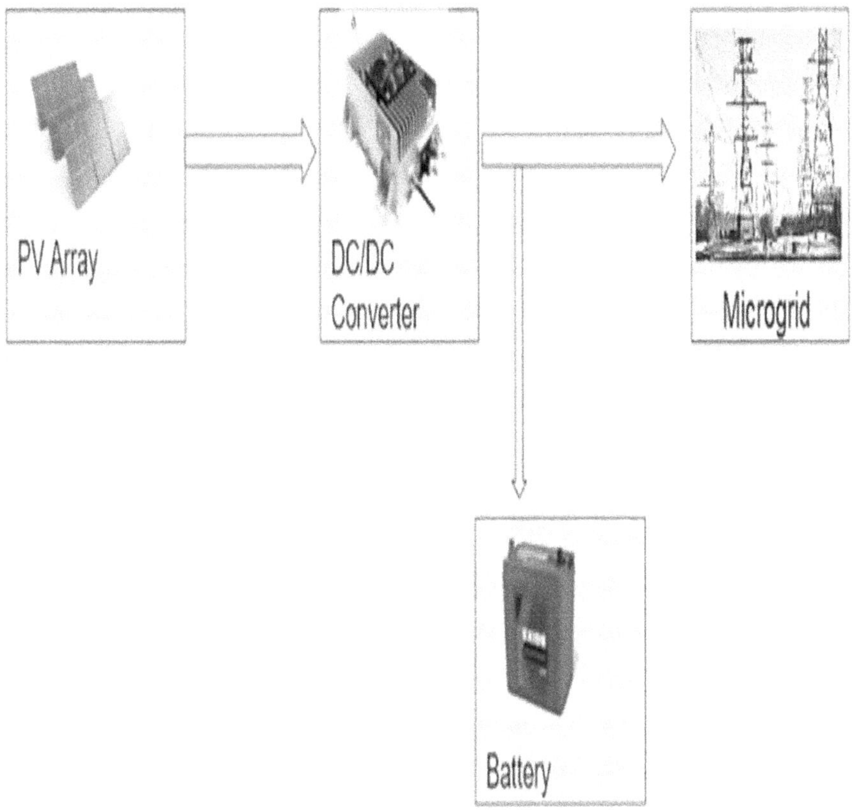

Figure 13.1 Overall block diagram

demand high efficiency, and they are compact and weigh less compared to isolated converters.

To minimize the cost of the magnetic elements, for instance, inductors and high-frequency transformers in converters, their size should be reduced, which is possible by increasing the switching frequency. However, operating at a high frequency increases the switching losses. Moreover, the forced commutation or hard switching of switches causes a sudden change in current and voltage that causes excessive electromagnetic interference (EMI). These concerns can be addressed by employing soft switching methods.

The converter topology is a bidirectional tapped inductor charge pump improved SEPIC converter [10–13]. The tapped inductor and charge pump technology are used to provide a high voltage conversion ratio. The topology encompasses a simple active regenerative snubber and a snubber capacitor. The leakage inductance is utilized as part of a snubber to enable zero-voltage (ZV) and zero-current (ZC) switching, thereby reducing losses associated with switching actions. The low voltage side enjoys a low ripple continuous current due to the filter inductor. The control of the four switches is simple and easy to implement.

A SEPIC-based converter working at a very high frequency (1 MHz) is suggested in [14]. The converter achieves ZV switching (ZVS) that reduces the switching losses. This allows the converter to work at high switching frequencies, empowering the design of the converters with smaller and lighter magnetic components and lower capacitor ratings. Apart from soft switching, the converter has increased static gain and the voltage stress in comparison to that of the existing SEPIC converter is reduced. Hence, this topology design has a high-power density. The converter is fairly efficient at full load. However, the converter lacks voltage control. The topology is suitable for applications that demand high-power-density converters, such as LED drivers and automotive headlights.

The converter is a reformed SEPIC-based improved gain and high-frequency converter [15]. The converter possesses features like low voltage stress, high step-up, and soft switching characteristics that allow it to function at high switching frequencies. A gallium nitride (GaN) switch is employed for this converter for better performance at high frequency and to lessen the conduction losses. The converter achieves a high-power density. To minimize the overall volume of the converter, the two magnetic components are confined in a single magnetic core. Even though the two components are physically coupled on a single magnetic core, they are magnetically decoupled so that the converter operates in the desired manner. By employing supplementary passive components, the converter achieves ZVS to reduce the switching losses. The converter is very efficient at full load, and appropriate for photovoltaic and fuel cell applications.

The converter in [16–18], suitable for RE applications, produces a high VG by employing a voltage tripler rectifier and coupled inductor (CI). The quasi-resonant function of the converter is to significantly decrease the losses incurred due to the switching action. The voltage strain across the semiconductor device is also less than that of the conventional SEPIC converter. The converter proposed in [19] is a unidirectional structure with large static gain and it functions with decreased voltage strain transversely for all switches. The converter proposed in [19] is a non-isolated SEPIC-based high VG converter. It is developed by combining a SEPIC and a boosting module to boost the voltage to higher levels than conventional SEPIC converters.

To meet high voltage requirements, CIs can be introduced in the converter topologies to meet the design requirements. This enables high VGs with the converter not operating at very high duty cycles. However, as an aid to the CI's leakage inductance, a snubber circuit should be integrated into the converter topology [20]. The secondary winding of the CI relates to a capacitor in series that ensures that the core does not become saturated from the DC currents. The converter achieves a much higher VG (> 10) than traditional CI DC/DC SEPIC-based topology.

For high-step-up applications, a non-isolated modified SEPIC DC/DC converter architecture is preferred [21–24]. The suggested converter suggests an increased VG and conversion efficacy, in addition to the features of the typical SEPIC converter. The converter commendably traps the leakage energy of the CIs via a passive (resistor-capacitor-diode (RCD)) clamp circuit. By adjusting the connected inductor's turn ratio, the converter gain may be modified. As an improvement, a converter topology with an active clamp is presented by adding a controlled active switch that further helps in enhancing the converter's efficiency.

13.2 CI SEPIC-BASED DC/DC TOPOLOGY

The suggested converter has a single input-output port as a result of integrating CIs, thereby increasing the converter's power density. Figure 13.2 depicts the CI-based SEPIC's higher gain DC/DC converter consisting of a CI (L_p, L_s), inductor (L_3), diodes (D_1,D_2, and D_3), and capacitors (C_1,C_2, and C_3) managed by a only one switch (S) working at 50kHz switching frequency. The suggested converter's primary properties include: (1) a single switch that decreases control circuitry complexity, (2) continuous supply current, (3) high VG, and (4) maximization of resources employed.

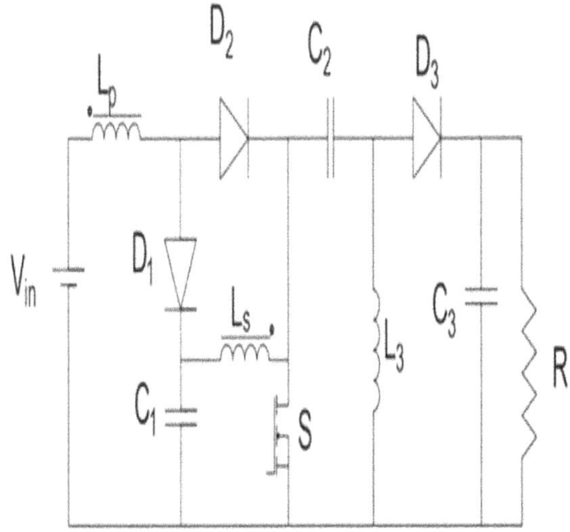

Figure 13.2 SEPIC-based converter with a CI

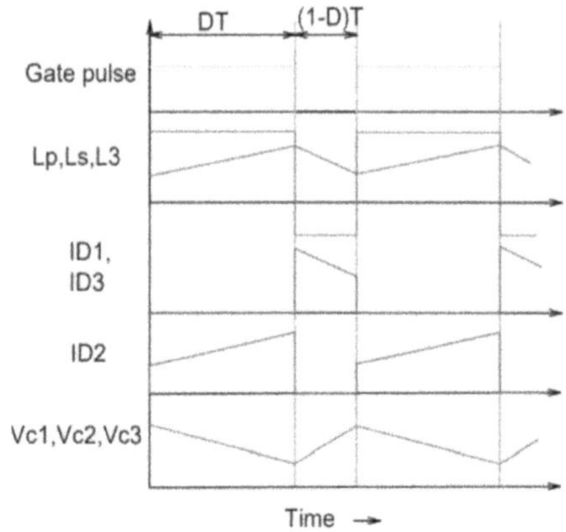

Figure 13.3 Theoretical waveform

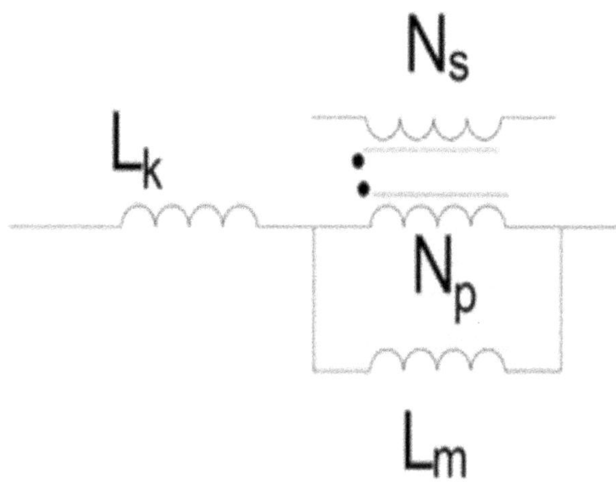

Figure 13.4 CI's model

13.3 WORKING OF THE CONVERTER

To understand steady-state functioning, various assumptions must be made, for example, that all components are ideally suited and capacitors used are high enough to sustain constant voltage. Figures 13.3 and 13.4 depict theoretical waveforms and the CI model. The converter function is discussed in two modes: Mode-a ($t0\,\text{to}\,t1$) and Mode-b ($t1\,\text{to}\,t2$), respectively. The relation between the turn's ratio, leakage inductance, mutual inductance, and coupling coefficient is defined by

$$n = \frac{N_2}{N_1} \tag{13.1}$$

$$k = \frac{L_m}{L_m + L_k} \tag{13.2}$$

where n = turns ratio, k = coefficient of coupling, L_m = magnetizing inductance, L_k = leakage inductance, N_2 = number of turns on the secondary part, and N_1 = number of turns on the primary part.

13.3.1 Mode-a [(*t*0 *to* *t*1)]

During Mode-a, the current path in the CI is L_p from the source ($V_{in} - V_{Lp} - D_2 - S - V_{in}$) and L_s from capacitor C_1 as ($VC_1 - VL_s - S - VC_1$). The current flows through inductor L_3 from capacitor C_2 as ($V_{C2} - S - V_{L3} - V_{C2}$). C_3 reverse biases the diode D_3 and allocates energy to the load simultaneously, as depicted in Figure 13.5.

Additionally,

$$V_{LM} = kV_{in} \tag{13.3}$$

$$V_{LK} = (1-k)V_{in} \tag{13.4}$$

$$V_{Ls} = V_C = nkV_{in} \tag{13.5}$$

$$V_{L3} = V_{C2}, \tag{13.6}$$

where V_{LM}, V_{LK}, and V_{Ls} are the magnetizing induced voltage and leakage inductance of the primary winding of the CIs, and voltage induced in the secondary winding of the CIs, respectively.

13.3.2 Mode-b [*t*1 *to* *t*2]

The CI and inductor L_3 are demagnetized in Mode-b in the following way: Capacitor C_1 is excited by L_p and the supply voltage (V_{in}) along the path ($Vin - VLs - D_1 - C_1 - V_{in}$). Through the path $V_{C1} - V_{Ls} - V_{C2} - D_3 - V_0 - V_{C1}$, the inductor $LY - C_1$ network charges the capacitor. Inductor L_3 discharges through the load at the same time, succeeding the pathway ($V_{L3} - D_3 - Vo$) as illustrated in Figure 13.5(b).

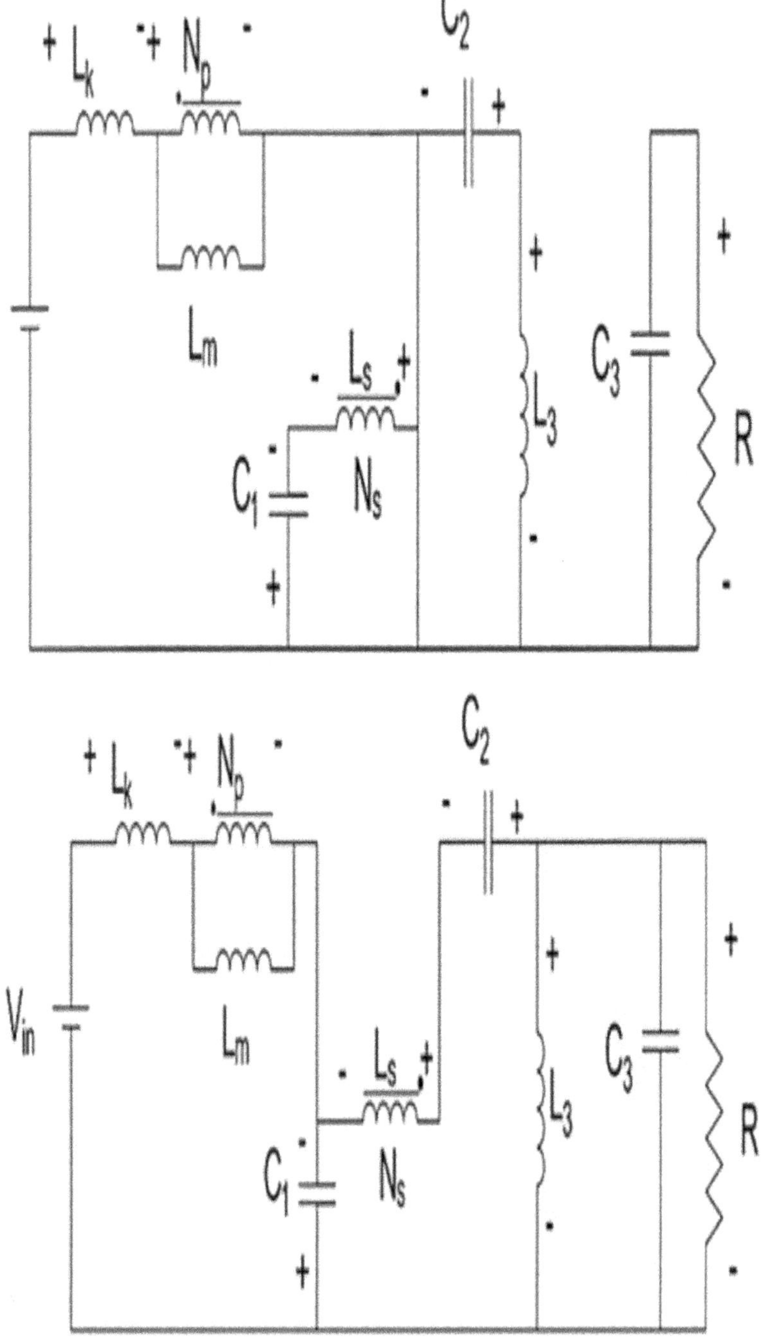

Figure 13.5 (a) Mode-a and (b) Mode-b

Figure 13.3 portrays the typical waveforms of each component during this interval:

$$V_{LM} = k\left(V_{in} + V_{C1}\right) \tag{13.7}$$

$$V_{LK} = (1-k)\left(V_{in} + V_{C1}\right) \tag{13.8}$$

$$V_{Ls} = V_o + V_{C1} - V_{C2} = nk\left(V_{in} + V_{C1}\right) \tag{13.9}$$

$$V_{L3} = V_o. \tag{13.10}$$

From the volt-second balance theory, the following expressions are derived:

$$\frac{V_{C1}}{V_{in}} = \frac{1}{(D-1)} \tag{13.11}$$

$$\frac{V_{C2}}{V_o} = \frac{(D-1)}{D}. \tag{13.12}$$

Stress analysis is done to calculate the voltage stress across the MOSFET during the off state. The following expressions show the relation between the voltage stress with the load voltage and duty ratio:

$$V_T = V_{LS} + V_{C1} \tag{13.13}$$

$$V_T = \frac{V_0}{D}, \tag{13.14}$$

where V_T is the voltage stress across the power device.

13.4 LOSS DISTRIBUTION OF THE CONVERTER

In this subsection, losses contributed by semiconductor devices are discussed. The CI-modified SEPIC converter with non-idealities of power components in the circuit, that is, parasitic losses, on-state conduction losses, diode forward voltage loss, and switching losses, is considered. For the analysis, the magnetic core loss, on-state conduction losses in diodes, and leakage conduction loss are ignored because they are negligible.
The expression for the switching loss is provided by

$$E_{sw} = \int_0^T V_{DS}(t) I_D(t)\,dt \tag{13.15}$$

$$P_{switch} = f_s E_{sw}, \tag{13.16}$$

where E_{sw} is the energy loss for each switching cycle of the MOSFET, $V_{DS}(t)$ is the time function of MOSFET's drain-to-source voltage, $I_D(t)$ is the time function of the drain current of the MOSFET, P_{switch} is the resultant switching loss, and f_s is the switching frequency at which the MOSFET operates. It depends on the magnitude of the blocking voltage, the ON state conduction current, the switching time, and the MOSFET switching characteristics. The switching loss made the biggest contribution to the total losses in a power electronic converter

The following equations represent the on-state conduction loss of MOSFET, and parasitic ohmic losses in the inductive and capacitive components, respectively:

$$P_{s,cond} = R_{DS,on} I_s D \tag{13.17}$$

$$P_{L,cond} = I_{Lp}{}^2 R_{Lp} + I_{Ls}{}^2 R_{Ls} + I_{L3}{}^2 R_{L3} \tag{13.18}$$

$$P_{C,cond} = \frac{V_{C1}{}^2}{R_{C1}} + \frac{V_{C2}{}^2}{R_{C2}} + \frac{V_{C3}{}^2}{R_{C3}}, \tag{13.19}$$

where $R_{DS,on}$ is the on-state resistance of the power device and I_s is the on-state average current through the MOSFET. I_{Lp} and I_{Ls} are the average currents through the primary side and secondary side of the CIs, while R_{Lp} and R_{Ls} are their parasitic resistances, respectively. I_{L3} is the mean value of the primary current of inductor L_3 and R_{L3} is parasitic resistance.

The following equations represent the power loss due to the diode's forward voltage drop:

$$P_{D1} = 0.7 I_{D1}(1 - D) \tag{13.20}$$

$$P_{D3} = 0.7 I_{D3}(1 - D) \tag{13.21}$$

$$P_{D2} = 0.7 I_{D2} D, \tag{13.22}$$

where I_{D1}, I_{D2}, and I_{D3} are the diode currents during the forward conduction mode of the diodesD_1, D_2, and D_3, respectively.

13.5 SIMULATION AND HARDWARE RESULTS

13.5.1 Simulation

The suggested converter is simulated using the specifications in Table 13.1 to ensure that the theoretical analysis performed on the topology is appropriate.

Table 13.1 Prototype specifications

Parameters	*Specifications*	*Parameters*	*Specifications*
Supply voltage	24 V	Load voltage	380 V
Switching frequency	50 kHz	Inductors (L_3)	1 mH
Duty ratio	70%	Capacitors (C_1, C_2, C_3)	220 µF
Output power	400 W	CI turns ratio	1:1

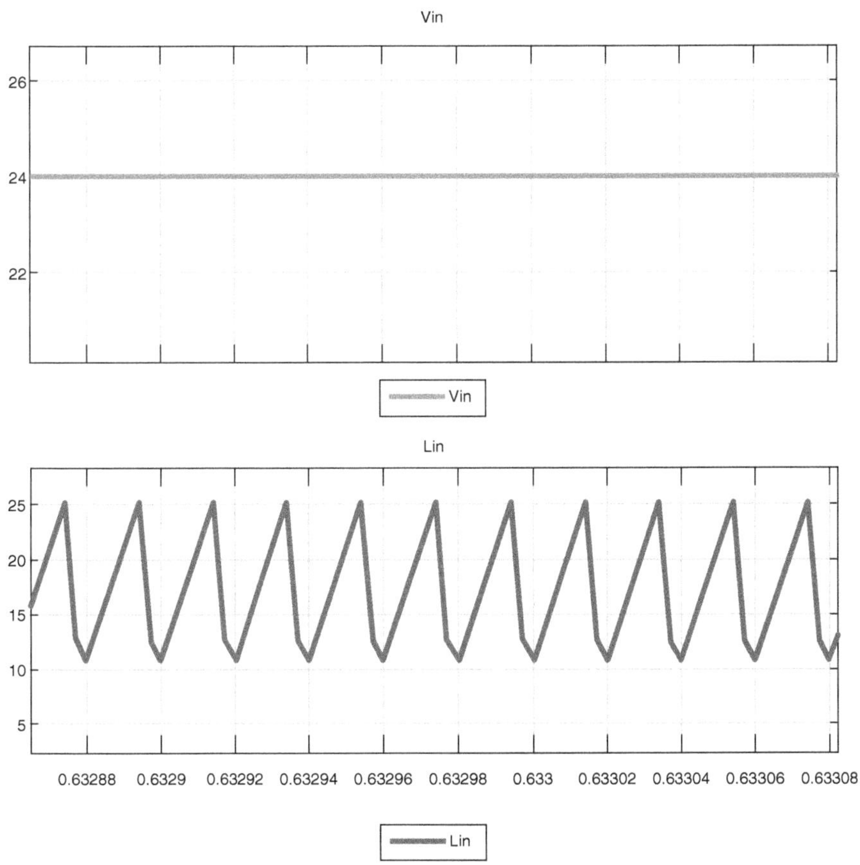

Figure 13.6 (a) DC source voltage and current waveform; Load voltage and load current; Gating pulses; Inductor currents

The converter is powered by a 24 V DC voltage source and the average current consumed from it is 18.33 A. The supply voltage waveform and supply current waveform are depicted in Figure 13.6 (a).

The 380 V load voltage waveform at 70% duty ratio and the load current waveform of 1.085 A are presented in Figure 13.6(b). As shown in

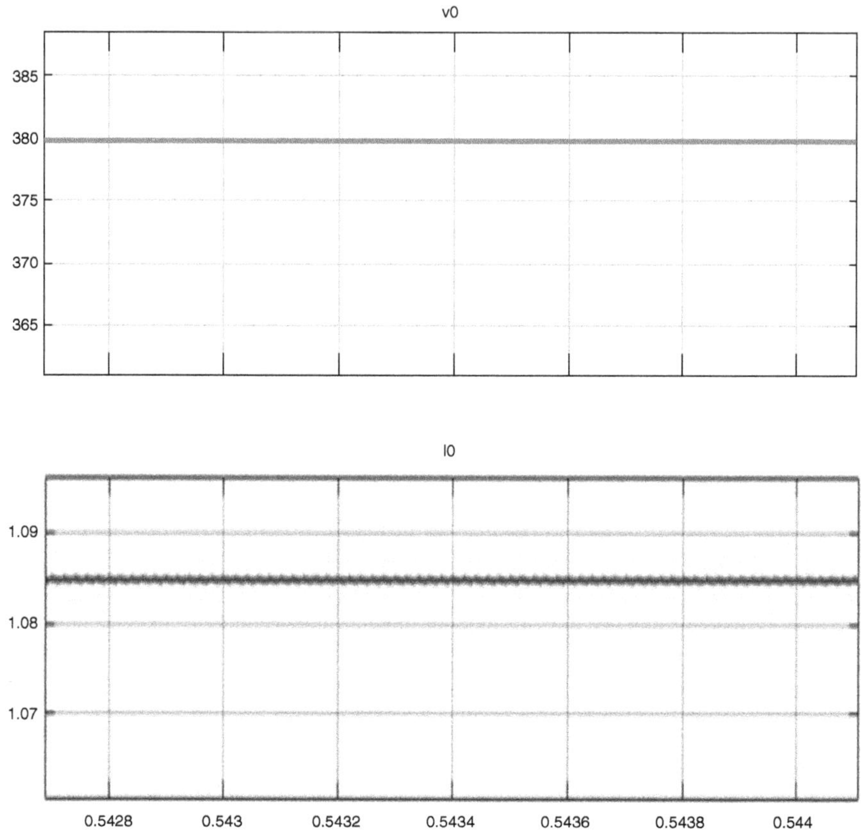

Figure 13.6 (Continued)

Figure 13.6(c), the switch is operated at a switching frequency fs of 50 kHz with a 70% duty ratio.

It is determined that 1A is the mean current through the L_3 inductors. The inductor's ripple current is within acceptable limits. The inductor current waveform and CI current waveform are presented in Figure 13.7. Figure 13.7 clearly illustrates the magnetization and demagnetization of the CI. The ripple current of the CI is within acceptable limits.

Capacitors' C_1 and C_2 voltage stresses are both observed to be 80.5 V. The capacitors' voltage ripple is within acceptable limits, as displayed in Figure 13.8.

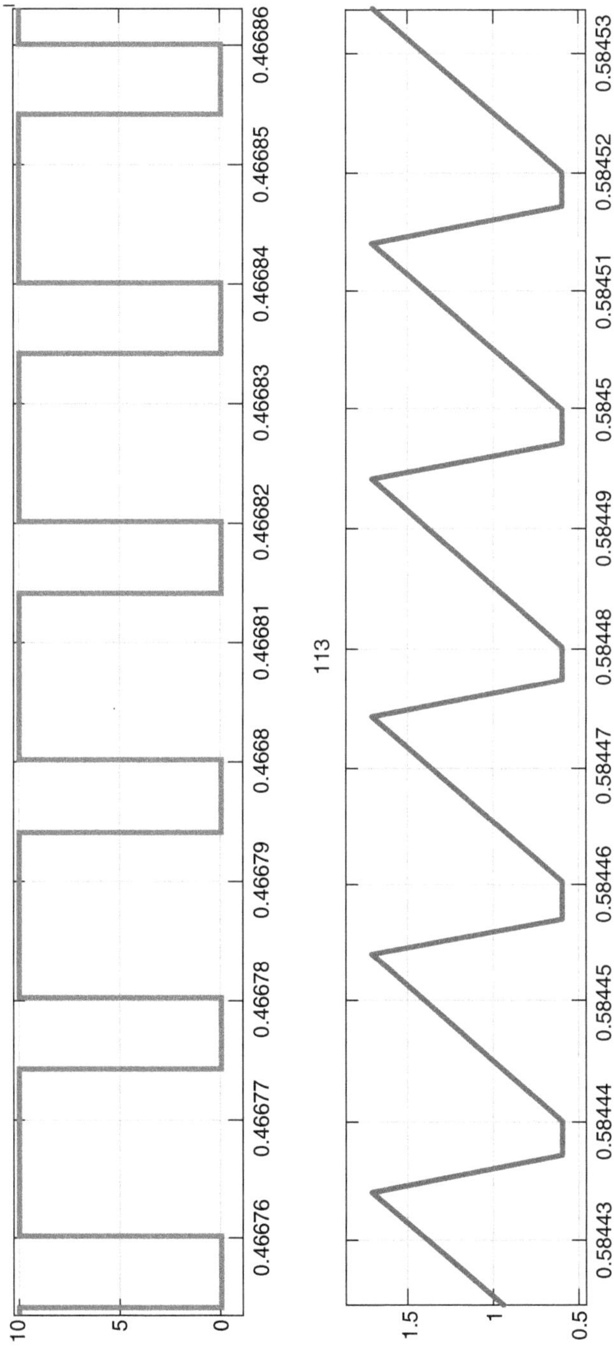

Figure 13.6 (Continued)

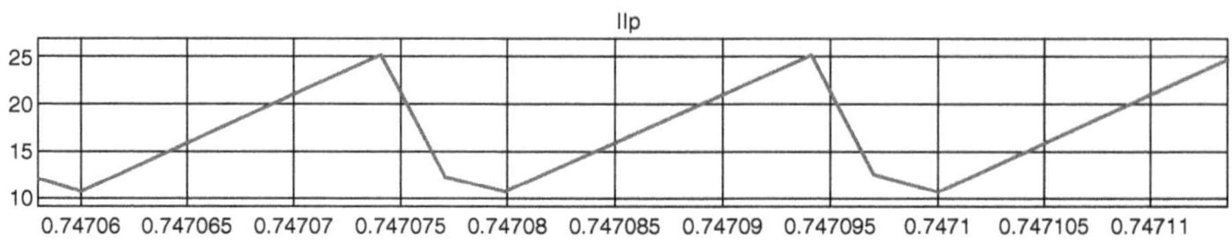

Figure 13.7 CI current

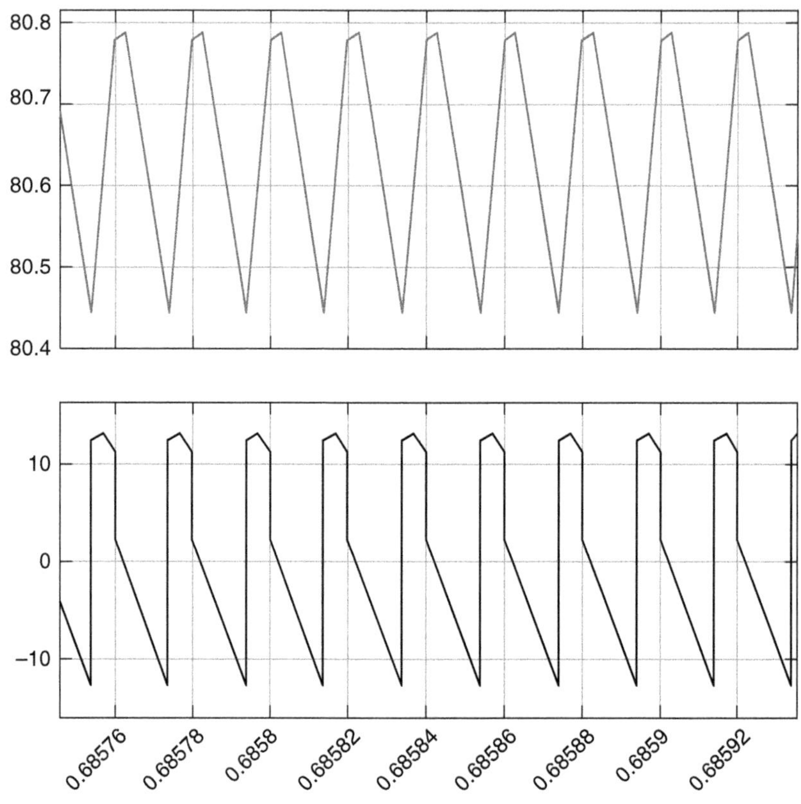

Figure 13.8 Capacitor waveforms

Table 13.2 Hardware prototype specifications

Parameters	*Specifications*	*Parameters*	*Specifications*
Input voltage	24 V	Output voltage	380 V
Switching frequency	50 kHz	Inductors (L_3)	220 uH
Diodes (D1, D2)	GC2X15MPS12-247	Capacitors (C_1, C_2, C_3)	220 μF/450 V
Switch (S)	SPW47N60C3 600 V/47 A	CI turns ratio	1:1

13.5.2 Hardware Results

The converter prototype is designed for 200 W. The specifications are given in Table 13.2.

Figure 13.9(a) demonstrates the loss contribution to the total converter losses. The switching losses comprise 55%, the forward voltage loss in the diode contributes 34%, and the conduction losses include 11% of the total losses. The converter's efficacy at 100 W output power at 70% duty ratio is found to be 93%.

13.6 COMPARISON

The suggested CI-modified SEPIC converter is compared to high-gain converters that have recently been addressed. The comparison is based on the output voltage for different voltage levels. It is evident from the plot shown in Figure 13.9(b) that the suggested topology has a much higher VG than the modified SEPIC topology. Higher VG allows the application of converter operation at lower duty ratios with increased efficiency. Figure 13.10 displays the comparison of the two converters and their efficiency at various power outputs.

13.7 CONCLUSION

It is recommended that a CI-based SEPICisolated high-gain DC/DC converter with critical qualities, such as isolated operation, continuous supply current, substantial VG, and non-inverted output voltage, be used. The recommended converter's operation methods are described in this chapter, as well as a derivation of the formulation of VG. The relevant equations for the converter design are supplied. To evaluate the performance of the suggested converter, MATLAB/Simulink software is used to simulate the conversion procedure. The results attained from simulation and hardware prototype are consistent with the analysis presented. To determine the converter's efficiency, the converter is further examined considering the non-idealities of the components. As a consequence, the converter is an excellent

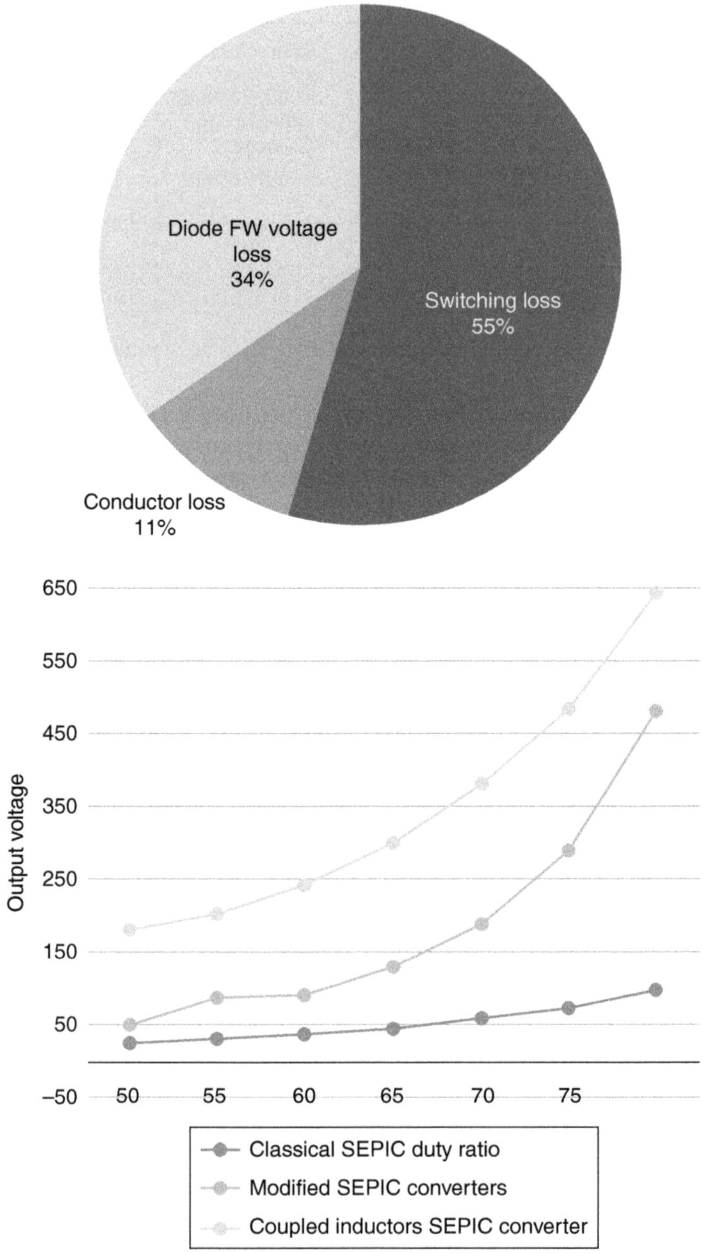

Figure 13.9 (a) Power loss distribution; Output voltage vs duty ratio

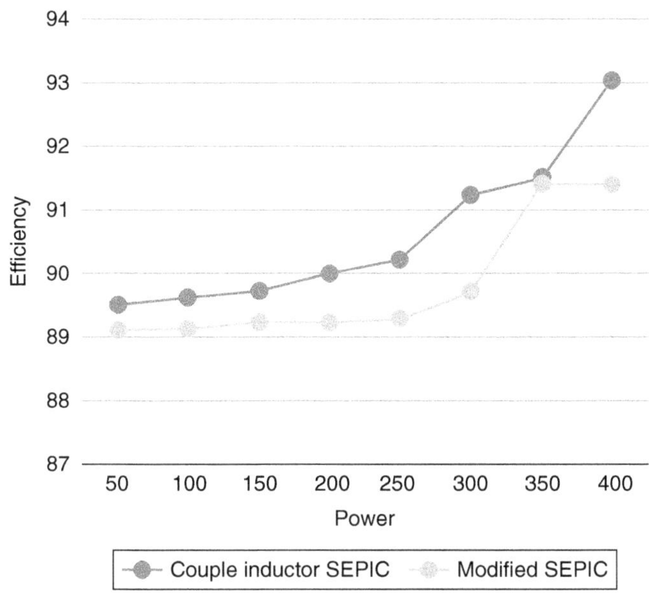

Figure 13.10 Efficiency vs power

choice for high-voltage applications that include critical loads and need galvanic isolation.

REFERENCES

1. G. N. Reddy, D. Guna-Shekhar, S. Choudhari and M. Ademola, "A Statistical Analysis Package for DC/DC Boost-Converter Design," *2011 IEEE 54th International Midwest Symposium on Circuits and Systems (MWSCAS)*, Seoul, 2011, pp. 1–4, doi: 10.1109/MWSCAS.2011.6026555.
2. A. K. Singh and M. K. Pathak, "Single-Stage ZETA-SEPIC-Based Multifunctional Integrated Converter for Plug-in Electric Vehicles," in *IET Electrical Systems in Transportation*, vol. 8, no. 2, pp. 101–111, June 2018, doi: 10.1049/iet-est.2017.0063.
3. R. Gules, W. M. dos Santos, R. C. Annunziato, E. F. R. Romaneli and C. Q. Andrea, "A Modified SEPIC Converter with High Static Gain for Renewable Applications," *XI Brazilian Power Electronics Conference*, Praiamar, 2011, pp. 162–167, doi: 10.1109/COBEP.2011.6085178.
4. M. Kesler, M. C. Kisacikoglu and L. M. Tolbert, "Vehicle-to-Grid Reactive Power Operation Using Plug-In Electric Vehicle Bidirectional Off board Charger," in *IEEE Transactions on Industrial Electronics*, vol. 61, no. 12, pp. 6778–6784, Dec. 2014, doi: 10.1109/TIE.2014.2314065.

5. C. Shi, A. Khaligh and H. Wang, "Interleaved SEPIC Power Factor Pre regulator Using CIs In Discontinuous Conduction Mode with Wide Output Voltage," in *IEEE Transactions on Industry Applications*, vol. 52, no. 4, pp. 3461–3471, July-Aug. 2016, doi: 10.1109/TIA.2016.2553650.
6. S. Bhagyashri and G.Kanimozhi, "Soft Switching Bridgeless Rectifier for NEV Battery Charger," *International Journal of Renewable Energy Research*, vol. 10, no. 2, 2020, pp. 988–999.
7. T. Gopalarathnam and H. A. Toliyat, "A New Topology for Unipolar Brushless DC Motor Drive with High Power Factor," in *IEEE Transactions on Power Electronics*, vol. 18, no. 6, pp. 1397–1404, Nov. 2003, doi: 10.1109/TPEL.2003.818873.
8. Y. Mochizuki and T. Yachi, "Relationship between Power Generated and Series/Parallel Solar Panel Configurations for 3D Fibonacci PV Modules," *2017 IEEE 6th International Conference on Renewable Energy Research and Applications (ICRERA)*, San Diego, CA, 2017, pp. 126–130, doi: 10.1109/ICRERA.2017.8191253.
9. P. K. Maroti, S. Padmanaban, J. B. Holm-Nielsen, M. Sagar Bhaskar, M. Meraj and A. Iqbal, "A New Structure of High VG SEPIC Converter for Renewable Energy Applications," in *IEEE Access*, vol. 7, pp. 89857–89868, 2019, doi: 10.1109/ACCESS.2019.2925564.
10. Y. R. Kafle, S. U. Hasan and G. E. Town, "Quasi-Z-Source Based Bidirectional DC/DC Converter and Its Control Strategy," in *Chinese Journal of Electrical Engineering*, vol. 5, no. 1, pp. 1–9, Mar. 2019, doi: 10.23919/CJEE.2019.000001.
11. S. Lee and H. Do, "Isolated SEPIC DC–DC Converter with Ripple-Free Input Current and Lossless Snubber," in *IEEE Transactions on Industrial Electronics*, vol. 65, no. 2, pp. 1254–1262, Feb. 2018, doi: 10.1109/TIE.2017.2733440.
12. H. Zhang, Y. Chen, D.-H. Kim, S.-J. Park and S.-M. Park, "Bidirectional DC/DC Converter Based on Quasi-Sepic for Battery Charging System," *Journal of the Korean Industrial Convergence Society*, vol. 23, no. 2_1, pp. 139–147, Apr. 2020.
13. S. Jaisudha, S. Srinivasan and G. Kanimozhi, "Bidirectional Resonant DC-DC converter for Microgrid Applications," *International Journal of Power Electronics and Drives Systems*, vol. 8, no. 4, Dec. 2017, pp. 1548–1561.
14. Y. Wang, S. Gao and D. Xu, "A 1-MHz-Modified SEPIC With ZVS Characteristic and Low-Voltage Stress," in *IEEE Transactions on Industrial Electronics*, vol. 66, no. 5, pp. 3422–3426, May 2019, doi: 10.1109/TIE.2018.2851974.
15. S. Hasanpour, A. Baghramian and H. Mojallali, "A Modified SEPIC-Based High Step-Up DC–DC Converter with Quasi-Resonant Operation for Renewable Energy Applications," in *IEEE Transactions on Industrial Electronics*, vol. 66, no. 5, pp. 3539–3549, May 2019, doi: 10.1109/TIE.2018.2851952.
16. L. Gu and K. Jin, "MHz Zero-Voltage-Switched Isolated Resonant Converter for Wide-Output-Voltage Application," in *IET Power Electronics*, vol. 10, no. 15, pp. 2170–2179, 15 Dec. 2017, doi: 10.1049/iet-pel.2016.0665.

17. M. P. Hirth, R. Gules and C. H. I. Font, "A Wide Conversion Ratio Bidirectional Modified SEPIC Converter with Non- Dissipative Current Snubber," in *IEEE Journal of Emerging and Selected Topics in Power Electronics*, vol. 9, no. 2, pp. 1350–1360, 2020, doi: 10.1109/JESTPE.2020.2966410.
18. P. F. de Melo, R. Gules, E. F. R. Romaneli and R. C. Annunziato, "A Modified SEPIC Converter for High-Power-Factor Rectifier and Universal Input Voltage Applications," in *IEEE Transactions on Power Electronics*, vol. 25, no. 2, pp. 310–321, Feb. 2010, doi: 10.1109/TPEL.2009.2027323.
19. P. K. Maroti, S. Padmanaban, J. B. Holm-Nielsen, M. Sagar Bhaskar, M. Meraj and A. Iqbal, "A New Structure of High VG SEPIC Converter for Renewable Energy Applications," in *IEEE Access*, vol. 7, pp. 89857–89868, 2019, doi: 10.1109/ACCESS.2019.2925564.
20. S. Gao, Y. Wang, Y. Guan and D. Xu, "A High Step-Up SEPIC-Based Converter Based on Partly Interleaved Transformer," in *IEEE Transactions on Industrial Electronics*, vol. 67, no. 2, pp. 1455–1465, Feb. 2020, doi: 10.1109/TIE.2019.2910044.
21. S. A. Ansari and J. S. Moghani, "A Novel High VG Non-CI SEPIC Converter," in *IEEE Transactions on Industrial Electronics*, vol. 66, no. 9, pp. 7099–7108, Sept. 2019, doi: 10.1109/TIE.2018.2878127.
22. G. Kanimozhi and V. T. Sreedevi, "Semibridgeless Interleaved PFC Boost Rectifier for PHEV Battery Chargers," *IETE Journal of Research, Taylor and Francis*, vol 65, no. 1, pp. 128–138, 2019.
23. Y. P. Siwakoti, M. Soltani, F. Blaabjerg and A. Mostaan, "A Novel Quasi-SEPIC High-Voltage Boost DC/DC Converter," *2017 IEEE Applied Power Electronics Conference and Exposition (APEC)*, Tampa, FL, 2017, pp. 2213–2216, doi: 10.1109/APEC.2017.7931006.
24. M. Premkumar, U. Subramaniam, H. Haes Alhelou, P. Siano, "Design and Development of Non-Isolated Modified SEPIC DC/DC Converter Topology for High-Step-Up Applications: Investigation and Hardware Implementation." *Energies,* vol. 13, p. 3960, 2020.

Chapter 14

IoT-Based Power Theft Detection

Ameesh Singh, Harsh Gupta, Aayush Karthikeyan, Akhtar Kalam, O.V. Gnana Swathika

14.1 INTRODUCTION

Electricity theft is an exceptionally regular issue in the nation, where the populace is extremely high and the utilization of electricity is enormous. In India, there is a consistently expanding number of electricity thefts. The ways in which theft occurs are so numerous that it is hard to keep track and the stakes are too high to not pay attention as soon as feasible given the conditions.

14.1.1 Objectives

The objectives are as follows:

- to identify whether power theft has occurred or not on the basis of the power consumed by different types of loads with the help of machine learning (ML) algorithms, implement algorithm-based learning for identifying theft using ML, and display the results on a user-friendly website
- to build a hardware prototype to detect power theft in the network and notify the user in the case of theft using a buzzer
- to set up a message alert system for the user whenever power theft is detected and integrate this with ThingSpeak
- to implement remote live monitoring for the proposed hardware prototype and integrate a voice and notification alert system.

14.1.2 Literature Survey

In today's linked world, the development of intelligent gadgets has ushered in a new era of innovation. This development has opened new gateways of machine-to-machine communication with the internet linked to it. In a study, researchers describe the design and implementation of an energy meter utilizing the Internet of Things (IoT) concept. Controls can be maintained from

 DOI: 10.1201/9781003537878-14

a distance and any theft, if detected, will result in automatically switching off power [1].

The IoT plays an important and significant role in the smart grid. Researchers have thrown light on the deepening of IoT technology and its application in the power system. They proposed a status monitoring and early warning system for the power distribution network based on IoT technology. The system can be used to display the status of all equipment in the distribution network and on-line theft monitoring can be performed [2].

In earlier times when analog meters were used, permanent magnets were used to slow down its rotation. Later, for digital meters, power was looted through bypassing the meter. Smart meters with communication networks and a database management system (DBMS), as discussed in a study, can help to locate and thus minimize power theft [3].

Electric power is the route to digitalization. With increasing demand, many people are undergoing the illegal practices of stealing power. This results in a massive amount of losses. In a paper, researchers described how these losses can be worked on if IoT technology is incorporated [4]. Electricity theft is a key source of concern for utilities because it has a significant impact on the grid. To work on it in the best way possible, researchers described combinations of various ML algorithms plus data handling techniques. They implemented different ML algorithms, for example, logistic regression (LR) and random forest, and then computed the results, and compared the area under the receiver operating characteristic (ROC) curve (AUC) and F1 score to check which combination provided the best results [5].

The distribution network undergoes many challenges, with power theft being the major one because it results in unbalanced loads and thus impacts utilities. An IoT-based solution was developed and described to deal with the power theft issue. It senses power theft in a line by comparing the sending and receiving end current values and thus alerts the authority to take action if theft is detected. It also keeps a check on energy consumption and alerts the authority if it is exceeded [6].

Researchers have described a strategy for locating power theft and taking appropriate action both under normal circumstances and when power theft is identified. When the controller detects theft, it turns off the power when it receives a signal and re-checks for theft. If theft is not detected after four or more attempts, the system is reset, otherwise an alert message with the stolen location is issued [7].

The combination of smart grid technologies and smart meters with the information communication technology provides a reliable way to identify power theft. The use of IoT in the detection of power theft, as well as the real-time monitoring of smart meters, were discussed in a study. The Android application is used to track consumer consumption and to inform the authorities in the case of power theft [8].

Arduino and Global System for Mobile Communications (GSM)-based smart energy meters have been proposed. These smart energy meters can read and send billing and metering data via a wireless network utilizing GSM technology and a GSM modem. Smart energy meters store data in the cloud and thus the user can access them using a mobile app. These smart meters will become cost friendly and, at the same time, with advanced technology can also monitor consumption and power theft [9].

Researchers described the digitalization of the load energy usage and power theft using various hardware and software tools. Using this smart energy monitoring, human intervention will be reduced. Using a channel identifier (ID), the user can monitor energy consumption using a webpage. Power theft detection is monitored using an infrared (IR) sensor and data are fed into a Raspberry Pi, which then relays them to a webpage [10].

In a study, researchers measured the current values at the shaft and at the meter using current curl and loop. If the current at both points is found to be equivalent, then there is no robbery, otherwise power robbery is detected. The difference in units is sent to the predefined site page using a Wi-Fi module. At the theft point, a Short Message Service (SMS) is sent to the concerned specialist with the house number [11]. Through data analysis techniques, smart grid adaption can considerably reduce this loss. The smart grid system creates a large amount of data, including individual user power usage. With the help of this data, ML and deep learning (DL) techniques may rightly identify electricity thieves. In a study, researchers proposed an electrical theft detection system based on convolutional neural networks (CNN) and long short-term memory (LSTM) architecture. CNNs are frequently used in technologies that automate power, such as feature extraction and classification. A novel data preprocessing strategy was also used in the study to find and evaluate missing chunks in the dataset [12].

The mechanism employed by businesses and academia to identify electricity theft is called electricity theft detection (ETD). However, ETD cannot be utilized effectively due to reasons such as unbalanced data, processing of high-dimensional data, and overfitting difficulties. As a result, a remedy to the aforementioned restrictions has been presented. To detect aberrant patterns in electricity consumption data, an LSTM algorithm is used, along with the random undersampling boosting (RUSBoost) algorithm, to optimize parameters. The support vector machine (SVM), CNN, and LR are examples of state-of-the-art approaches that outperform the proposed method (LR) [13].

In a study, researchers proposed a deep residual neural network-based model for power theft that is, in turn, based on the timing and periodicity trait of smart meter data to mine the enormous amount of data offered by smart meters in a better way and improve the accuracy of power theft detection models. The proposed method has a high detection rate; different studies on real datasets have validated that the proposed model has a higher detection rate compared to others, as well as a low false-positive rate [14].

A brief analysis of ML research in power theft detection using the data of smart meters has been offered in a study. It then summarizes and compares different algorithms utilized in terms of indicators, simulation and analytic environment, and sets of data used. Finally, it emphasizes the difficulties in detecting power theft. It also acknowledges that researchers have not yet paid sufficient attention to or explored these issues in prior studies and that they must be solved to enhance power theft detection [15].

The goal of a study was to use IoT to measure power usage in a family unit and automatically generate a bill. Information on electricity usage can be updated automatically, reducing the amount of effort required by humans. The microcontroller is used to enable workouts with a sophisticated metering system, and to connect the system to the internet and server. A latent infrared sensor is integrated into the framework to detect any unauthorized changes in the metering system. The proposed framework is capable of accurately detecting the amount of energy utilized and improves smart home security [16].

The goal of another study was to develop a system for monitoring load consumption, and detecting and eliminating power theft in transmission lines and energy meters. This initiative also focuses on using IoT to communicate theft information to the Electricity Board (EB). A network of connected devices, such as sensors, has the potential to communicate real-time information via the internet. A Raspberry Pi is used in this project to detect power theft and send commands to the GSM module, which then sends the theft information message to the EB [17].

With the transformation in the electric industry, the dependability of power supply and the quality of energy are receiving more attention. Electricity providers and users are both concerned about reliable electricity, regardless of whether the focus is on intrusion, aggravation, or protracted power outages. The primary hurdle to delivering quality control to valued subscribers is power theft [18].

The main purpose of a research project was to design a system that can track how much electricity is utilized per load, as well as trace and eliminate electricity theft in the current line and meter. This includes telling officials from the EB about the theft that occurred as a result of the IoT. The transmission of real-time data through the internet is aided by a network of connected devices, such as sensors. The Arduino Uno detects energy theft and sends information to the GSM module, which then sends it to the EB [19].

Due to physical and human constraints, detecting and eliminating power theft in developing countries such as India is extremely challenging. In a study, researchers described an IoT-based solution for monitoring and eliminating power theft, identifying distribution system faults, and upgrading existing grids to smart grids that can be added to existing power grid and distribution company (DISCOM) infrastructure [20].

14.2 PROPOSED METHODOLOGY

14.2.1 Overview

A complete flowchart of the project is shown in Figure 14.1. ML analytics is performed on a large dataset and different models are deployed successfully to detect power theft in the system.

A hardware prototype is created to detect power theft in the system. It consists of a consumer end and distributor end. Then this hardware prototype is integrated with software after interfacing with ThingSpeak. The simulation results are presented on a user-friendly website.

Figure 14.1 Complete flowchart of the project

Various modes are established to detect power theft, including a buzzer in the hardware prototype, a notification and alarm system on the live monitoring website, and an SMS alert to the concerned user in the case of power theft.

14.2.2 Modules

The following modules represent the different phases of the project in chronological order.

14.2.2.1 Module 1

On a big complex dataset, different ML algorithms are employed to train an ideal system for ML prediction. Because it is not possible to construct such datasets using a hardware prototype, the model is later successfully employed for complex predictions, whose results are displayed on the website as well.

14.2.2.2 Module 2

A hardware prototype to detect power theft in real time is created. Two energy meters, one at the distributor end and one at the consumer end, two loads, two light-dependent resistor (LDR) sensors, and a microcontroller are used. Counter values are in sync with the LDR sensors and are used to detect power theft. If the counter values differ significantly, a subsequent alarm and notification of power theft are raised.

14.2.2.3 Module 3

The hardware prototype is interfaced with software through ThingSpeak and a remote live monitoring system is deployed, which shows the status of the blinks detected by the LDR sensor at the distributor and consumer end energy meters, and also notifies the user in the case that power theft has occurred. An SMS alert system is also set to notify the user in the case of power theft.

14.3 DESIGN SPECIFICATIONS

14.3.1 Design Approach (Machine Learning Part)

Three ML algorithms are deployed on a complex dataset, which is found on a government website that shows the power consumption of different households at different points in time.

A dataset is considered that contains the consumption of various electrical appliances for different households: 80 per cent of the dataset is used for training the two models and 20 per cent is used for testing purposes. Based on the training of the ML models, various points in the dataset are found where theft is detected. A website is created using a Python-based Flask framework. The ML models are dumped into the framework and then tested for certain values against all the columns of the dataset on which they were trained. The entered values are then passed to the corresponding variables using the POST method. Then the ML models check for theft in the given condition of the variables. Based on the output, it is concluded that power theft might or might not exist in the system and the corresponding results are displayed on the website.

14.3.2 Design Approach (Hardware Prototype and its Integration with Software Platforms)

The design approach is as follows:

- An IoT-based hardware prototype is created with a consumer side and distributor side.
- Two analog energy meters are deployed, one acting as a distributor end energy meter and the other as a consumer end energy meter.
- An LDR sensor is attached to the energy meters, which senses and matches the amount of light coming out of both the meters in the form of blinks.
- Two types of loads (which are two 100 W bulbs in this prototype) are used. The load connected on the consumer side is the legal load and the load connected on the distribution line connecting the distributor and consumer is the illegal load.
- In the case of no power theft, that is, when only the legal load is switched on, the power reading at the distributor side is the same as the power consumed on the consumer side; hence, there is no difference in the amount of light from both meters, resulting in the same number of blinks sensed by the LDR sensor.
- Now, in the case of power theft, that is, when an illegal load is switched on (which is connected on the distribution line connecting the distributor and the consumer sides), there is a difference in the power reading at the distributor end and the power consumed at the consumer end, resulting in a difference in the amount of light produced by both meters and hence, a different number of blinks is sensed by the LDR sensor, which is communicated to the microcontroller.
- An SMS alert system is also set up, which triggers a message to the mobile number of the concerned user in case electricity theft has occurred. The message triggered is displayed.

- The phone number to which the power theft message needs to be sent is provided in the Arduino code.
- Hardware is interfaced with ThingSpeak and thus the live status of the counters corresponding to the blinks sensed by the respective LDR sensors is observed as a graph pattern on ThingSpeak.
- The ThingSpeak application programming interface (API) is used in Node-RED to establish a remote live monitoring system.
- A user-friendly website is created, which comes with login authorization and a username to be entered on the login page to log in the user, who has to enter the same number that appears in the SMS alert, which is verified with the help of the microcontroller code.
- In the case in which the user enters a wrong username, the user is unable to log in and the message "wrong username entered" pops up on the login page.
- Once logged in successfully, the user is able to see a dashboard with two gauges corresponding to the consumer end and distributor end energy meters. These gauges work on the blinks detected by the LDR sensor connected at both the consumer and distributor ends.
- In the case of no power theft, that is, when only the legal load is switched on, the blinks detected by the LDR sensor connected at both the consumer and distributor ends are the same, so the reading of the gauge increases at the same rate.
- Now, in case of theft, that is, when the illegal load connected to the distribution line is also switched on, the blinks vary and thus the gauge reading also varies. Once this variation reaches a significant difference, a voice alarm and notification pop-up notifying the user that power theft has occurred are deployed.

14.3.3 Design Requirements

The hardware and software requirements are as indicated in Tables 14.1 and 14.2, respectively.

14.4 RESULTS AND DISCUSSION

14.4.1 Analytical Results

ML-based analytics is performed on a complex large dataset using different ML algorithms. The models succeed in detecting power theft. Of all the three models used, random forest obtains the most accurate results.

Figures 14.2–14.4 show the accuracy results for all three ML models.

The user inputs the values of the power consumed by different appliances on the webpage created and clicks the submit button. This directs the user to a page that shows whether power theft has occurred or not with the help

Table 14.1 Hardware

Components	*Quantity*	*Specification*
Arduino UNO	1	R3 Atmega328P
Energy meter	2	Single phase 240 V, 05-30A
LDR sensors	2	150 V DC, 100 mW
LCD screen	1	16 × 2
Wi-Fi module	1	ESP8266
GSM module	1	SIM900A
Buzzer	1	Voltage: 5 V Battery: 9 V
Zero PCB	1	10 × 7.5 cm
Bulbs	2	100 W each
Female-to-female jumper wires	50	N/A
Switch	2	N/A

Table 14.2 Software

Software used	*Requirement*
Arduino IDE 1.8.14	For microcontroller coding
Node-RED	For live monitoring dashboard
Web server: ThingSpeak	For integration with software
VS code	For website coding

of the best algorithm working on the user-provided values at the backend, which is the random forest algorithm. Figure 14.5 shows the interface of the final webpage.

14.4.2 Simulation Results

Node-RED is used for the voice alert system when power theft is detected and for creating the dashboard of the website. The ThingSpeak API is fetched and the status of the LDR sensor readings is passed to the webpage via Node-RED. The flow of the Node-RED screen is shown in Figure 14.6.

A screenshot of the dashboard on which the user is directed for live monitoring after logging in, which contains the distributor end and consumer end blink status sensed by the LDR, is shown in Figure 14.7.

Screenshots of the ThingSpeak dashboard with values being updated live in an Excel sheet are shown in Figures 14.8 and 14.9.

14.4.3 Hardware Results

The hardware prototype is shown in Figure 14.10.

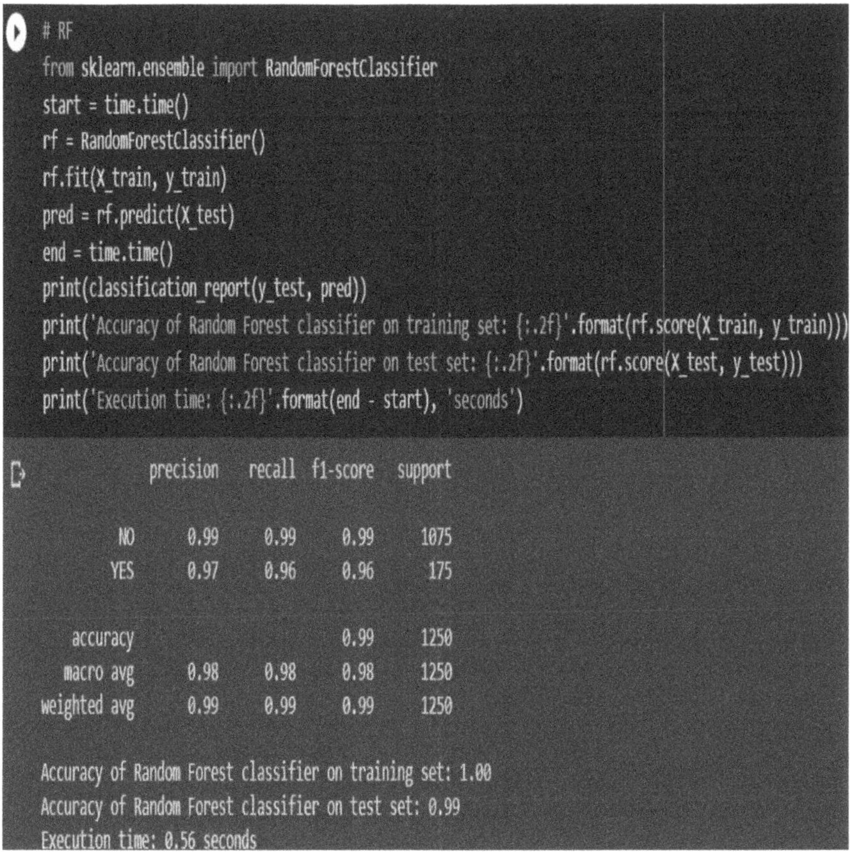

Figure 14.2 Random forest classifier results

Figure 14.11 shows the case of no power theft, when only the legal load is switched on, and Figure 14.12 shows the zoomed in picture of the liquid crystal display (LCD) screen corresponding to this case, in which the counter values, that is, C1 and C2, are equal.

Figure 14.13 represents the power theft condition in which the legal load at the consumer end and the illegal load on the distribution line are switched on at the same time. Figure 14.14 is a zoomed picture of the LCD screen. In this case, both counter values differ significantly, which is sensed by the microcontroller, and then the user is alerted accordingly.

14.4.4 Website and Integration Results

Figure 14.15 shows the website created with the help of Flask, HTML, CSS, and JavaScript. The website has a home page and login authorization

```
# Naive Bayes
from sklearn.naive_bayes import GaussianNB
start = time.time()
gb = GaussianNB()
gb.fit(X_train, y_train)
pred = gb.predict(X_test)
end = time.time()
print(classification_report(y_test, pred))
print('Accuracy of Random Forest classifier on training set: {:.2f}'.format(gb.score(X_train, y_train)))
print('Accuracy of Random Forest classifier on test set: {:.2f}'.format(gb.score(X_test, y_test)))
print('Execution time: {:.2f}'.format(end - start), 'seconds')
```

```
              precision    recall  f1-score   support

          NO       0.99      0.90      0.94      1075
         YES       0.60      0.94      0.73       175

    accuracy                           0.90      1250
   macro avg       0.80      0.92      0.84      1250
weighted avg       0.93      0.90      0.91      1250

Accuracy of Random Forest classifier on training set: 0.91
Accuracy of Random Forest classifier on test set: 0.90
Execution time: 0.01 seconds
```

Figure 14.3 Naive Bayes result

page, as shown in Figures 14.16 and 14.17, respectively, and the final live monitoring dashboard, as shown in Figure 14.18.

14.4.4.1 Power Theft Case

The whole setup includes the hardware prototype shown in Figures 14.19 and 14.20, integrated with ThingSpeak and a remote live monitoring system. As explained, when both the loads are switched on, the counter values are not equal and differ substantially. This in turn is displayed on the live monitoring website shown in Figure 14.21, which is monitored from a different device (laptop). As soon as the illegal load is switched on, the user is notified through a pop-up on the website, as well as a voice alert.

```
from sklearn.metrics import classification_report, confusion_matrix
print(classification_report(ytest2, y_pred2, target_names=['No', 'Yes'])) #SVC

              precision    recall  f1-score   support

          No       1.00      0.99      0.99       861
         Yes       0.95      0.97      0.96       139

    accuracy                           0.99      1000
   macro avg       0.97      0.98      0.98      1000
weighted avg       0.99      0.99      0.99      1000
```

Figure 14.4 SVM results

Figure 14.22 shows the message received by the user in the aforementioned scenario of power theft using the number entered by the user, which is the same as the number in the microcontroller coding.

14.5 CONCLUSION

A big dataset collected from a government website was used to train and create an ML-based system, which was then successfully integrated with a website that allowed the user to enter values and check if power theft had occurred or not based on the dataset provided. Because it is not possible to generate such a versatile dataset using a prototype model, this system was successfully deployed for large and complex already-available datasets.

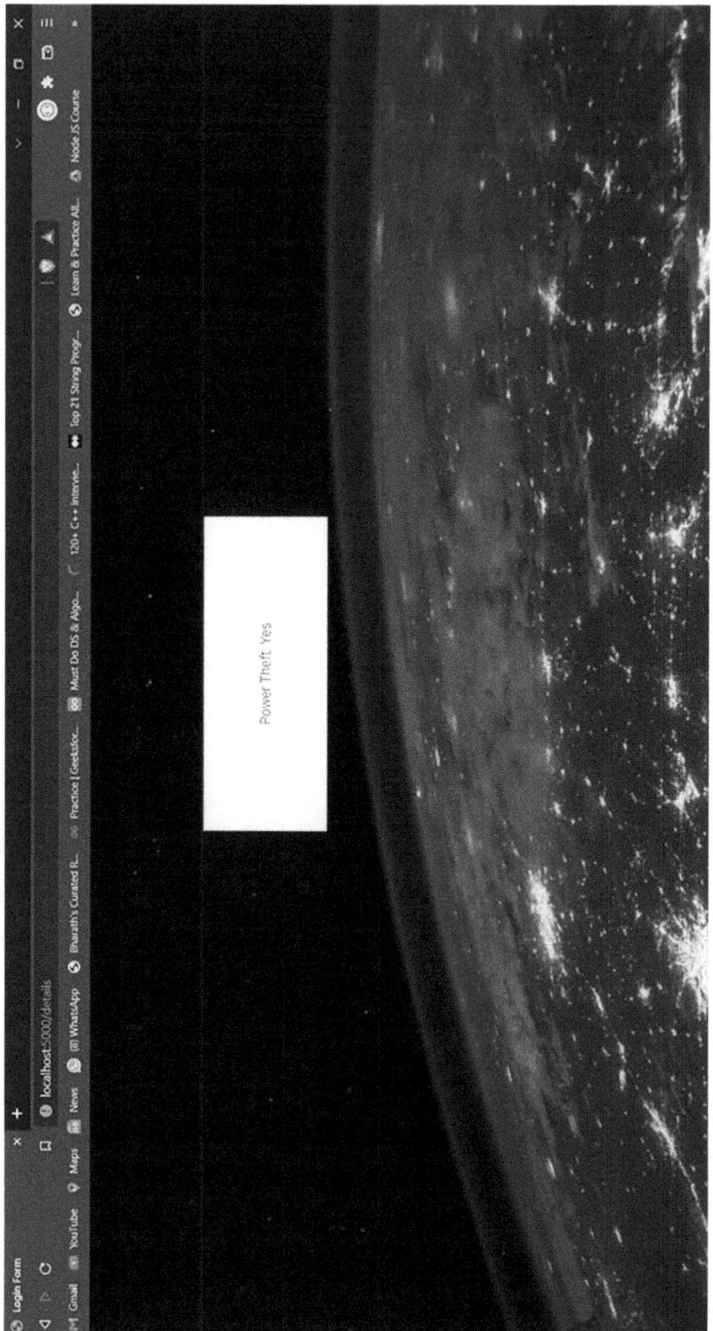

Figure 14.5 Website showing whether power theft has occurred or not

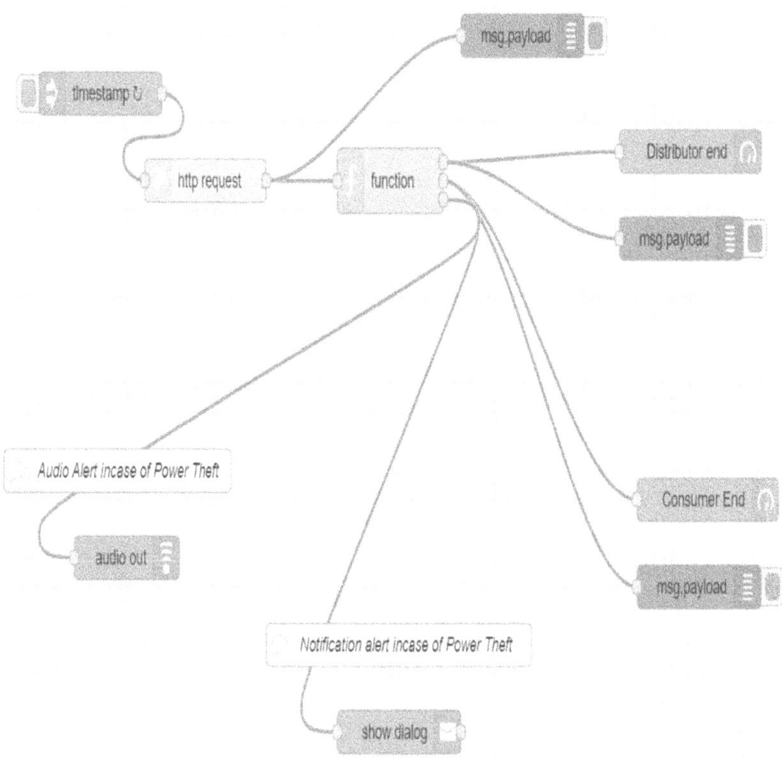

Figure 14.6 Node-RED flow

A hardware model was successfully implemented to detect power theft and alert the user. The hardware was interfaced with ThingSpeak, which also plotted the counter values. Additionally, the API of ThingSpeak was used in Node-RED to provide a user-friendly dashboard incorporating a login system consisting of a website.

An SMS alert system with the help of a GSM module was also set up, which sent a message to the authorized user in the case in which power theft was detected.

This way multiple user-friendly platforms were incorporated to alert the user if power theft occurred.

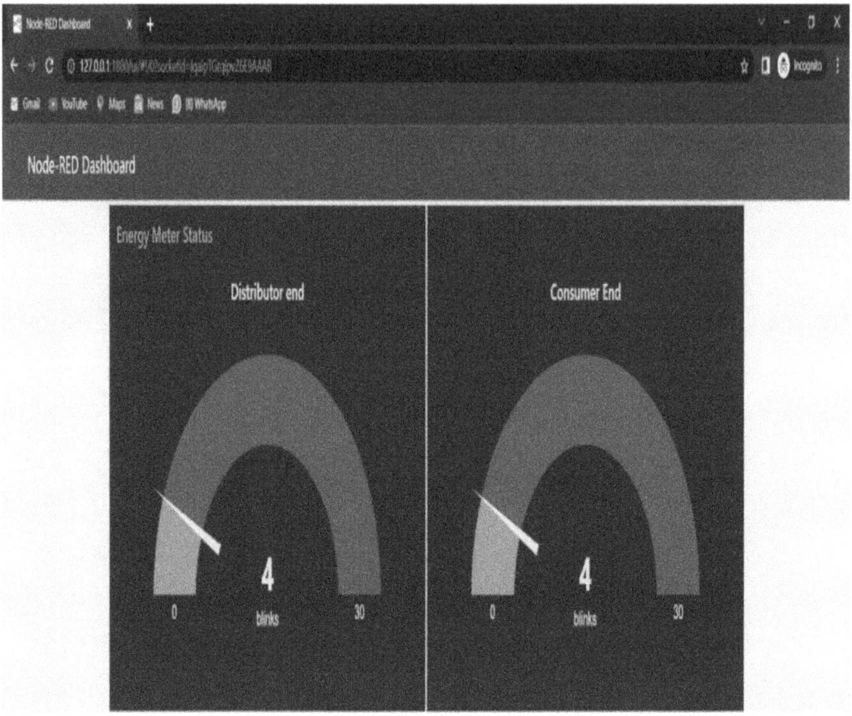

Figure 14.7 Node-RED dashboard

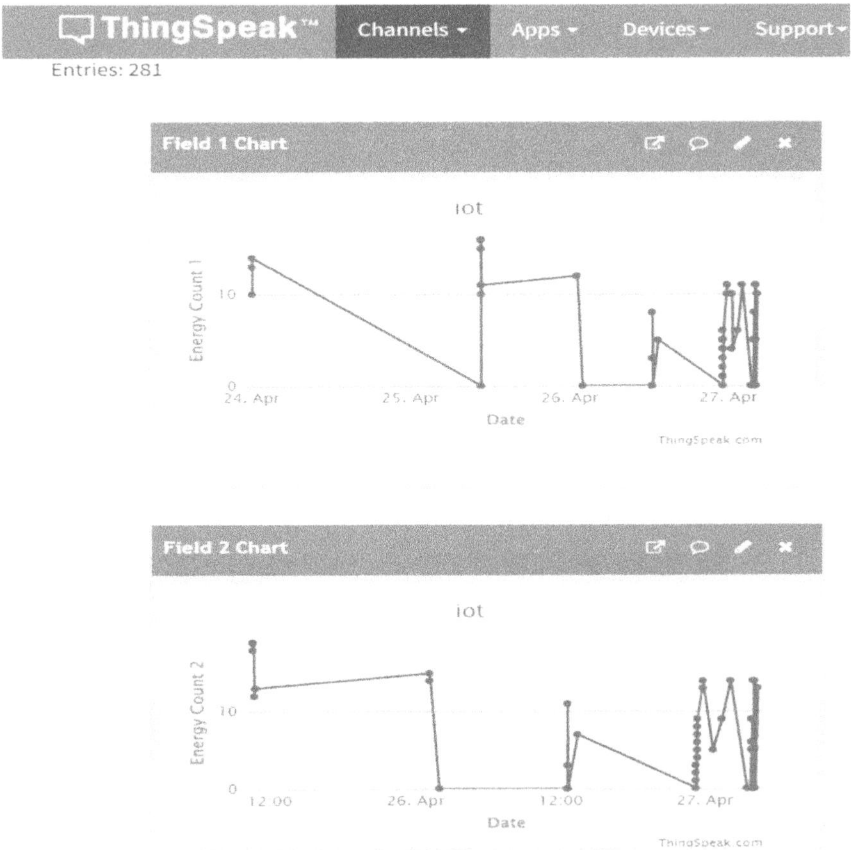

Figure 14.8 ThingSpeak dashboard and chart

	A	B	C	D	E
1	created_at	entry_id	field1	field2	
2	2022-03-06 12:58:26 UTC	143	0	0	
3	2022-03-13 15:12:22 UTC	144	1	1	
4	2022-03-13 15:12:40 UTC	145	2	2	
5	2022-03-13 15:12:59 UTC	146	2	3	
6	2022-03-13 15:13:18 UTC	147	2	4	
7	2022-03-13 15:13:35 UTC	148	2	5	

2022-04-26 17:50:28 UTC	243	0	0
2022-04-26 17:51:20 UTC	244	1	1
2022-04-26 17:51:45 UTC	245	2	2
2022-04-26 17:52:06 UTC	246	3	3
2022-04-26 17:52:22 UTC	247	4	4
2022-04-26 17:52:59 UTC	248	4	5
2022-04-26 17:53:37 UTC	249	5	6
2022-04-26 17:54:13 UTC	250	5	7
2022-04-26 17:54:44 UTC	251	5	8
2022-04-26 17:55:00 UTC	252	6	9

Figure 14.9 Excel timestamp for ThingSpeak

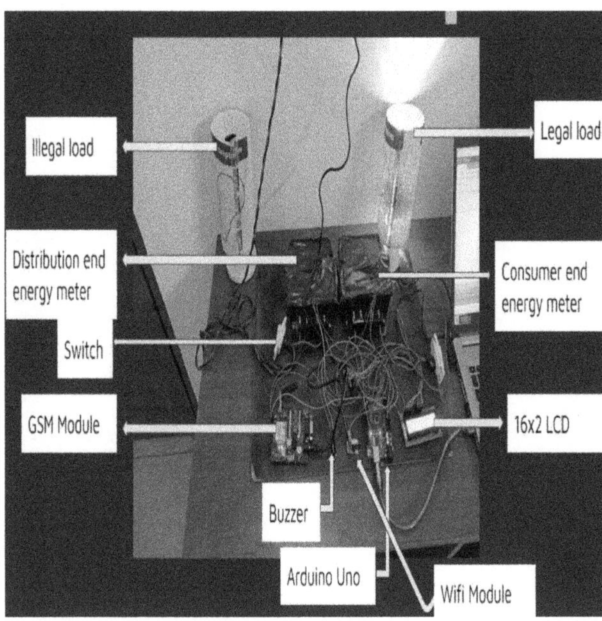

Figure 14.10 Labeled image of the hardware model

Figure 14.11 Only the legal load is switched on

Figure 14.12 Zoomed in image of the LCD screen in the case of no power theft

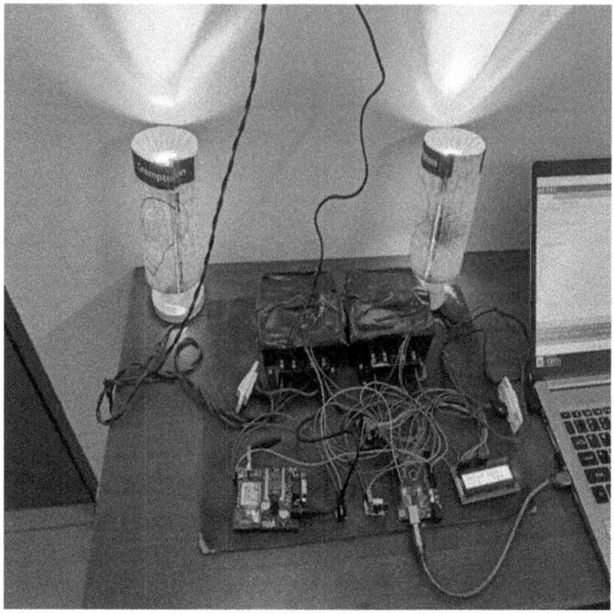

Figure 14.13 Both legal and illegal loads are switched on

Figure 14.14 Zoomed in image of the LCD screen in the case of power theft

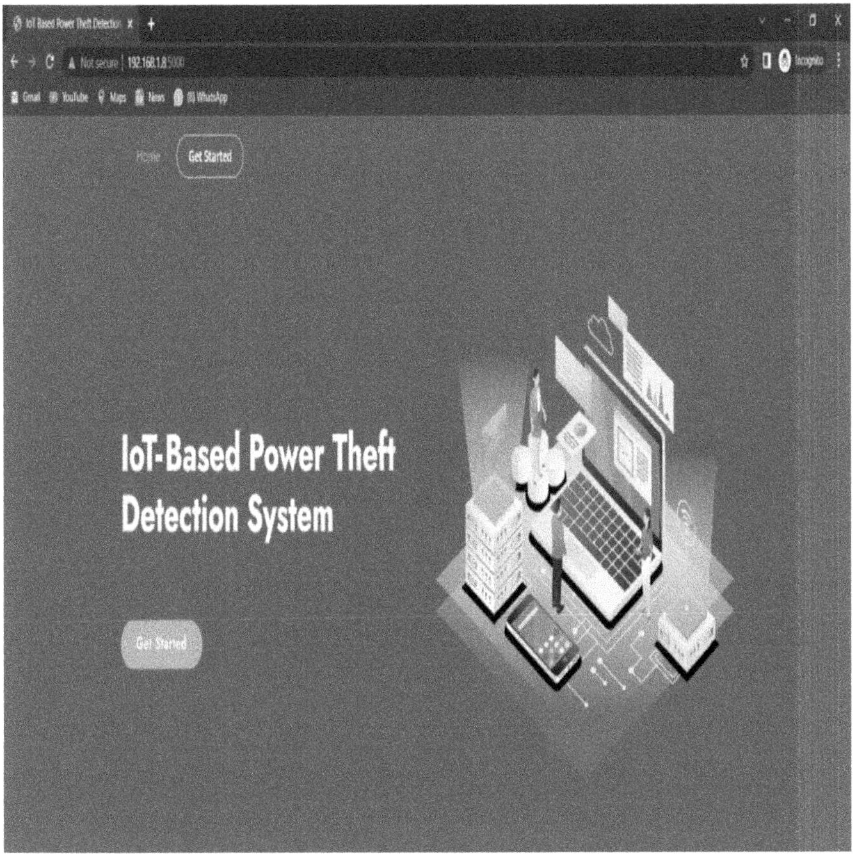

Figure 14.15 Home page of the website

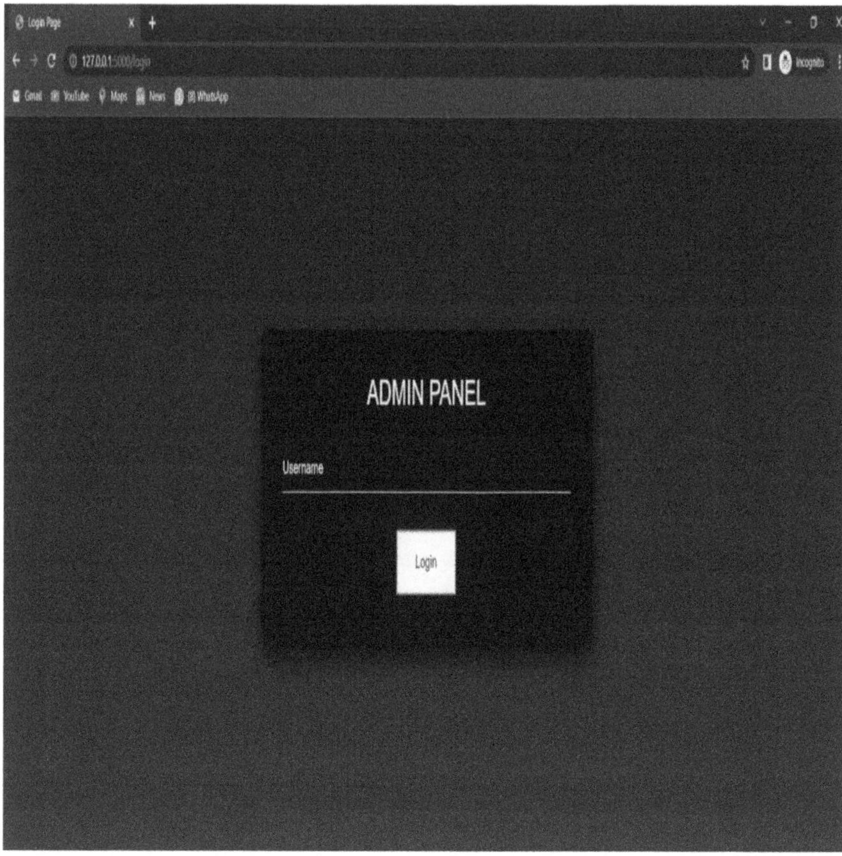

Figure 14.16 Admin login page

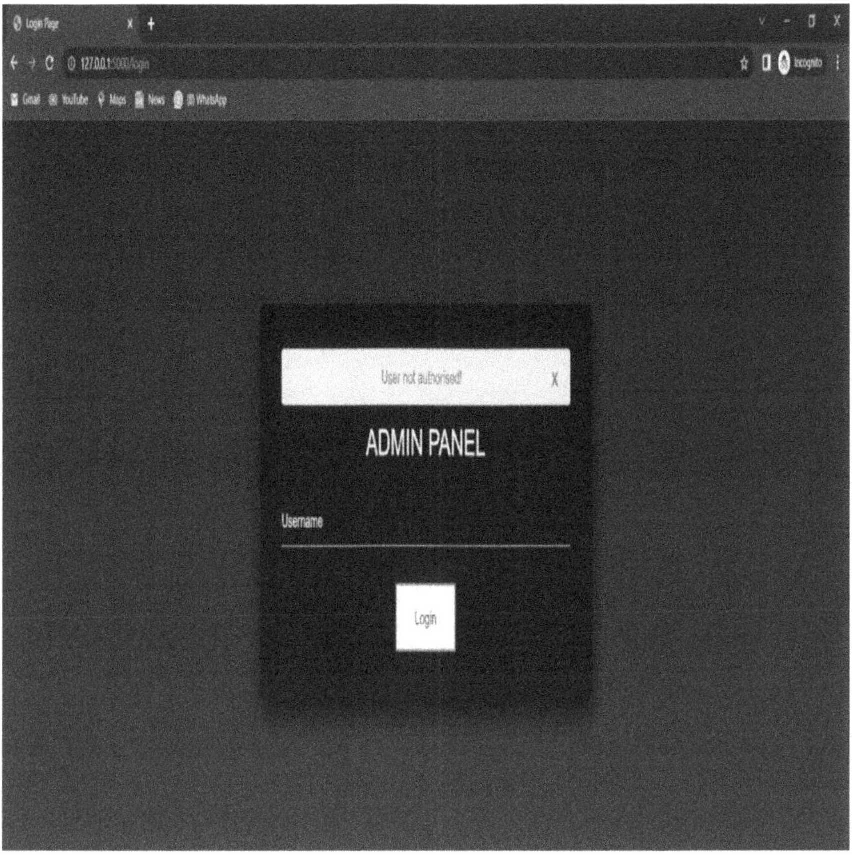

Figure 14.17 Login screen in the case that the wrong phone number is entered

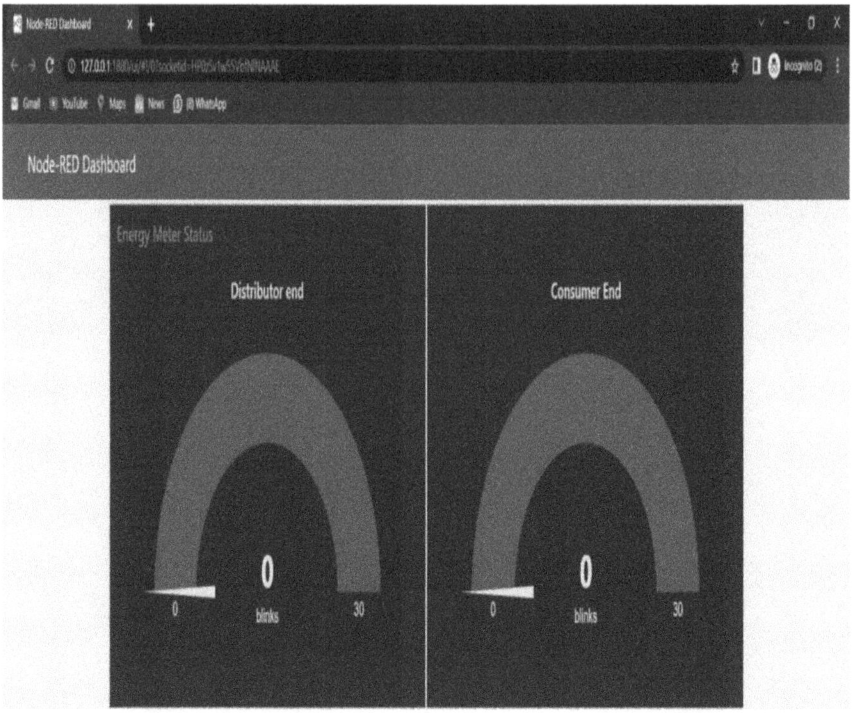

Figure 14.18 Dashboard visible to the user when the user logs in with the right credentials

Figure 14.19 Hardware connected with Laptop 1

Figure 14.20 Zoomed image of the LCD screen in the power theft case

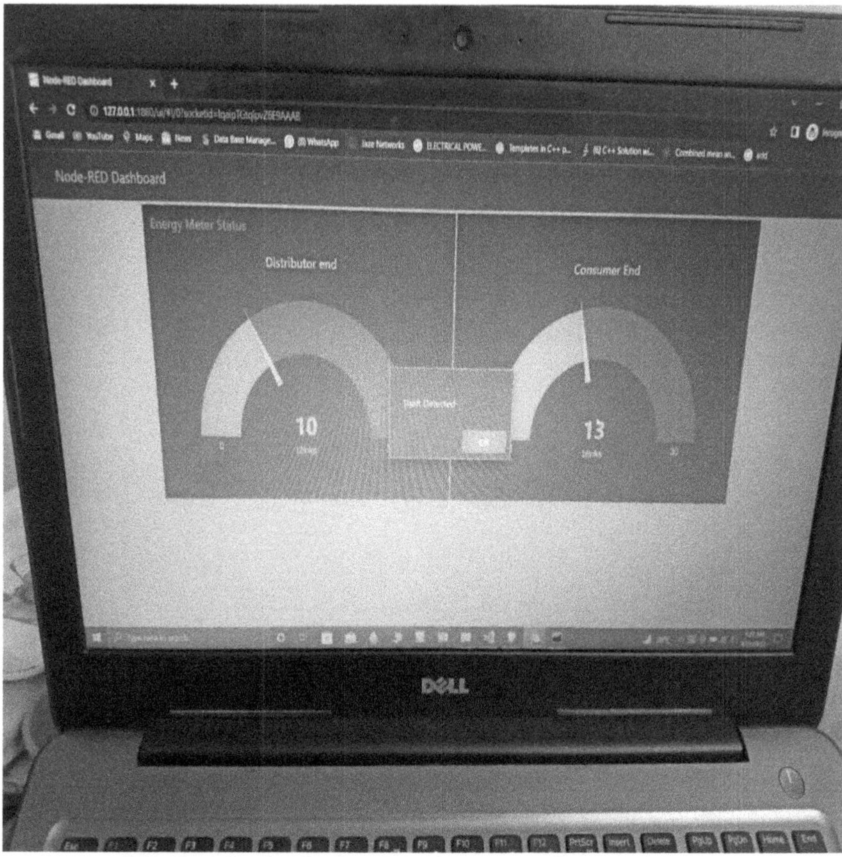

Figure 14.21 Live monitoring of the status on Laptop 2 with a power theft pop-up

Figure 14.22 SMS received using the user's mobile number

REFERENCES

[1] Uddanti, S., Joseph, C. and Kishoreraja, P.C., 2017. IoT Based Energy Metering and Theft Detection. *International Journal of Pure and Applied Mathematics*, *117*(9), pp.47–51.

[2] Qing-Hai, O., Zheng, W., Yan, Z., Xiang-Zhen, L. and Si, Z., 2013, October. Status Monitoring and Early Warning System for Power Distribution Network Based on IoT Technology. In *Proceedings of 2013 3rd International Conference on Computer Science and Network Technology* (pp. 641–645). IEEE.

[3] Sathyapriya, R. and Jeyalakshmi, V., 2020. Hardware Implementation of IOT Based Energy Management Theft Detection and Disconnection Using Smart Meter. *Malaya Journal of Matematik*, *5*(2), pp.4177–4180.

[4] Sirisha, B.L., 2020. Minimizing Electricity Theft Using IOT. *SAMRIDDHI: A Journal of Physical Sciences, Engineering and Technology*, *12*(SUP 3), pp.49–52.

[5] Pereira, J. and Saraiva, F., 2020, July. A Comparative Analysis of Unbalanced Data Handling Techniques for Machine Learning Algorithms to Electricity Theft Detection. In *2020 IEEE Congress on Evolutionary Computation (CEC)* (pp. 1–8). IEEE.

[6] Rajagiri, A.K., Ajitha, A. and Thalluri, A.K., 2021, January. Development of an IoT Based Solution for Smart Distribution Systems. In *2021*

International Conference on Sustainable Energy and Future Electric Transportation (SEFET) (pp. 1–6). IEEE.

[7] Uvais, M., 2020, February. Controller Based Power Theft Location Detection System. In *2020 International Conference on Electrical and Electronics Engineering (ICE3)* (pp. 111–114). IEEE.

[8] Jeffin, M.J., Madhu, G.M., Rao, A., Singh, G. and Vyjayanthi, C., 2020, July. Internet of Things Enabled Power Theft Detection and Smart Meter Monitoring System. In *2020 International Conference on Communication and Signal Processing (ICCSP)* (pp. 0262–0267). IEEE.

[9] Mir, S.H., Ashruf, S., Bhat, Y. and Beigh, N., 2019. Review on Smart Electric Metering System Based on GSM/IOT. *Asian Journal of Electrical Sciences*, *8*(1), pp.1–6.

[10] Saritha, K. and Murthy, A.S.N., 2020. Energy monitoring and theft system using IoT. *International Journal for Research in Applied Science and Engineering Technology*, *8*, pp.1705–1711.

[11] Shravani, K. and Reddy, G.S., 2018. Implementation of Embedded Web Server Based Power Theft Detection and Smart Monitoring System. *International Journal of Advanced Technology and Innovative Research*, *10*(1), pp. 68–70.

[12] Hasan, M., Toma, R.N., Nahid, A.A., Islam, M. and Kim, J.M., 2019. Electricity Theft Detection in Smart Grid Systems: A CNN-LSTM Based Approach. *Energies*, *12*(17), p.3310.

[13] Adil, M., Javaid, N., Qasim, U., Ullah, I., Shafiq, M. and Choi, J.G., 2020. LSTM and Bat-Based RUSBoost Approach for Electricity Theft Detection. *Applied Sciences*, *10*(12), p.4378.

[14] Chen, Y., Hua, G., Feng, D., Zang, H., Wei, Z. and Sun, G., 2020, September. Electricity Theft Detection Model for Smart Meter Based on Residual Neural Network. In *2020 12th IEEE PES Asia-Pacific Power and Energy Engineering Conference (APPEEC)* (pp. 1–5). IEEE.

[15] Maamar, A. and Benahmed, K., 2018, January. Machine learning techniques for energy theft detection in AMI. In *Proceedings of the 2018 International Conference on Software Engineering and Information Management* (pp. 57–62).

[16] Vineeth, V.V., Ambrish, V., Haricharann, D.V., Harshini, V. and Abilash, C., 2021, May. Power Theft Recognition and Data Security in Smart Meter Reading of a Smart Grid. In *Journal of Physics: Conference Series* (Vol. 1916, No. 1, p. 012216). IOP Publishing.

[17] Meenal, R., Kuruvilla, K.M., Denny, A., Jose, R.V. and Roy, R., 2019, November. Power Monitoring and Theft Detection System Using IoT. In *Journal of Physics: Conference Series* (Vol. 1362, No. 1, p. 012027). IOP Publishing.

[18] Rajashekar, M., Govardan Kumar, R.R. and Rashmi, K., 2021. Monitoring the Power Theft Using IoT. *Journal of Advances in Computing and Information Technology*, *1*(1), pp.46–53.

[19] Kumaran, K., 2021. Power Theft Detection and Alert System Using IOT. *Turkish Journal of Computer and Mathematics Education (TURCOMAT)*, *12*(10), pp.1135–1139.

[20] Nalinaksh, K., Pathak, L. and Rishiwal, V., 2018, February. An Internet of Things Solution for Real-Time Identification of Electricity Theft and Power Outrages Caused by Fault in Distribution Systems (Converting Existing Electrical Infrastructure of Third World Countries to Smart Grids). In *2018 3rd International Conference On Internet of Things: Smart Innovation and Usages (IoT-SIU)* (pp. 1–8). IEEE.

Chapter 15

IoT-Based Prioritized Load Management Technique for PV Battery-Powered Building

Mohit Sharan, Prantika Das, Apurv Malhotra, Aayush Karthikeyan, Ramani Kannan, O.V. Gnana Swathika

15.1 INTRODUCTION

The need of the hour is to come up with sustainable power supply systems. An example is a demand-side load management strategy that keeps costs at a minimum while regulating loads in a residential building in a way that the satisfaction of consumers is maximized. The prime aim is to provide an energy-oriented production control method that treats electric energy as a restricted production capacity with specified load profiles for manufacturing. This control as mentioned tries to match the manufacturing demand of energy with a limited supply of energy, lowering energy expenditures. It considers the significance of adaptive power utilization coupled with Internet of Things (IoT) technology and data analytics. The adaptive power system paves the way for an intelligent real-time interaction platform of power utilization. IoT technology, with its reliable communication and strong data processing, can effectively enhance the adaptive utilization of power. Photovoltaics (PVs) are playing a major role in the electricity industry around the world. At the same time, battery energy storage systems (BESS) are becoming cheap and are being deployed by PV users. Apartments offer a chance to adapt central battery storage and solar generation that is common to aggregated apartments, in addition to loads that are common through a microgrid (MG). The purpose is to have a hybrid consumption setup that smartly couples the existing power consumption setup with demand and a reliable PV energy-based power consumption setup to ensure power for all.

15.1.1 Literature Review

15.1.1.1 Load Management in Residential Buildings

One of the ways to achieve sustainable power supply systems is to follow the incentive-based pricing system. The method can be based on "model predictive control," which works in line with the thermal load management

 DOI: 10.1201/9781003537878-15

of buildings [1]. Demand-side management (DSM) programs, which are implemented through residential energy management systems (EMSs) for smart cities, have a number of advantages: consumers save money on electricity and the utility works at lower peak demand. The DSM model for scheduling residential user's appliances is provided in this study, which is based on evolutionary techniques (binary particle swarm optimization (PSO), genetic algorithm, and cuckoo search) [2].

In hybrid renewable energy system (HRES) sizing, the load is split into major and minor importance sections, and based on those categorizations, the load is managed. The sizing is determined using a smart PSO algorithm implemented in MATLAB. The results reveal that the cost of generated energy is inversely proportional to the load division percentage [3].

In a demand-side load management strategy, loads are managed within a residential complex by means of maximum user satisfaction by keeping costs at the minimum possible value [4],[5]. A smart grid setting has energy management units (EMUs). It is essentially aided by an artificial neural network (ANN) applied alongside a DSM method that helps in peak cutting and load prioritization to incorporate both primitive and modern day (renewable) energy sources [6]. In the production control method, energy is produced as per the capacity mentioned in the load profiles. This "energy-oriented" production control tries to match energy production with a set energy supply, lowering energy costs [7].

Researchers have considered demand response (DR) for economic and system dependability reasons. The concentration is on a reward-based program for consumers with the goal of matching a set consumption unit to the total requirement to provide peak control and lower market costs [8]. Sensor technologies, communications, and control methods can be relied on for proper energy usage. The use of reinforcement learning in building energy management and control has been an aid in proper energy utilization [9]. A DR program (DRP) enables a dual targeted optimization system for the appropriate siting and scaling of a storage system in an MG [10].

In a study, researchers outline a broad framework for evaluating prospects for effective load management in houses where renewable energy sources and energy storage facilities are provided in response to changing utility incentives [11]. Electrical distribution systems can have distributed energy resources (DERs). Control methods are based on a primitive operational setup that allows overlapping control layers to be integrated smoothly with local control methods [12–16]. The EMS has the latest criterion called the criticality level, which is the threshold value integrated in apartments based on their energy consumption. The model in the study maximizes the power consumed simultaneously with cost minimization [17].

15.1.1.2 IoT

The importance of flexible power consumption associated with IoT technology has been considered. The flexible power system paves the way for a real-time interactive energy field. This improves the viability, stability, and reliability of the power supply. IoT technology, and reliable communication and robust data processing, can effectively improve flexible power consumption [4]. Researchers have discussed developments in the energy sector. With the growing adoption of smart housing technologies and smart construction, there is an urgent need to establish an energy-efficient ecosystem. The solution has been recognized as environmentally friendly and as an important global aim to upgrade planetary stability [14]. The sudden surge in availability of electricity and the development of smart grids are a convenience in the emergence of local home EMSs (HEMS) for improving energy efficiency. This includes a DR tool that changes the need to improve home energy use considering power bills, environmental constraints, load profile, and consumer reassurance.

Clever home technologies and load scheduling controllers embedded with artificial intelligence have been studied [18]. The goal of the study is to create a home power control and monitoring system. It consists of ESP8266, Arduino, the current sensor ACS 712, and the cloud platform as a data storage and analysis service. The solution focuses on reducing energy consumption by repeatedly informing electrical appliances and providing optimized information to users, which helps the user to reduce the use of unnecessary energy, and thus save more resources and money [19].

A theoretical and practical review was conducted that provided a solution to the problem of the need for electricity supply in Zimbabwe during peak hours. The concept model was developed using C++, directed by the load management algorithm. It continues to monitor the feeder, open the automatic controller, and perform the load-sharing process based on important settings that ensure the availability of electricity to be used, which are critical in times of increasing demand. The equipment is attached to the power grid with the help of smart plugs [20].

In a study, an intelligent IoT-focused system for power regulation in buildings was introduced. The combination of power and the construction industry will ultimately be an important part of building a smart city. The proposed system extends to building data, power generation, energy bills, climate data, and consumer behavior, in order to produce day-to-day consumer applications [21].

Researchers have conducted a detailed review of the literature on strategies for the efficient use of energy and planning in smart households. An intricate description is provided of various materials that contribute to thermal relief, visual comfort, and good air quality. Computer systems used in smart homes have also been updated [22]. The goal of this project is to build an

Arduino and IoT-focused hierarchical multi-agent system (HMAS) for small grid loading using a promotional strategy. The effectiveness of the proposed algorithm on the small grid is confirmed in this study [23]. The project proposes the construction of a virtual web (IoT) based on HRES systems, along with a wind turbine, PV setup, battery storehouse equipment, and diesel generator. The parameters to check are labeled as various categories, which include electrical values, status, and environment [24].

15.1.1.3 Photovoltaic Power Systems

In a study, the optimal EMS is provided for reducing residential building electricity bills. The goal is to acquire peak shaving along with lowering the price of electrical energy for the consumer using a dynamic energy pricing model [12]. The study demonstrates how to utilize a device with stored energy such as a battery on an MG with PV along with load generating energy in the most efficient way possible. The method forecasts daily loads and generation power, then looks for the best storage action strategy for the energy storage structure based on those forecasts [13]. A system is more difficult to govern because of the fluctuation and uncertainty of output and consumption. In reality, weather conditions have an impact on renewable energy output, whereas occupancy has an impact on power consumption. As a result, reliable temporary projections are required for the perfect unification of renewable energy systems (RESs), such as solar systems and wind turbines, alongside the conventional power grid in MG systems. Researchers proposed a method that anticipates power generation and usage in MG systems, along with the state of charge (SoC) of the batteries [15]. The study focuses on boosting the usage of solar energy, which is often pushed as a viable alternative to fossil fuels due to its ease of use and environmental benefits. To compensate for its erratic and unpredictable qualities, electrical energy storage solutions are being developed to synchronize generated power with building demand. The paper presents an overview of current research, scope of study, and optimization of the construction of hybrid PV-electrical energy storage systems (ESSs) required for power supply in buildings, which can be utilized like a roadmap for future studies in the subject [25].

Distributed PVs are becoming more common in the world's electrical businesses, whereas BESS are becoming more affordable and being utilized by users of energy with PVs. By an embedded network or MG, apartment complexes can integrate main battery storage along with PV generation to apartments and communal loads [26].

Traditional electricity users are being transformed into electricity prosumers as a result of the smart grid. Research has examined how the people in a current household in Sweden were converted into electricity prosumers. For the coupled PV-heat pump-thermal storage-electric vehicle

(EV) system, an objective function built on genetic algorithms is formed to optimize the capabilities and stances of PV elements at the cluster level, with the goal of increasing self-consumed electricity under a non-negative net present value during the economic lifecycle [27].

As the utilization of renewable energy along with DC-based loads has increased dramatically in recent years, DC MGs are being examined as a probable solution to India's electrification. Researchers have explained how to manage energy in DC MGs and how to overcome common design issues when putting off-grid power systems in place. The rule of conservation of energy serves as the basis for the system's operation. The algorithm's scalability was demonstrated using a prototype hardware configuration at Vellore Institute of Technology's (VIT's) Solar Energy Research Center [28].

The influence of PV systems in residential and commercial settings has been examined. Its significance in these places has grown as a result of daily electricity usage during solar hours. The study presents a controller-based management solution for power circulation in the electrical grid and PV system, and EV batteries. It demonstrates the advantages of deploying EVs as a vital component in balancing energy, resulting in lower energy expenditures [29]. PV systems do not produce electricity continuously and are mostly dependent on weather conditions, resulting in lesser energy production than other renewable alternatives. As a result, the ESS may prove to be a viable answer to the aforementioned issue.

The practicality in terms of the cost of lead-acid ESS paired with PV panels in apartments has been proposed in a study [30]. The efficiency of a PV plant placed on the roof of an educational structure is evaluated and a strategy for enhancing it using only feeder connections is provided. The demand of the electrical system of the structure is then reduced by substituting obsolete lights and ventilation fans with more energy efficient types [31].

Researchers have investigated the influence of four distinct strength management techniques (EMSs) on the dependability of PV-powered MGs, which are projected to play a significant part in future distribution networks [32]. The control of the EMS for MGs using PV-based distribution generation systems is discussed in the study. During cloudy situations, the maximum power point tracking (MPPT) algorithm, in this case, the hill climbing technique, is utilized to harvest the maximum power provided by the PV source, which is assisted by a battery-based ESS [33].

15.2 PROPOSED METHODOLOGY

15.2.1 Basic Working of the Project

A hospital building is considered for the demonstration of the hybrid operating system design. Regarding prioritizing the loads in a building setup, in this case for a hospital environment, energy can be supplied by prioritizing a certain set of loads. Therefore, a mechanism is needed to define priority and non-priority loads in a hospital.

As observed in a hospital building, rooms such as operation theaters, incubation centers, and intensive care units (ICUs), where the power demand is high, are concentrated on a few floors. These floors are labeled as high-priority sections or critical floors. On the other hand, rooms such as the reception area, waiting area, and canteen or mess area, where the energy demand is low, are labeled as low-priority sections.

As mentioned above, the prime focus is to design a hybrid operating system to achieve the set goal; hence, the RES and the conventional source of energy function together.

Considering a scenario of a power cut, the building has a diesel generator to support building operations at the time of outages. As mentioned, running a diesel generator is a very expensive process. Hence, the solution to cutting down costs is a hybrid model that ensures power for everyone.

The working of the system shown in Figure 15.1 is coupled with a solar panel along with a conventional diesel generator.

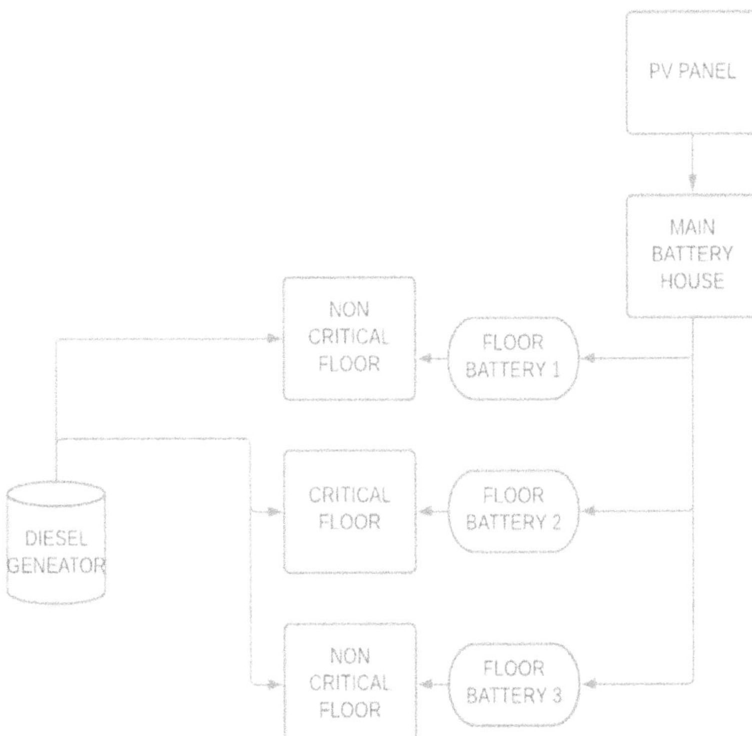

Figure 15.1 Building setup

The power generated by the PV system is stored in the main battery pack when there is no power cut. When there is a power cut, the diesel generator comes into action and takes over the responsibility of providing power to the building for $0.4T$, where T is the total power cut time.

Meanwhile, the main battery starts the process of battery-to-battery charging of the floor batteries labeled as critical and non-critical. The state of the main battery is monitored by the control center on the control and monitoring interface. Once the main battery voltage falls to a certain threshold level, a notification for this is triggered on the interface, and from the interface, the controller or supervisor selects the mode that focuses on charging the critical floor batteries. Thus, power is provided to all, along with prioritization of the loads.

After $0.4T$, the dependency on the diesel generator ends and the building critical loads are fed by the topped-up floor batteries.

15.2.2 Modules of the Project

To achieve the set goal, our system has three major modules:

- solar power generation prediction model
- battery-to-battery charging simulation
- power generation monitoring and control interface.

15.2.2.1 Solar Power Generation Prediction Model

Solar power prediction is a forecasting model that tests and predicts solar power generation. Because solar power is a renewable source of energy and is irregular throughout the length of the day, it is not possible to be completely dependent on the equipment used to gather and record the solar power readings. Hence, to eradicate the effects of the intermittency of solar power generation and for monitoring and controlling the parameters of the PV power panel for future reference, which is an essential component of the objective of the project, accurate readings of solar power are required for subsequent days or weeks. Forecasting also helps to increase the management efficiency of power grids, thus providing a way for grid operators to predict and balance energy generation and consumption.

15.2.2.2 Battery-to-Battery Charging Simulation

Proteus simulation was performed to depict the charging of load batteries using the main battery. It is a scaled down version of a charging system in a building where the main battery will charge the floor batteries. The values in the simulation are provided in accordance with the parameters of the hardware in the project.

15.2.2.3 Power Generation Monitoring and Control Interface (IoT)

The power generation monitoring and control interface works in coherence with the hardware module developed. The entire functioning of this module is dependent on the hardware module developed. Hence, this particular module has two parts.

Hardware Setup

The hardware consists of various sensor modules, along with a microcontroller governing the operational status. A voltage sensor is embedded along with a temperature sensor, which senses the voltage of the battery pack, which is charged by the solar panels and the ambient temperature of the entire unit.

Based on the voltage-to-battery capacity ratio, a rough estimated SoC of the battery is extracted. All the parameters are sent to the control interface developed by the means of serial communication. This partnership of sensor and communication technology makes the module an IoT-based smart system.

Monitoring and Control Interface

A control and monitoring interface helps to keep track of the live operational feed and guide the operator with correct decisions as per the system response.

These real-time responses ultimately pave the way to strengthen the building power management sector by coming up with smart and hybrid operational features.

15.2.3 Tasks and Milestones

15.2.3.1 Machine Learning for Solar Power Prediction

Machine learning (ML) is the best way to forecast solar power because prediction is a time series problem using historical data. Various ML algorithms are tested and compared on the solar power dataset, which is obtained using virtual instrumentation, where captured data are logged straight into a spreadsheet from the microcontroller. The flowchart of the data logger system is shown in Figure 15.2.

In order to compare different ML algorithms on the dataset obtained, an open-source ML library is used, known as PyCaret. The caret package automates all the major steps for comparing ML models and evaluating the best ML model suited for the prediction of solar power, thus providing accurate results.

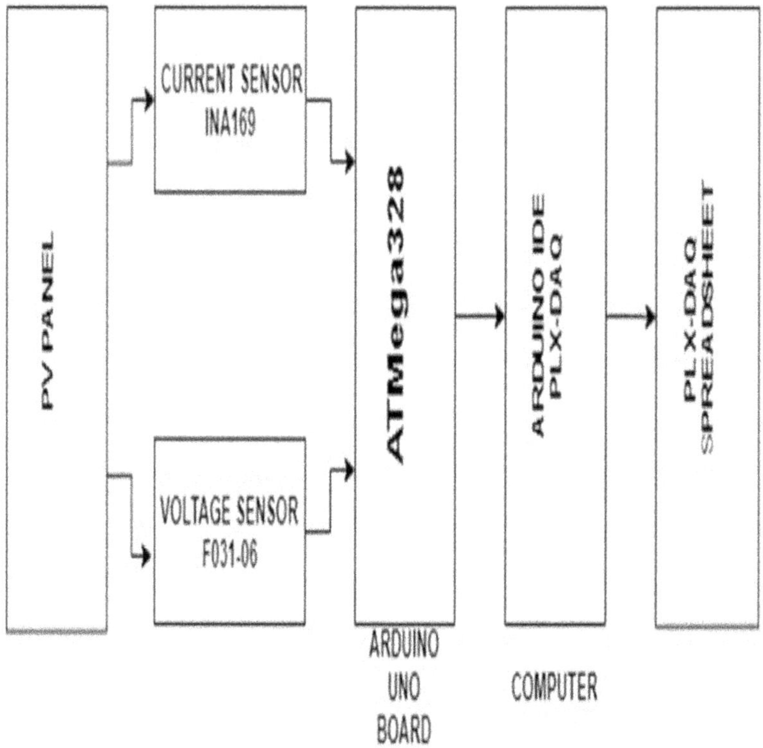

Figure 15.2 Flowchart of the data logger system

15.2.3.2 Proteus Simulation

The main battery is used for charging all the floor batteries. Each circuit to the floor battery consists of the following:

- Diode: It allows unidirectional current, that is, from the main battery to the floor battery.
- Zener diode: It starts conducting when the voltage goes above a certain voltage.
- Transistor switch: The base of the transistor switch starts getting current. When the current exceeds a certain limit, the transistor switch also starts conducting.
- Relay: One end of the relay gets grounded. Hence, the relay gets activated. Once the relay gets activated the current flows through the relay, charging the battery. Each relay also has a cutoff voltage, so when the voltage through the relay exceeds a certain limit, the relay switches off, thus protecting the circuit from overvoltage.

15.2.3.3 Control and Monitoring Interface

All the data gathered from the sensors are channeled via the means of serial communication to the monitoring interface developed on the Node-RED IoT visualization dashboard. All the parameters are monitored in real time by the site operator. If there is any discrepancy found in the operations, certain operative measures are taken to ensure power to all. The control interface allows priority-based switching to be conducted based on the real-time feedback displayed on the interface. By only the click of a button, charging of the critical load batteries is enabled by cutting off charging from the rest of the low-priority labeled loads.

15.3 DESIGN METHODOLOGY

15.3.1 Design Approach

The logical operation design of the system is shown in Figure 15.3.

The control and monitoring IoT-based interface works on the logic design presented above. All the sensors coupled in the hardware system push the captured real-time data to the interface via the means of serial communications. The data gathered trigger the logical operations required for the smooth operation of the building. The state of the main battery can be monitored in terms of voltage and SoC, and once they drop to a certain threshold level, the system triggers a message that it is necessary to focus on charging the critical loads.

15.3.2 Codes and Standards

15.3.2.1 Arduino

Arduino boards interface with many boards and systems. The regulations for the interfacing of Arduino are known as Arduino communication protocols. There are three protocols: Universal Asynchronous Reception Transmission (UART), Serial Protocol Interface (SPI), and Inter-Integrated Circuit (I2C). UART permits Arduino to communicate with various serial devices. SPI allows Arduino to send data to, for example, shift registers, sensors, and SD cards. I2C is used for long-distance communication. To read the sensor value, Arduino is used as the microcontroller.

15.3.2.2 Node-RED

Node-RED is an online flow editor that makes it trouble-free to connect various flows using the different nodes in the palette. The outputs can then be viewed in the user interface (UI) during runtime in a single click. Node-RED is built on Node.js, which is a JavaScript environment. JavaScript

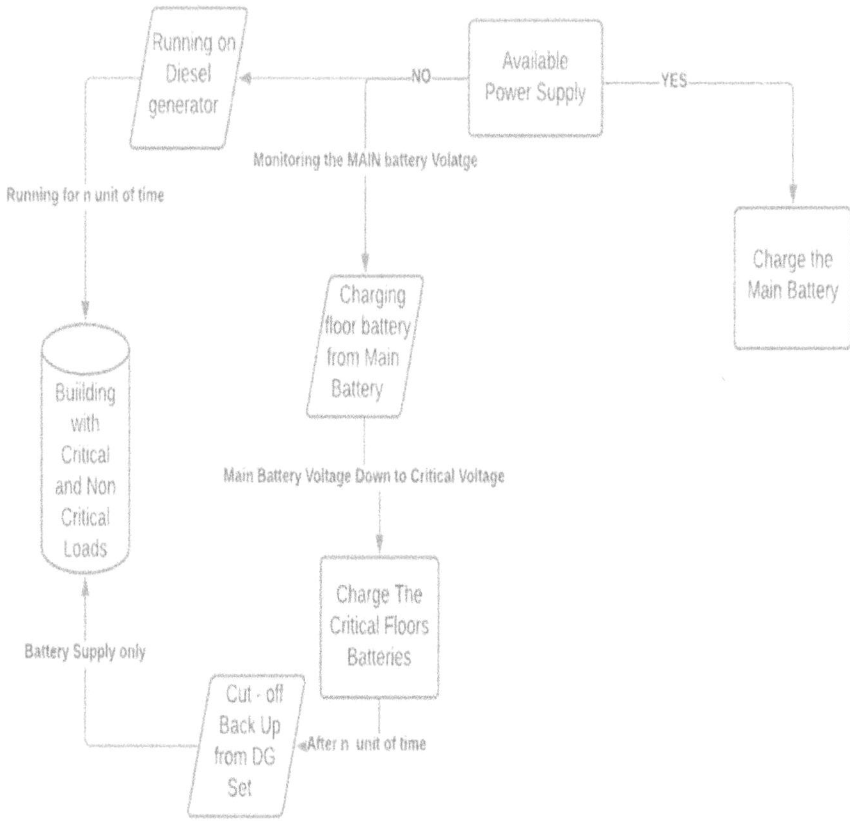

Figure 15.3 Logic of the operation

is the language of the web. Node.js allows the use of JavaScript as server (back-end) language. The flows that are made on Node-RED can be easily imported and exported because they are stored in the form of JSON code. An inbuilt online flow library is available that permits the sharing of flows with people around the globe.

In the project, Node-RED is used as the programming tool for the control and monitoring of the UI for performing operations smoothly.

15.3.2.3 PyCaret for ML

PyCaret is an open-source low-code ML library in Python whose objective is to lessen the time taken for testing a dataset with various ML models. It helps in data preprocessing. It is an end-to-end ML and model management tool that increasingly speeds up the experiment cycle. It tests and trains various models at the same time, compares all these models, and provides a comparison table as output based on parameters such as precision, recall, and F1 score. Python code for the ML model is developed in Jupyter Notebook. It consists of various ML algorithms that are tested on the dataset. Comparison analysis is performed to determine the best ML algorithm and solar power prediction is performed.

15.3.2.4 Realistic Constraints

The realistic constraints are as follows:

- The collection of reliable solar power data used in PV buildings is not possible due to the unavailability of a solar panel used in an actual building.
- A particular value of power consumption in a building is determined.
- A secure but fast data collection and transmission network is established.
- Hardware for battery-to-battery charging is dangerous to operate as it leads to overheating of the equipment.

15.3.2.5 Alternatives and Tradeoffs

The alternatives and tradeoffs are as follows:

- To obtain the dataset for the power prediction model, a solar data logger system previously developed has to be relied on. The data are gathered and used in solar power prediction.
- To develop the prototype, an average expected power consumption value for a building is taken and then scaled down for the prototype accordingly.
- To achieve secure data transmission, line communication is performed via the means of serial communication, which is slower than a wireless Wi-Fi connection.
- Proteus simulation of battery-to-battery charging is performed as it is a safer alternative to working on the hardware.

15.3.3 Design Specifications

15.3.3.1 Software Design Specifications

Node-RED

Node-RED is a programming tool developed by IBM. It paves the way for hardware devices to communicate along with application programming interfaces (APIs), and web services as part of the IoT. In Node-RED, flows are designed that construct JavaScript functions in a web browser.

ThingSpeak

ThingSpeak is software that makes communication with internet-enabled devices possible. It allows data access and retrieval, alongside the logging of data, by the means of API to both devices and web interfaces for analysis purposes.

Jupyter Notebook

The original web application for producing and sharing computational documents is Jupyter Notebook. It provides a straightforward, simplified, and document-focused environment.

Proteus

The Proteus Design Suite is a software tool used for electronic design simulation studies. The created schematic can be used to study the working of the circuit and get a feel of how various actions will take place in it.

15.3.3.2 Hardware Design Specifications

Arduino

Arduino contains a physical programmable circuit board and a piece of software, alongside an integrated development environment (IDE). On the IDE, a certain chunk of code is written to work with other peripherals, such as sensors. Then the computer code is uploaded to the physical board to perform the necessary actions.

Voltage Sensor (0–25 V)

The Voltage Sensor Module works on the potential divider rule. This helps to reduce the input voltage by a factor of 5. This module allows analog input from a microcontroller. The input is then monitored at voltages much higher than it is capable of sensing. The sensing range of this module is set between 0 V and 25V.

Temperature Sensor (LM35)

The LM35 is an analogue temperature sensor. Electricity is output in relation to the degrees Celsius of the ambient environment where the sensor

Table 15.1 Specifications of the solar panel used in the data logger system

Specifications	*Rating*
Rated power	50 W
Maximum rated voltage	20 V
Current at maximum power	2.5 A
Open circuit voltage	22.5 V
Short circuit current	0.69 A
Tolerance	1.5%

Table 15.2 Specifications of the hardware prototype

Specifications	*Rating*
Arduino UNO	3.3 V/150 mA
Solar panel and battery pack	3.7 V/300 mAh
Voltage sensor	(0–25) V
Temperature sensor	LM35
Jumper set	-
Breadboard	-

has been put to use. These sensors are highly accurate; hence, to achieve the usual accuracy, the LM35 sensor does not require any extra calibration or trimming.

15.3.3.3 Design Specification for the Data Logger for Generating an ML Dataset

Table 15.1 shows the design specification for the data logger.

15.3.3.4 Design Specification for the Hardware Prototype

Table 15.2 shows the design specification for the hardware prototype.

15.4 RESULTS AND DISCUSSION

15.4.1 Analytical Results

The dataset in Figure 15.4. shows the readings that were retrieved in the spreadsheet through the data logger system. The dataset was then imported to the Jupyter Notebook for further analysis.

PyCaret was installed in the Jupyter Notebook, as shown in Figure 15.5, in order to get the best ML model suited for the dataset.

t	voltage	current	power
12:15:35 PM	19.94	2.66	53.02
12:15:40 PM	19.92	2.54	50.52
12:15:45 PM	19.89	2.53	50.36
12:15:50 PM	19.82	2.53	50.18
12:15:55 PM	19.82	2.53	50.08
12:16:00 PM	19.79	2.53	50.12
12:16:05 PM	19.79	2.53	50.02
12:16:10 PM	19.79	2.56	50.6
12:16:15 PM	19.77	2.46	48.6
12:16:20 PM	19.67	2.49	48.94
12:16:25 PM	19.67	2.5	49.13
12:16:30 PM	19.72	2.51	49.45
12:16:35 PM	19.75	2.55	50.38
12:16:40 PM	19.72	2.49	49.06
12:16:45 PM	19.72	2.52	49.64
12:16:50 PM	19.75	2.52	49.8
12:16:55 PM	19.75	2.52	49.7
12:17:00 PM	19.77	2.52	49.76
12:17:05 PM	19.77	2.52	49.86
12:17:10 PM	19.79	2.52	49.83
12:17:15 PM	19.77	2.52	49.76

Figure 15.4 Dataset obtained

The comparison shown in Figure 15.6 occurred based on various parameters. The ranking in the table goes from best ML model to worst ML model suited for the imported solar power dataset.

The best ML algorithm for the imported dataset was Orthogonal Matching Pursuit, as shown in Figure 15.7.

Power prediction was performed every 5 seconds using the best ML algorithm (Orthogonal Matching Pursuit), as shown in Figure 15.8.

15.4.2 Simulation Results

The Proteus simulation is depicted in Figure 15.9, where the main battery is used to charge the respective floor batteries. The specification of the main battery voltage matches with the solar panel parameter used in the

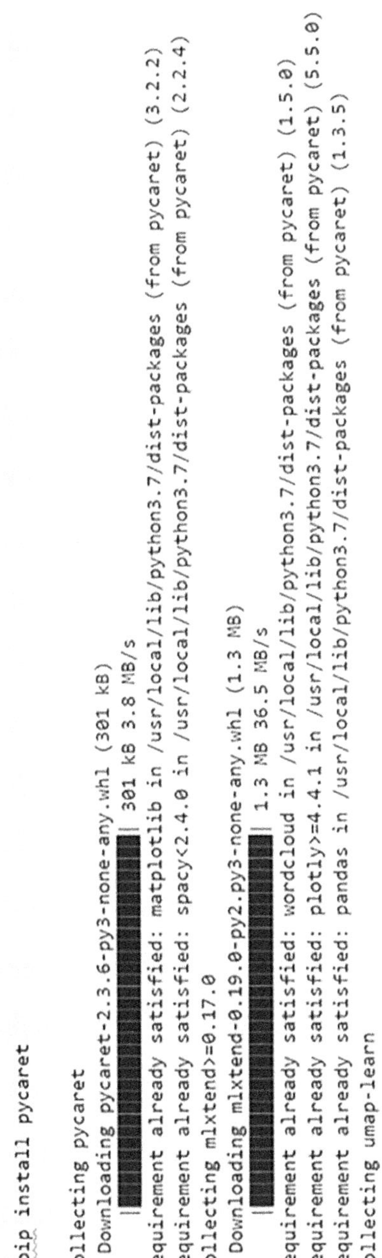

Figure 15.5 Installation of PyCaret in the Jupyter Notebook

```
best = compare_models()
```

	Model	MAE	MSE	RMSE	R2	RMSLE	MAPE	TT (Sec)
omp	Orthogonal Matching Pursuit	5.110000e-02	3.700000e-03	5.830000e-02	9.968000e-01	0.0011	1.000000e-03	0.032
et	Extra Trees Regressor	2.615000e-01	1.306300e+00	5.425000e-01	9.131000e-01	0.0098	4.800000e-03	0.492
gbr	Gradient Boosting Regressor	2.845000e-01	1.318800e+00	5.835000e-01	8.935000e-01	0.0107	5.300000e-03	0.091
rf	Random Forest Regressor	3.158000e-01	1.407500e+00	6.329000e-01	8.686000e-01	0.0116	5.900000e-03	0.545
dt	Decision Tree Regressor	3.371000e-01	1.246600e+00	6.398000e-01	8.512000e-01	0.0118	6.300000e-03	0.032
ada	AdaBoost Regressor	4.204000e-01	1.428200e+00	7.252000e-01	8.046000e-01	0.0133	7.900000e-03	0.146
knn	K Neighbors Regressor	4.339000e-01	1.787700e+00	8.095000e-01	7.929000e-01	0.0152	8.300000e-03	0.082
br	Bayesian Ridge	6.069000e-01	1.772600e+00	9.866000e-01	6.467000e-01	0.0188	1.190000e-02	0.034
huber	Huber Regressor	7.002000e-01	2.137200e+00	1.120800e+00	5.302000e-01	0.0214	1.370000e-02	0.091
lr	Linear Regression	7.219000e-01	1.969700e+00	1.101300e+00	4.871000e-01	0.0210	1.410000e-02	0.030
ridge	Ridge Regression	7.629000e-01	2.566200e+00	1.228300e+00	4.244000e-01	0.0235	1.490000e-02	0.031

Figure 15.6 Comparison table

hardware prototype as the power gathered by the solar panel is stored in the main battery to be distributed to various loads. The specifications in the project are scaled down versions of the parameters that are to be used in an actual building setup. The critical and non-critical loads are depicted as motor loads in the simulation.

15.4.3 Hardware Results

The hardware prototype developed in Figure 15.10 has a voltage sensor and temperature sensor coupled onto it, which helps in determining the estimated SoC of the battery, voltage, and the ambient temperatures of the battery unit. All this information is relayed to the control and monitoring

```
[ ] best

OrthogonalMatchingPursuit(fit_intercept=True, n_nonzero_coefs=None,
                          normalize=True, precompute='auto', tol=None)
```

Figure 15.7 Best ML model

```
predictions = predict_model(best, data = dataset)
predictions.head()
```

	voltage	current	power	t	Label
0	19.94	2.66	53.02	12:15:35	52.988058
1	19.92	2.54	50.52	12:15:40	50.593102
2	19.89	2.53	50.36	12:15:45	50.315463
3	19.82	2.53	50.18	12:15:50	50.122616
4	19.82	2.53	50.08	12:15:55	50.122616

Figure 15.8 Power prediction using the best ML algorithm

interface by the means of serial communication and logical conclusions are made based on the data received.

Figure 15.11 shows that the system is working under the normal operational state and the main battery is able to charge all the floor batteries, whether critical or non-critical.

In Figure 15.12, the system, in real time, notices that the voltage of the main battery has dropped below the critical level and responds accordingly by changing the operational state tab. When the voltage reaches the critical level, the operator is notified that the critical point is reached, so the system focuses on charging the critical floor batteries only.

In Figure 15.13, the operator responds on the basis of the feedback obtained by the system. As the feedback suggests that the critical point has been reached, the operator switches the battery charging switch from the control panel tab and focuses on charging the critical floor batteries only.

In Figure 15.14, by the means of the ThingSpeak IoT analysis platform, a real-time data observation and storing infrastructure is provided.

Runtime data are pushed to the cloud server. The same data are retrieved for further analysis when required, as shown in Figure 15.15.

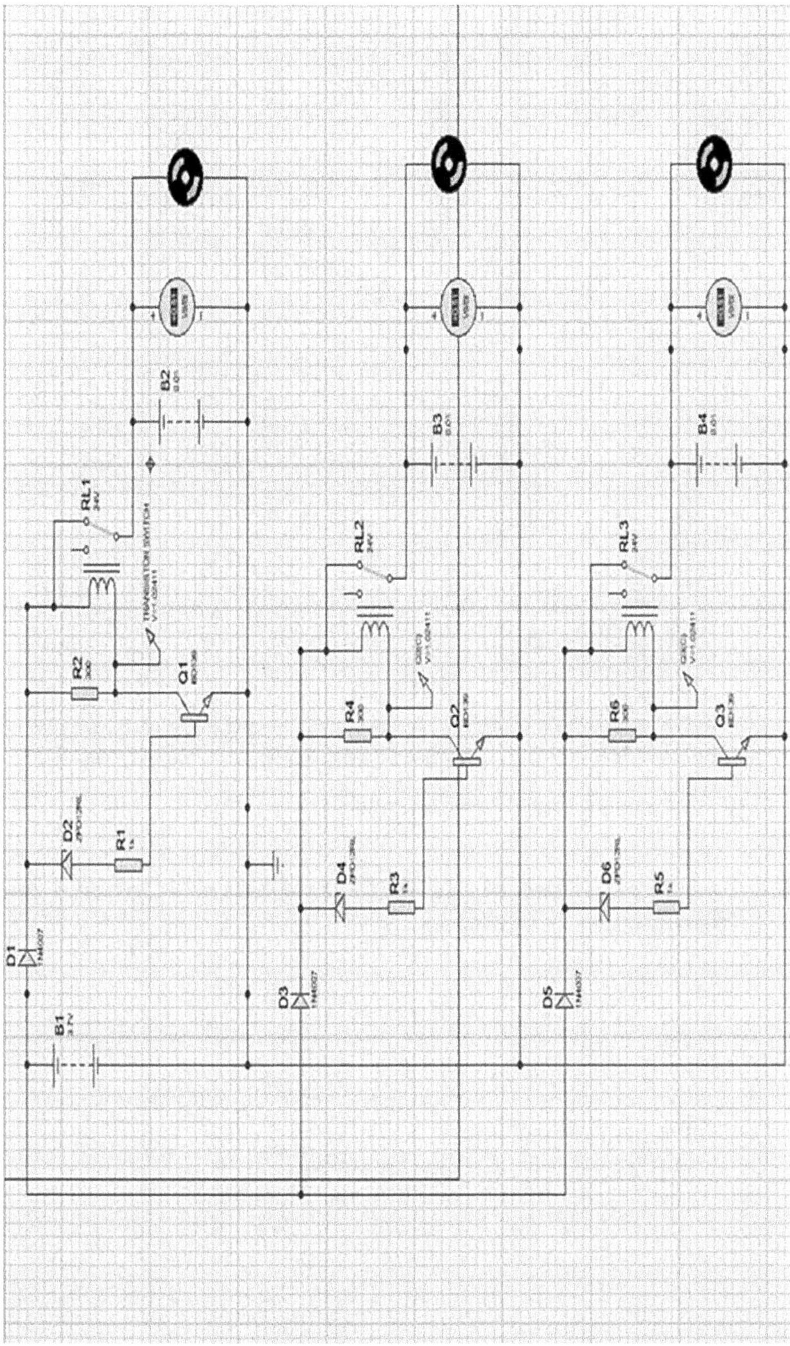

Figure 15.9 Main battery for floor battery charging (Proteus)

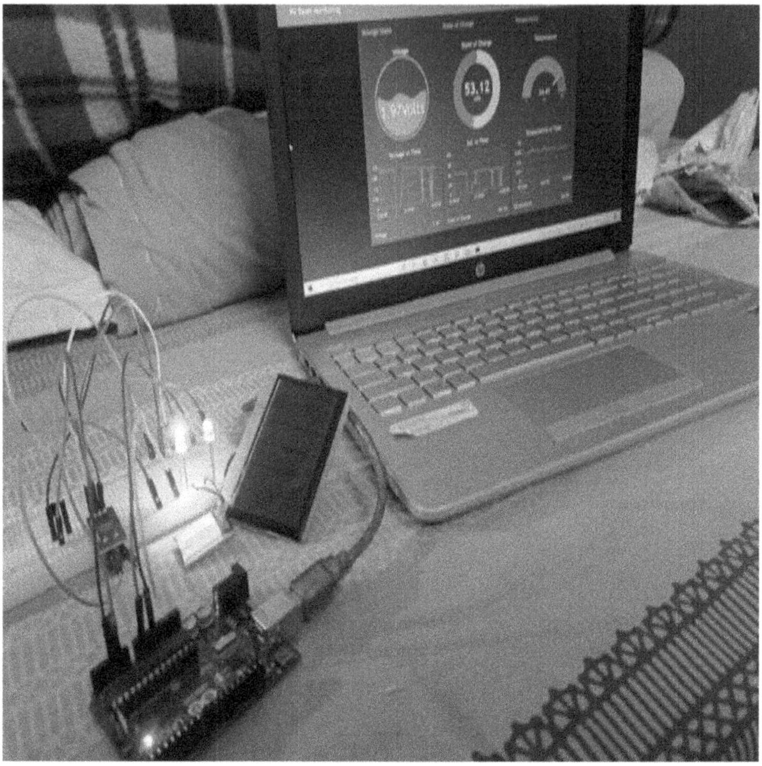

Figure 15.10 Hardware prototype under operation

15.4.4 Features of the Node-RED Dashboard (Control and Monitoring Interface)

The features are as follows:

- real-time monitoring of all the parameters, such as voltage, SoC of the battery, and ambient temperature
- graphical representation of the parameters in real time
- real-time operational status
- provision for remote charging switching
- real-time data are also logged into the ThingSpeak cloud.

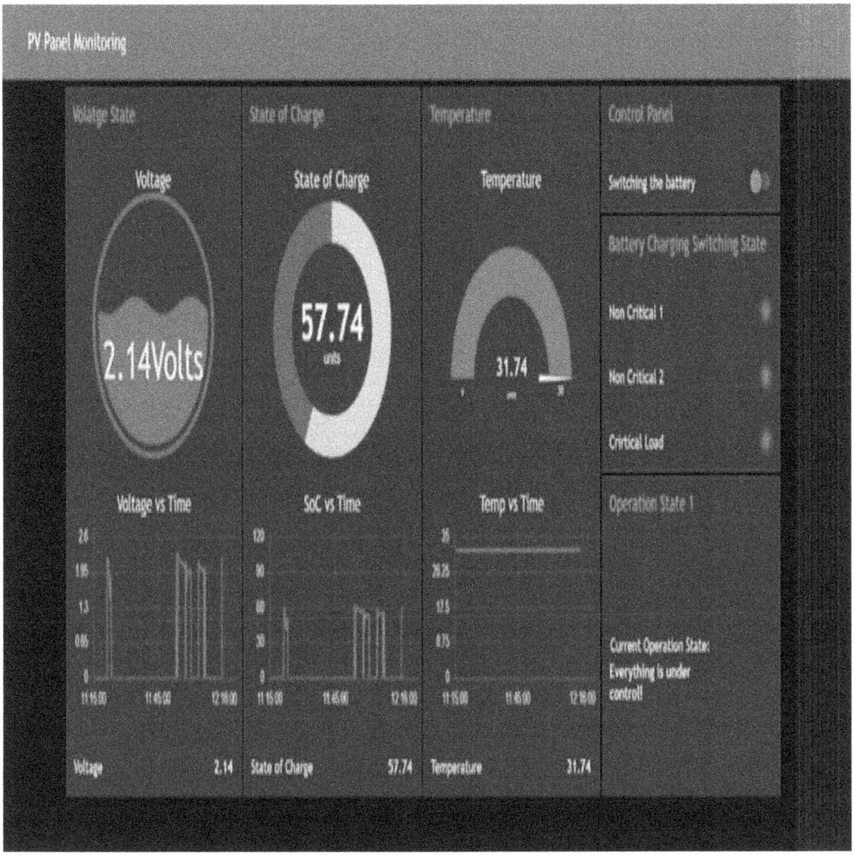

Figure 15.11 Normal running condition

15.5 CONCLUSION

15.5.1 Cost Analysis

The cost of this project directly relates to the operating building. If the power demand is high, then the one-time installation cost would be high or vice versa. However, the post-installation maintenance and running cost would roughly be the same as per general estimates.

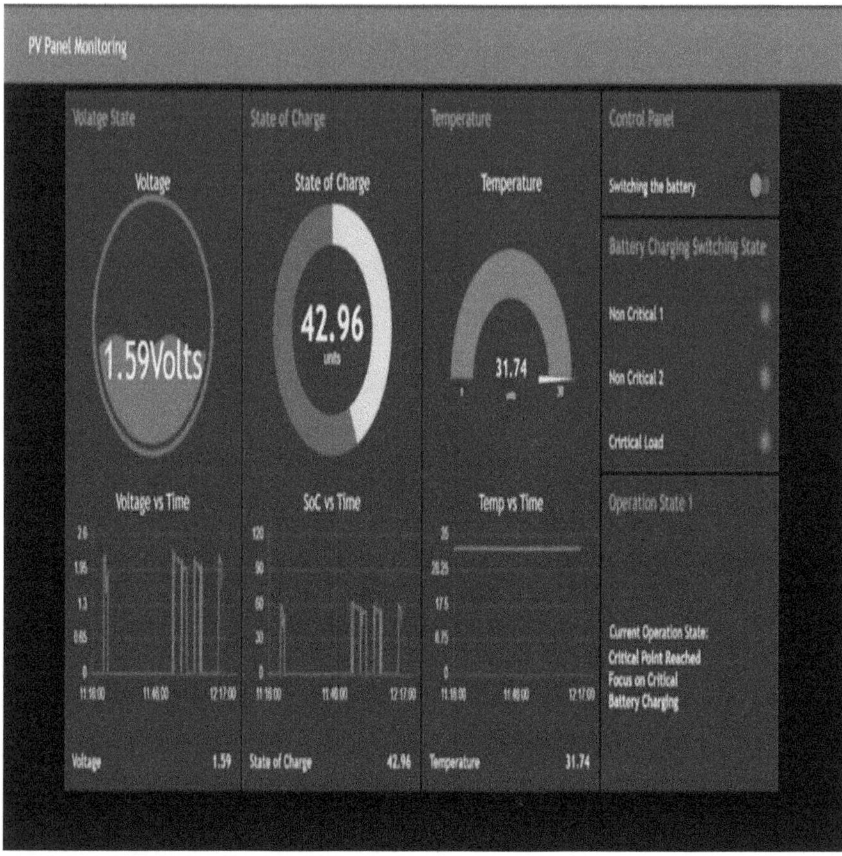

Figure 15.12 Change of state notification triggered

15.5.2 Scope of the Study

The scope of study is as follows:

- In the future, a shift to a Wi-Fi-based protocol is advisable rather than serial communication.
- The current system only monitors the state of the main battery; a system that monitors every battery is more efficient.
- Monitoring the current drawn from the system also strengthens operations.

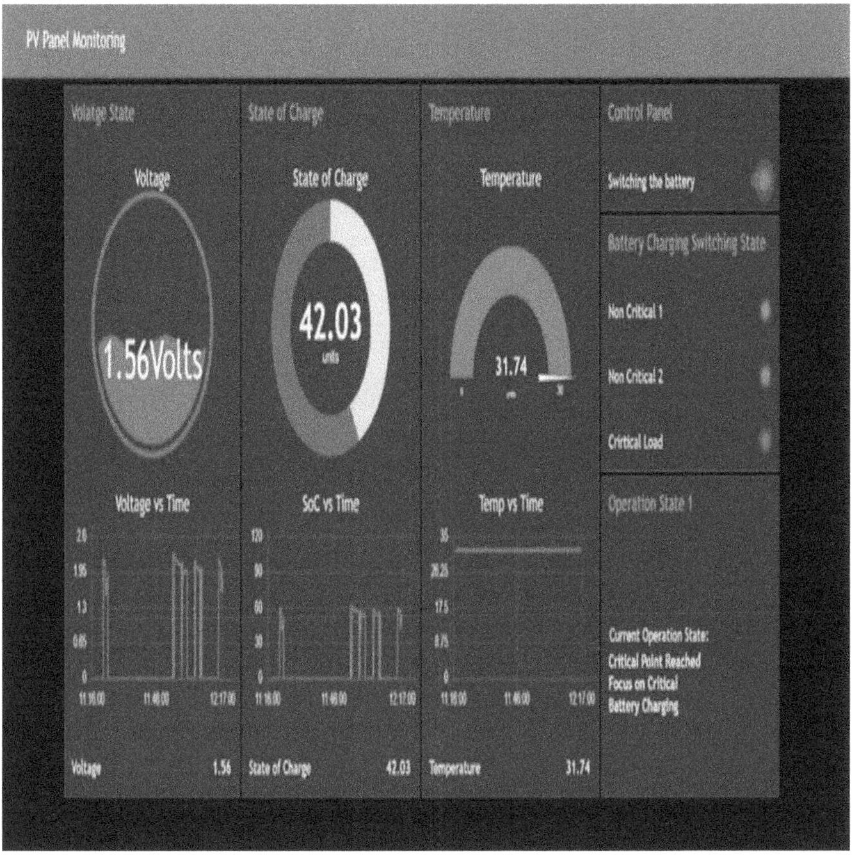

Figure 15.13 Battery charging state changed

15.5.3 Summary

By implementing the system suggested, we could achieve the goal of efficiently managing power consumption and supply from the existing source, and ensure power to all. An HRES coupled with an already existing conventional energy source, cuts down building power management costs by manifolds and proves to be a boon in the building power management segment.

Figure 15.14 ThingSpeak cloud

```
created_at,entry_id,field1,field2,field3,latitude,longitude,elev
2022-03-11T13:28:35+05:30,1,,20.70,,,,,
2022-03-11T13:28:50+05:30,2,0.00,,,,,,
2022-03-11T13:29:05+05:30,3,,33.30,,,,,
2022-03-11T13:29:20+05:30,4,,0.00,,,,,
2022-03-11T13:29:40+05:30,5,24.98,,,,,,
2022-03-11T13:29:55+05:30,6,,0.59,,,,,
2022-03-11T13:30:10+05:30,7,,0.00,,,,,
2022-03-11T13:30:25+05:30,8,,33.30,,,,,
2022-03-11T13:30:40+05:30,9,,0.00,,,,,
2022-03-11T13:30:55+05:30,10,19.92,,,,,,
2022-03-11T13:31:15+05:30,11,24.98,,,,,,
2022-03-11T13:31:30+05:30,12,0.00,,,,,,
2022-03-11T13:31:45+05:30,13,0.00,,,,,,
2022-03-11T13:32:00+05:30,14,24.98,,,,,,
2022-03-11T13:32:15+05:30,15,0.00,,,,,,
2022-03-11T13:32:35+05:30,16,,33.30,,,,,
2022-03-11T13:32:50+05:30,17,,26.82,,,,,
2022-03-11T13:33:10+05:30,18,1.17,,,,,,
2022-03-11T13:33:25+05:30,19,,33.30,,,,,
2022-03-11T13:33:40+05:30,20,19.60,,,,,,
2022-03-11T13:33:55+05:30,21,1.54,,,,,,
2022-03-11T13:34:10+05:30,22,,0.00,,,,,
2022-03-11T13:34:30+05:30,23,,0.00,,,,,
2022-03-11T13:34:45+05:30,24,24.98,,,,,,
2022-03-11T13:35:00+05:30,25,,0.00,,,,,
```

Figure 15.15 Runtime data

REFERENCES

[1] Goy, S. and Sancho-Tomás, A., 2019. Load management in buildings. In *Urban Energy Systems for Low-Carbon Cities* (pp. 137–179). Academic Press.

[2] Javaid, N., Ullah, I., Akbar, M., Iqbal, Z., Khan, F.A., Alrajeh, N. and Alabed, M.S., 2017. An intelligent load management system with renewable energy integration for smart homes. *IEEE Access*, 5, pp.13587–13600.

[3] Eltamaly, A.M., Mohamed, M.A., Al-Saud, M.S. and Alolah, A.I., 2017. Load management as a smart grid concept for sizing and designing of hybrid renewable energy systems. *Engineering Optimization*, 49(10), pp.1813–1828.

[4] Abou Emira, S.S., Youssef, K.Y. and Abouelatta, M., 2019, December. Adaptive Power System for IoT-Based Smart Agriculture Applications. In

2019 15th International Computer Engineering Conference (ICENCO) (pp. 126–131). IEEE.

[5] Ogunjuyigbe, A.S.O., Ayodele, T.R. and Akinola, O.A., 2017. User satisfaction-induced demand side load management in residential buildings with user budget constraint. *Applied Energy*, 187, pp.352–366.

[6] Raju, M.P. and Laxmi, A.J., 2017, November. A novel load management algorithm for EMU by implementing demand side management techniques using ANN. In *2017 International Conference on Electrical and Computing Technologies and Applications (ICECTA)* (pp. 1–6). IEEE.

[7] Schultz, C., Braun, S., Braunreuther, S. and Reinhart, G., 2017. Integration of load management into an energy-oriented production control. *Procedia Manufacturing*, 8, pp.144–151.

[8] Ahmed, S. and Bouffard, F., 2017, October. An online framework for integration of demand response in residential load management. In *2017 IEEE Electrical Power and Energy Conference (EPEC)* (pp. 1–6). IEEE.

[9] Mason, K. and Grijalva, S., 2019. A review of reinforcement learning for autonomous building energy management. *Computers & Electrical Engineering*, 78, pp.300–312.

[10] Nojavan, S., Majidi, M. and Esfetanaj, N.N., 2017. An efficient cost-reliability optimization model for optimal siting and sizing of energy storage systems in a microgrid in the presence of responsible load management. *Energy*, 139, pp.89–97.

[11] Georges, E., Braun, J.E. and Lemort, V., 2017. A general methodology for optimal load management with distributed renewable energy generation and storage in residential housing. *Journal of Building Performance Simulation*, 10(2), pp.224–241.

[12] Bonthu, R.K., Pham, H., Aguilera, R.P. and Ha, Q.P., 2017, August. Minimization of building energy cost by optimally managing PV and battery energy storage systems. In *2017 20th International Conference on Electrical Machines and Systems (ICEMS)* (pp. 1–6). IEEE.

[13] Galván, L., Navarro, J.M., Galván, E., Carrasco, J.M. and Alcántara, A., 2019. Optimal scheduling of energy storage using a new priority-based smart grid control method. *Energies*, 12(4), p.579.

[14] Mangla, M., Akhare, R. and Ambarkar, S., 2019. Context-aware automation based energy conservation techniques for the IoT ecosystem. In *Energy Conservation for IoT Devices* (pp. 129–153). Springer, Singapore.

[15] Elmouatamid, A., Ouladsine, R., Bakhouya, M., Zine-Dine, K. and Khaidar, M., 2019, October. A control strategy based on power forecasting for micro-grid systems. In *2019 IEEE International Smart Cities Conference (ISC2)* (pp. 735–740). IEEE.

[16] Ghaffar, M., Naseer, N., Sheikh, S.R., Naved, M., Aziz, U. and Koreshi, Z.U., 2019, October. Electrical energy management of buildings using fuzzy control. In *2019 International Conference on Robotics and Automation in Industry (ICRAI)* (pp. 1–8). IEEE.

[17] Mahlangu, G., Musungwini, S. and Nleya, S., 2018. An algorithm for using internet of things (IoTs) to improve load management in electric power grid. *International Journal of Computer Applications, 179*(35), pp.10–17.

[18] Marinakis, V. and Doukas, H., 2018. An advanced IoT-based system for intelligent energy management in buildings. *Sensors*, 18, p.610, doi:10.3390/s18020610.
[19] Liu, J., Chen, X., Cao, S. and Yang, H., 2019. Overview on hybrid solar photovoltaic-electrical energy storage technologies for power supply to buildings. *Energy Conversion and Management*, 187, pp.103–121.
[20] Shah, A.S., Nasir, H., Fayaz, M., Lajis, A. and Shah, A., 2019. A review on energy consumption optimization techniques in IoT based smart building environments. *Information*, 10, p.108.
[21] Cucchiella, F., D'Adamo, I., Gastaldi, M. and Stornelli, V., 2018. Solar photovoltaic panels combined with energy storage in a residential building: An economic analysis. *Sustainability*, 10, p.3117, doi:10.3390/su10093117.
[22] Roberts, M.B., Bruce, A. and MacGill, I., 2019. Impact of shared battery energy storage systems on photovoltaic self-consumption and electricity bills in apartment buildings. *Applied Energy*, 245, pp.78–95.
[23] Paul, S. and Padhy, N.P., 2019. Real-time bilevel energy management of smart residential apartment building. *IEEE Transactions on Industrial Informatics*, 16(6), pp.3708–3720.
[24] Mozaffari Legha, M. and Farjah, E., 2020. IoT based load management of a micro-grid using Arduino and HMAS. *Iranian Journal of Electrical and Electronic Engineering*, 16(2), pp.228–234.
[25] Huang, P., Lovati, M., Zhang, X., Bales, C., Hallbeck, S., Becker, A., Bergqvist, H., Hedberg, J. and Maturi, L., 2019. Transforming a residential building cluster into electricity prosumers in Sweden: Optimal design of a coupled PV-heat pump-thermal storage-electric vehicle system. *Applied Energy*, 255, p.113864.
[26] Azharuddin Shamshuddin, M., Babu, T.S., Dragicevic, T., Miyatake, M. and Rajasekar, N., 2018. Priority-based energy management technique for integration of solar PV, battery, and fuel cell systems in an autonomous DC microgrid. *Electric Power Components and Systems*, 45(17), pp.1881-1891. DOI: 10.1080/15325008.2017.1378949.
[27] Shah, A.S., Nasir, H., Fayaz, M., Lajis, A. Shah, A., Shareef, H., Ahmed, M.S., Mohamed, A. and Al Hassan, E., 2019. Review on home energy management system considering demand responses, smart technologies, and intelligent controllers. *Information*, 10, p.108, doi: 10.3390/info10030108.
[28] Aneesh, R., Sivraj, P. and Kottayil, S.K., 2020, November. Load management and smart monitoring for Rroftop PV in academic building. In *2020 International Conference on Power Electronics and Renewable Energy Applications (PEREA)* (pp. 1–6). IEEE.
[29] Sharan, M., Das, P., Malhotra, A., Karthikeyan, A., Kannan, R. and Swathika, O.G., 2023. IoT-Based Prioritized Load Management Technique for PV Battery-Powered Building: Mini Review. *IoT and Analytics in Renewable Energy Systems (Volume 1)*, pp.275-280.
[30] Jiang, J., Peyghami, S., Coates, C. and Blaabjerg, F., 2021. A comprehensive study on reliability performance of Photovoltaic-battery-based microgrids under different energy management strategies. *Journal of Energy Storage*, 43, p. 103051.

[31] Tejas, R.N. and Vaikund, H., 2021. Design and control of PV connected microgrid. *Int J Eng Res Technol*, *10*(01), pp.66-72.

[32] Eltamaly, A.M., Alotaibi, M.A., Alolah, A.I., Ahmed, M.A., 2021. IoT-based hybrid renewable energy system for smart campus. *Sustainability*, 13, p.8555. https://doi.org/10.3390/su13158555.

[33] Torres-Moreno, J. L., Gimenez-Fernandez, A., Perez-Garcia, M. and Rodriguez, F., 2018. Energy management strategy for micro-grids with PV-battery systems and electric vehicles. *Energies*, 11, p.522, doi:10.3390/en11030522.

Chapter 16

A Predictive Analysis of Crop Yield for Sustainable Farming in Rajasthan

Savani M. Kulkarni, Puppala Anusha, Veeram Priyanka, G. Hannah Grace, Obeng-Denteh W., Revathi G. K.

16.1 INTRODUCTION

Hydrometric stations play a crucial role in monitoring the components of water quality, generating indispensable data for evaluating project feasibility and developing initiatives for water conservation. The assessment of water quality is of paramount significance in the planning of agricultural projects, influencing decisions related to cropping patterns, selecting irrigation systems, and establishing industrial water purification systems.

Freshwater is the most vital natural asset on Earth; it is essential for the sustenance of all forms of life. To effectively manage water resources, it is imperative to comprehend the cycles of river fluctuations in both physical and chemical concentrations, encompassing nutrients. Additionally, the ability to differentiate the impacts of physical and biogeochemical processes is crucial in ensuring effective water management.

Climate change is having a substantial impact on water resource projections, as well as their potential future availability and quality. As the world's population grows, so will the number of activities that demand huge amounts of water. Most of these methods cause significant pollution and push the boundaries of sustainable water resource use.

Human activities impact virtually all facets of river water management, encompassing evaporation, precipitation, and runoff. Altering the characteristics of river basins inherently brings about modifications in the hydrological dynamics of rivers, guaranteeing the system's endurance. Due to factors such as pollution, rising seawater, and other influences, these responses culminate in diminished water quality and alterations in the habitat of the ecosystem.

The condition of surface water is a sensitive matter crucial to sustained economic development, social well-being, and environmental stability. In recent times, global attention and apprehension regarding water pollution have grown significantly. Consequently, innovative strategies for attaining sustainable water resource management have emerged globally. Regular

DOI: 10.1201/9781003537878-16

monitoring of water storage facilities is still necessary to evaluate the quality of water for various uses.

16.2 OBJECTIVES

The objectives of this study are as follows:

- predict crop yields for different regions of Rajasthan, considering the variations in water quality, soil analysis, and irrigation methods, and their impact on agricultural productivity
- optimize the use of water, fertilizers, and other resources by tailoring agricultural practices according to predicted yields, minimizing wastage and ensuring resource efficiency.

16.3 LITERATURE REVIEW

In the study, it is highlighted that approximately one-third of the population relies on groundwater in residential and industrial areas. Large industrial companies, such as those in the fertilizer, electricity, oil, and gas industries, release their treated wastewater into unlined canals via drains. Some of these companies also hold their treated wastewater in slurry or ash ponds. The water quality index (WQI) serves as an unified metric representing the overall water quality at a specific location and time, considering multiple factors. Originally formulated by Horton in the early 1970s, the WQI functions as a mathematical approach for condensing numerous test results into a single value. The outcomes of the index convey the water quality within a specific water basin, such as ponds, lakes, rivers, or streams.

It has been mentioned that agriculture is a key economic sector in India, employing 60% of the population. More than 20 million private and government tube wells are active in India, and are used for drinking and irrigation. In India, 80% of groundwater is used for residential uses, with the remaining 45% used for agriculture. The majority of people contribute 45% of the gross national product. Groundwater quality data tells us about the geological history of the rocks, variations in groundwater level, groundwater recharge and outflow movement, and aquifer storage capacity.

In a study, diverse machine learning (ML) methods were explored for the estimation of water quality. Water quality was evaluated using a range of ML models with respect to their predictive accuracy and precision. Characteristics such as the pH level, total salts in the water, viscosity, phosphorus, nitrates, zinc, chemical oxygen demand, chlorine, and sodium were all included in the raw dataset.

Subsequent to building the models, they undergo testing and analysis to identify the optimal one. This involves a comparison and evaluation of prediction accuracy, precision rate, and execution time across all models.

Out of all the methods scrutinized, the random forest approach emerges with the highest accuracy at 95%, coupled with the swiftest execution time. According to the random forest model, water quality is assessed to be 84% contaminated, rendering it unsafe for consumption. This suggests a potential degradation in water quality attributable to anthropogenic activities and inadequate maintenance.

The escalation in global population growth, coupled with the expansion of industrial and agricultural development, has elevated water quality and supply to a significant global concern. Precipitation patterns are additionally shifting as a result of the ongoing effects of global warming.

It has been mentioned in studies that groundwater serves as the most significant and plentiful asset of nature, as well as a major source of water supply worldwide. Water quality is important for humans because it is linked to human well-being. Groundwater is important not just for its long-term sustainability but also for its high efficiency.

Water quality is being impacted by a variety of anthropogenic and natural activities, including irrigation systems and practices, geology, and regional climate. Rapid development and urbanization have a negative impact on surface and groundwater quality. A total of 65% of piped groundwater is used for agriculture globally. Groundwater supplies 50% of the world's drinking water and 45% of its irrigated agricultural land.

A study examined how well artificial intelligence systems, including the group method of data handling (GMDH), artificial neural network (ANN) algorithm, and support vector machine (SVM) algorithm, predict different elements of the water quality of the Tireh River in southwest Iran. Various transfer and kernel functions were explored for ANN and SVM development, and the radial basis function emerged as the most effective. The outcomes indicate that both the ANN and SVM exhibit commendable predictive capabilities for water quality components, with SVM identified as the most precise model based on error indices. While acceptable, GMDH demonstrates a somewhat lower accuracy compared to the ANN and SVM.

Control charts, commonly referred to as charts, are essential statistical instruments employed for overseeing and managing techniques. They indicate the existence of both typical and unique causes in a technique. Every process exhibits variation and common causes fall within the control limits. Conversely, special causes require heightened attention as they manifest values outside the control limits, necessitating prompt intervention.

Utilizing control charts for routine analysis monitoring enables the prompt identification of unfavorable trends in analytical work. This, in turn, facilitates immediate corrective measures, reduces instances of out-of-specification results, and thus prevents non-compliance. Control charts are widely applied globally. Employing control charts aided in uncovering anomalies in the daily analysis cycle, revealing similarities in the characteristics of Indonesia's Surabaya River between 2014 and 2015 across various metrics such as biochemical oxygen demand over five days (BOD5), chemical oxygen demand (COD), and total suspended solids (TSS).

Control charts serve the purpose of determining whether a process is under control. Traditionally, they have been integral in monitoring quality in the manufacturing process. When the process of production is affected only by random factors, it is regarded as natural or normal. The control charts in the balanced system of coordinates typically exhibit random fluctuations at this stage. While the standard control chart readily identifies irregularities beyond the limits, pinpointing the range of abnormality proves challenging, often requiring human judgment and influenced by various factors.

Control chart objectives are outlined in ISO 7870-1:2019, which include analyzing process stability, comparing sample data to control limits, calculating process variability, and detecting oscillations. It has been demonstrated that control charts are efficient in analyzing sewage treatment plant performance, allowing for timely issue diagnosis and resolution to preserve water quality. A preparatory phase is required to set control limits in sample analysis when using the commonly used Shewhart-type control chart.

Within accredited laboratories, quality management comprises a series of measures aligned with global regulatory standards, deemed crucial for preserving the adequacy of goods and amenities. More and more organizers are exploring avenues to enhance their processes and employ tools to fulfill their objectives. By employing statistical controls, fostering continuous improvement, providing training, and encouraging interaction, organizations can mitigate process unpredictability, consequently raising the efficiency and quality of work. A variety of instruments are used in quality control: check sheets of paper, histogram charts, and control charts are a few examples. Quality is upheld through the reduction of variation in both product and procedural characteristics.

A multidimensional control graph that makes use of the author's Gower distance is presented in a study. It aims to address handling difficulties that arise from high-dimensional processes that include both persistent and categorical data. Through a comprehensive overlapping investigation comparing it with existing charts, superior performance is observed, notably when the sheer number of categorical factors increases. Furthermore, a real-world case study reflects the versatility and effectiveness of the lately indicated control charts.

The essential assumptions underlying the development of quality control diagrams include the consistency and autonomy of subsequent samples. Although the conception of regularity is generally valid, many continuous-flow processes, like chemical processes, display some degree of autocorrelation among serial samples. Disregarding the autocorrelation structure during the construction of control charts diminishes the in-control run time, leading to an increase in false alarms. Additionally, when identifying slight variations in a process's mean vector, cumulative sum (CUSUM) control charts outperform Shewhart control charts.

In a project, the authors combined advanced data mining techniques with statistical process control (SPC) methodologies to properly monitor self-correlated and multivariate data in modern systems. The authors used ANN, support vector regression (SVR), and multivariable adaptive regression models to build total control charts with residuals from data mining methods. The simulation results demonstrate that data mining model-based control charts outperform traditional time-series model-based control charts.

Over the past two decades, scholars have shown interest in process parameter monitoring in a multidimensional routine. Nevertheless, the majority of previous attempts to achieve this goal have concentrated on creating statistical control charts. By contrast, the integration of ML techniques into control charts has been relatively infrequent. Three ML control charts in a study are grounded in the concepts of random forests, SVMs, and ANNs. These ML algorithms are trained to produce linear results and the process is categorized as in-control or out-of-control using memory-less control charts. The suggested ML structures make use of a couple of unique input situations and two training methodologies.

ML algorithms have been applied extensively in process monitoring through the utilization of control charts for various purposes, such as problem diagnosis, identification, reduction of dimensions, recognition of patterns, alteration of the estimation of points, and detection of signals. Among these techniques, the ANN stands out as the most commonly employed ML method in this context. Numerous scholars have made substantial contributions to this field.

Applying ML structures for creating control graphs is infrequently observed in the literature. A perceptron-based neural network is presented in a study that is intended to recognize and categorize mean shifts in processes with multiple attributes. Another study involved the development of three ANN control charts for tracking the parameters of a simple linear profile, such as the point of intersection, gradient, and variance residual. Among these control charts, one addressed both detection and identification, while the other two focused solely on detection. Building on the work in, scholars further extended the research by creating an SVR control chart tailored to keep an eye on an administrative task. Furthermore, to track generalized linear regression profiles, in, a variety of analytical and ML-based control graphs were developed with both fixed and variable design parameters.

The scope of social network surveillance was expanded through the introduction of an ANN-centered control chart for binary surgical outcomes, thereby enhancing ML controls customized for self-correlated profiles, while building on the groundwork laid in. While our approach shares similarities with theirs, we distinguish ourselves by addressing multivariate processes and concurrently monitoring parameters using a customized input set for ML structures. In contrast to their utilization of a single ML structure, we employ multiple ML techniques, subjecting them to extensive comparison across various scenarios. Additionally, unlike previous ML control charts that trained structures with a random shift size for out-of-control data representation, our method incorporates an additional technique.

In the first study in the literature to present a unique ANN, SVM, and random forest for the concurrent monitoring of multiple normal process parameters relying on control charts, two training approaches are investigated using two input circumstances, including the novel use of statistics, and dual control scenarios (detection and detection-identification). This involves developing ML structures with small shifts and training them for the first time with both big and small mean shifts. The statistical control charts and the ML control charts for two, three, and four characteristics of quality can all be compared with one another.

Through statistical analysis, the correlation coefficient between the WQI and physicochemical variables is examined in a study. The investigation reveals a noteworthy linear correlation and a strong coefficient of correlation among various measurements of water quality. Consequently, this study delves into the percolation of leachate and its influence on groundwater, employing WQI and analysis of correlation.

Various instruments and methodologies created for water quality evaluation, including the WQI model, have been highlighted. This paradigm streamlines the evaluation of enormous datasets into a single numerical result, and has gained appeal due to its simplicity. However, questions have been raised about aspects such as uncertainty, model dependability, transparency, and sensitivity. Recent research raises doubts about the accuracy of this model, implying that it may not accurately reflect the actual water quality situation.

Groundwater is significant due to its continuous high efficiency, as well as its sustainability. Water quality is declining due to a variety of man-made and natural factors, including geology, irrigation techniques, and regional climate. The quality of surface and groundwater is declining due to the rapid rise of industrialization and urbanization.

Utilizing the weighted arithmetic approach to calculate the WQI, it becomes evident that all bore wells exhibit a WQI exceeding 100, signaling a concerning classification of "totally unsuitable for drinking." This concerning circumstance suggests that the groundwater is impure, posing a hazard for drinking and domestic purposes.

Providing a strategy for estimating the process area of proportions from a sample of size n taken at regular intervals using an adequately prepared control chart, the technique proposed in is unusual as it does not require any assumptions about the latent variable.

It has been mentioned that the p-chart is the most basic method for manipulating attribute data. Using a distinct p-chart for each category may thus be one approach for dealing with data with many categories. Dealing with many p-charts may be challenging to implement in practice.

A log-linear model that generates a control chart for regulating ordinal variables improves the partial adjacent category logit model by incorporating both unconstrained and restricted variants. Furthermore, the model is modified by loosening the proportionate odds assumption only for a subset of independent variables.

16.4 DESCRIPTION OF THE DATASET

Due to a lack of water, the Indian state of Rajasthan has been implementing trendy farming practices. As a result, the state has developed into a special hub for data scientists and ML engineers to apply their predictive analysis techniques to the field of agriculture in the future. The dataset covers agricultural data from 2018 to 2019 in the Indian state of Rajasthan. It consists of three CSV files: Crop_Production_Data.csv, Water_Usage_Data.csv, Soil_Analysis_Data.csv.Researchers. Data scientists working in the agricultural sector can assess and simulate multiple aspects of market dynamics and agricultural practices in Rajasthan with the help of these datasets.

Information on crop production, including district, crop type, season, area (hectares), yield (quintals), and production (tonnes), is contained in the Crop_Production_Data.csv file. With the use of this data, agricultural yield, seasonal fluctuations, and production trends in various districts may all be studied.

Information about water usage in agriculture, such as district, crop type, irrigation method, water consumption (liters/ha), and water availability (liters/ha), can be found in the Water_Usage_Data.csv file. This data can be used to examine crop productivity effects of water availability, evaluate irrigation methods' effectiveness, and investigate water management strategies.

Information about soil quality, including area, soil type, pH, percentage of organic matter, and amount of nitrogen, phosphorus, and potassium is present in the Soil_Analysis_Data.csv file. This information is useful for researching soil characteristics, determining soil fertility, and comprehending the amount of nutrients in various Rajasthani soil types.

16.5 METHODOLOGY

16.5.1 Data Collection

Our main purpose is to predict crop yield in different regions of Rajasthan. Hence, we consider various datasets that are useful in our research. The analysis incorporates a comprehensive set of data encompassing crop production, various districts, types of crops, seasonal variations, area in hectares, yield in quintals, production in tonnes, soil quality, soil type, pH levels, and organic matter percentage, as well as nitrogen, phosphorus, and potassium content measured in kilograms per hectare. This diverse dataset is employed to examine and draw insights from information pertaining to crop production, soil analysis, and water usage. Additionally, an effort is made to offer insights on water usage in agriculture utilizing data on irrigation methods, districts, crop varieties, and water availability (in liters per hectare). The effectiveness of different irrigation techniques can be assessed and practices for managing water resources can be looked into using this data.

16.5.2 Exploratory Data Analysis

Our culture is visual, including everything from television and movies to advertisements and artwork. Because our eyes are able to perceive colors and patterns, we can instantly distinguish red from blue and squares from circles. Thus, data visualization is an additional visual art approach that sparks our interest and maintains our primary attention on the message that is visually represented. A crucial stage in every data science or data analysis study is exploratory data analysis (EDA). The process of analyzing a dataset to find patterns and anomalies (outliers), and generate hypotheses based on our comprehension of the dataset is known as EDA. EDA includes creating several graphical representations to enable a better understanding of the data and providing summary statistics for the numerical data in the dataset. If researchers have some topic expertise, they may use data visualizations to create plots and charts that emphasize and illustrate significant links that are more useful to them and their stakeholders than measures of association or significance. In our research, we generate various visual graphs, such as a "histogram," "scatter plot," "heat map," "count map," and "box plot." Different graphs are used in order to understand our data in a better way.

16.5.2.1 Histograms

Histograms of various numerical values are shown in Figure 16.1, with frequencies on the *Y*-axis and various independent variables on *X*-axis.

16.5.2.2 Scatter Plot

A scatter plot serves as a visual depiction that showcases the values of two variables in a dataset using Cartesian coordinates. Each data point in the plot is positioned based on its corresponding values along the *X*-axis and *Y*-axis. Typically, the *X*-axis represents the independent variable, and in the

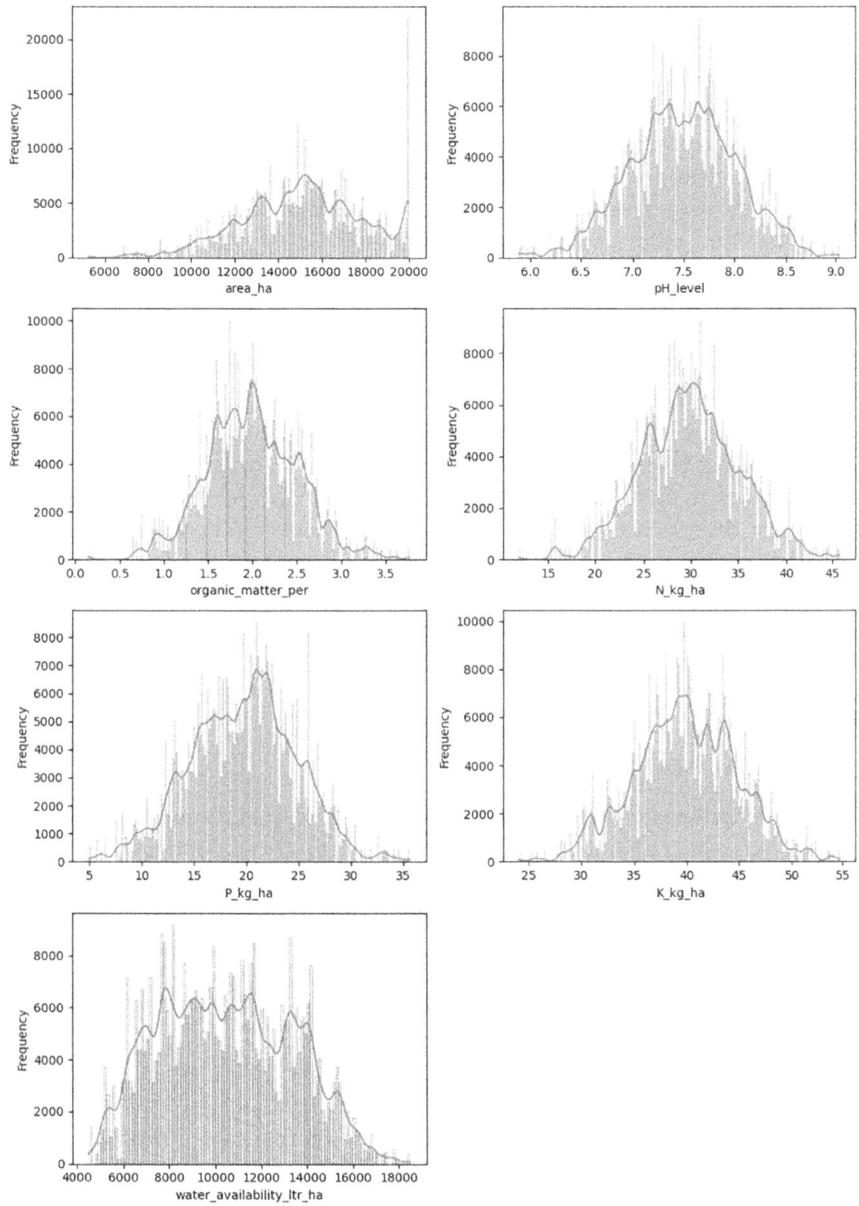

Figure 16.1 Histograms representing the data's frequency distribution

context of this representation, the *Y*-axis signifies the dependent variable, which, in this case, is the yield. The remaining variables are considered independent and are notated accordingly. A stronger correlation between two variables is evident in a scatter plot when data points closely approximate a straight line, indicating a more pronounced relationship. Conversely, a weaker association is observed when data points are more widely dispersed. On examination, it is apparent that nearly all scatter plots exhibit a broad distribution, suggesting a lack of significant linear correlation between yield in quintals and other features. See Figure 16.2.

16.5.2.3 Heat Map

A heatmap is a graphical data representation where various values are represented by color coding. Warmer colors, such as red or orange, indicate higher degrees of intensity, while cooler colors, such as blue or green, indicate lower levels. This is how the "heat" in "heatmap" relates to the use of colors to describe the intensity. The correlation matrix helps us to find out if there is a correlation between two features. It can be a positive (toward +1.00) or negative (toward -1.00) correlation. As can be seen in the confusion matrix in Figure 16.3, no two features have correlation values more than 0.1. This indicates that there is no direct correlation between any two features, so we consider training most of the features from the dataset.

16.5.2.4 Bar Plot

Data are represented by rectangular bars in a bar plot, sometimes referred to as a bar chart. Each bar's height and width correspond to a different data point's value and category, respectively. A numeric and a categorical variable's relationship can be seen using bar graphs. The plot's other axis displays a measured value, while the first axis displays the particular categories under comparison. The bar plot is used to keep a check on the count of the variables. See Figures 16.4–16.8.

16.5.2.5 Box Plot

A box-and-whisker plot is a visual representation that offers a succinct summary of a dataset. Designed to showcase the central tendency, variability, and skewness of numerical datasets, box plots provide a graphical depiction of the distribution of data. They are effective in revealing the presence of outliers and facilitating comparisons between different datasets. See Figures 16.9–16.13.

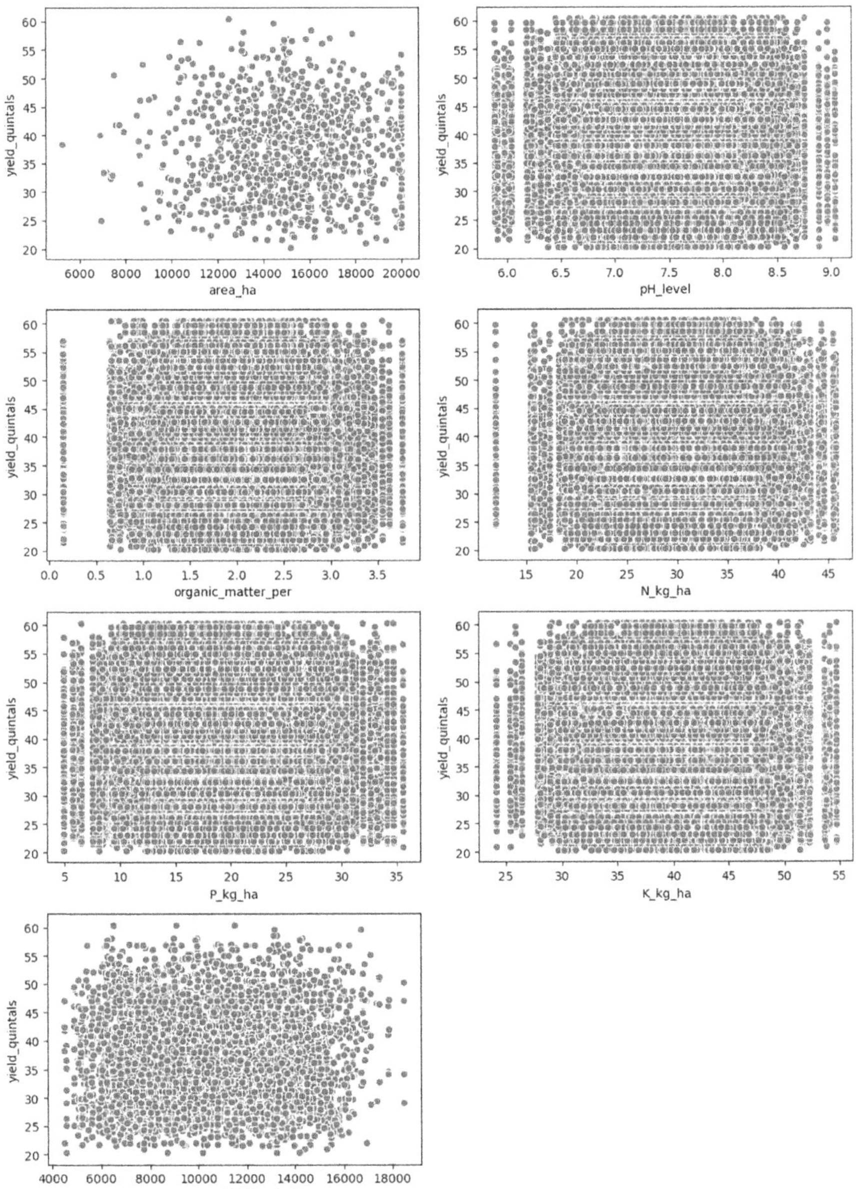

Figure 16.2 Scatter plots representing the relationship between two variables plotted on the *X* and *Y*-axes

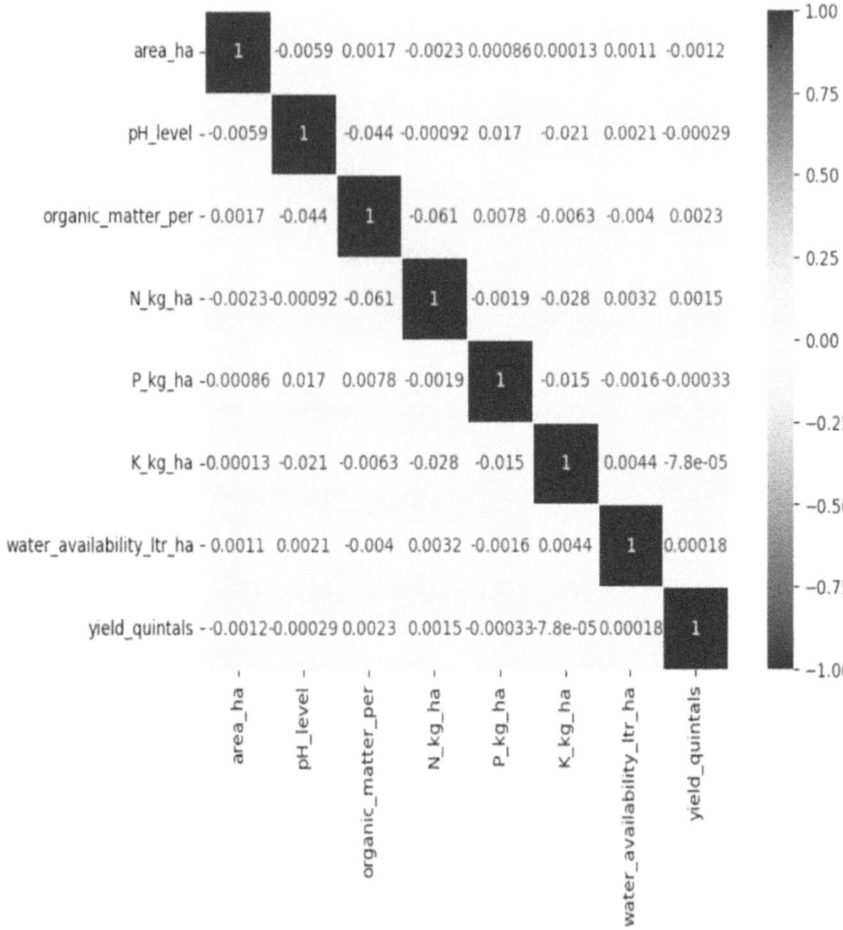

Figure 16.3 Correlation matrix represented as a heatmap

16.5.3 Data Preprocessing

16.5.3.1 Data Cleaning and Integrating

Integrating and cleaning data are essential phases in the data analysis process. They are crucial for guaranteeing the quality, accuracy, and reliability of the data utilized in analysis, research, and decision-making. The data cleaning process involves removing outliers and erroneous entries from the dataset to ensure accuracy and reliability. The datasets that are taken into consideration have extreme outliers that don't affect any further process.

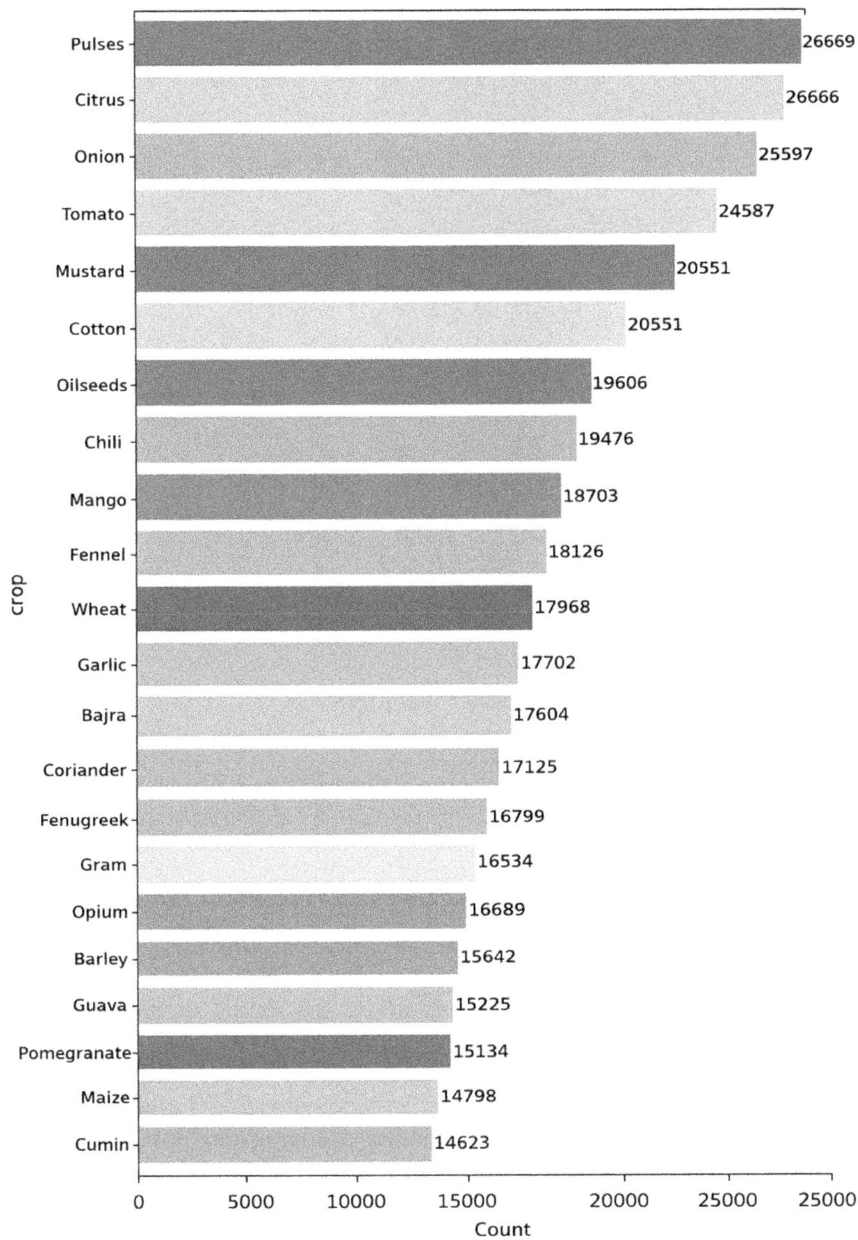

Figure 16.4 Bar graph of crops

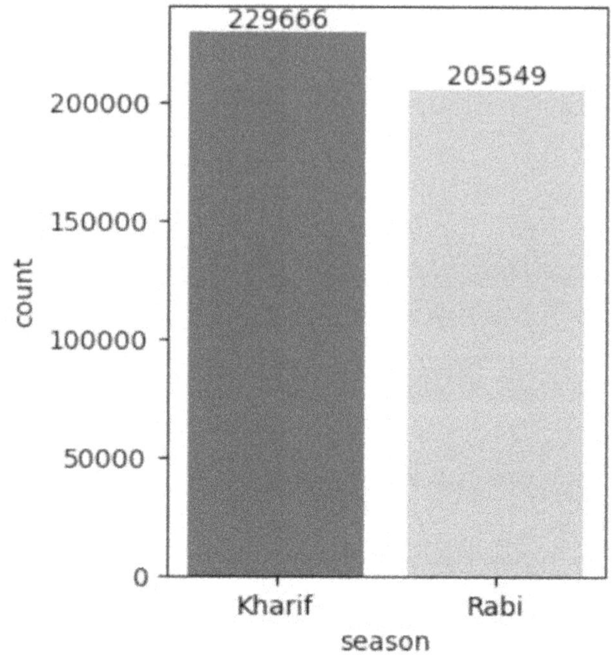

Figure 16.5 Bar graph of seasons ("Kharif " and "Rabi")

However, it is recommended to remove those outliers in order to achieve good accuracy for the models.

16.5.3.2 Converting Categorical Values into Numerical Values

Datasets have various categorical variables, such as "Districts," "Soil type," "Crop type," "Season," and "Irrigation method." These categorical values are converted into numeric values to allow the model to recognize and extract appropriate data. Many methods are used to convert categorical data into numeric data, including "one-hot encoding," "label encoding," "binary encoding," "ordinal encoding," "mean encoding," and "frequency encoding." In our research, we use "one-hot encoding and "label encoding" to deal with categorical variables.

16.5.3.3 Normalization

The normalization process is the act of appropriately arranging data. Its goal is to decrease the number of relationships that are exactly the same in the database. Furthermore, it's used for troubleshooting anomalies such

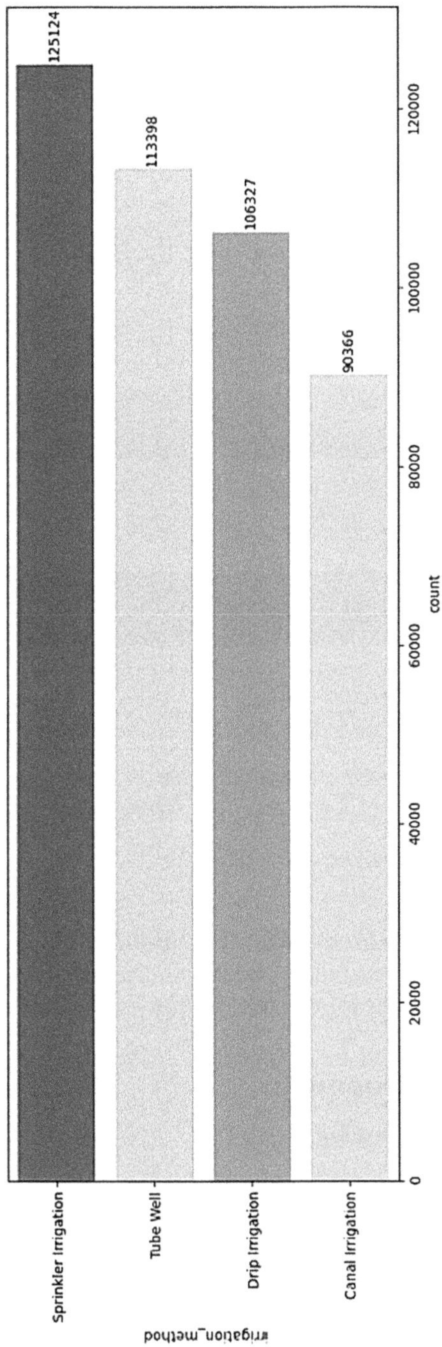

Figure 16.6 Bar graph of irrigation methods

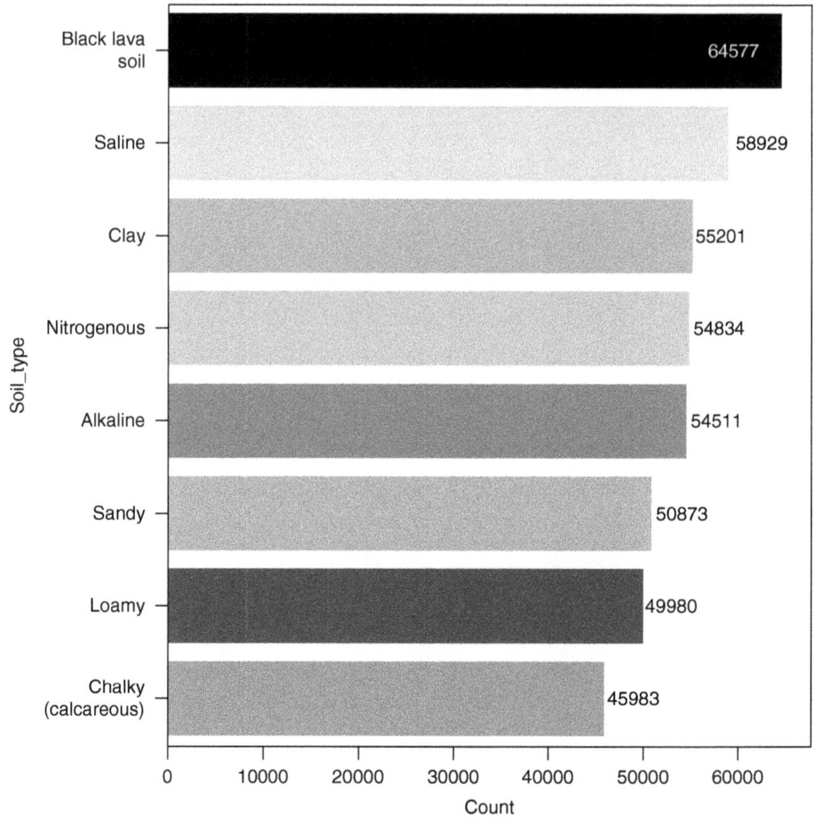

Figure 16.7 Bar graph of soil types

as inserts, deletes, and table changes. It is useful to divide a large table into several smaller, normalized tables. We normalize soil data, water usage data, and crop production data for the interpretation of control charts.

16.5.4 Model Development

16.5.4.1 Training and Testing

There are training and testing sets inside the dataset. The ML model is trained using the training set, which consists of 60% of the total data. The model gains an understanding of the relationships and patterns in the dataset during the training phase. The testing set, which is made up of the remaining 40% of the data, is used to assess the model's efficacy and determine how well it generalizes to new, unseen data. Predictions are produced

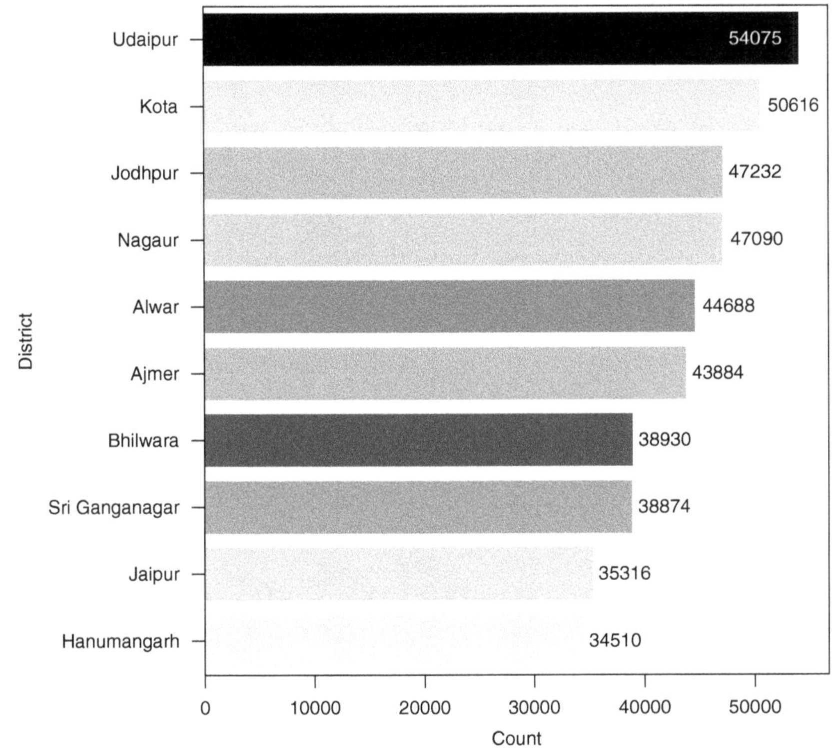

Figure 16.8 Bar graph of districts

by applying the trained model to the testing data. Then these are compared to the actual labels found in the testing dataset. On the normalized data, ANN, SVR, and random forest regression (RFR) models are trained, and control charts for each dataset are interpreted to check if the process is in control.

16.5.4.2 Machine Learning Algorithms Integrated with Control Charts

The correlation between dependent and independent variables in a dataset can be assessed using regression analysis and other predictive modeling techniques. This method is commonly used when the dependent and independent variables have a linear or nonlinear relationship, and the objective variable has a set of continuous values. Thus, time series modeling, forecasting, and the identification of causal relationships between variables are supported by regression analysis approaches. Regression analysis, for

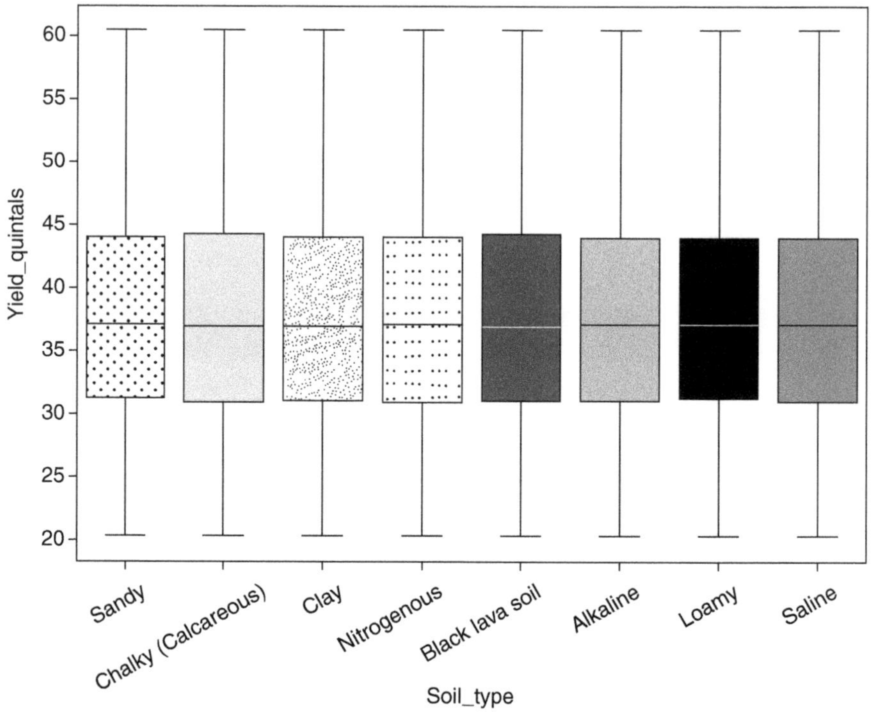

Figure 16.9 Boxplot of soil types

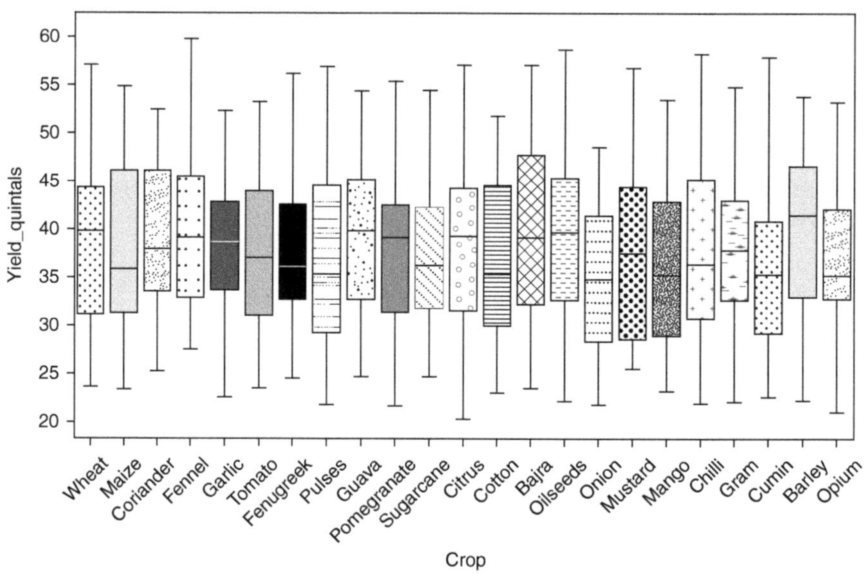

Figure 16.10 Boxplot of crops

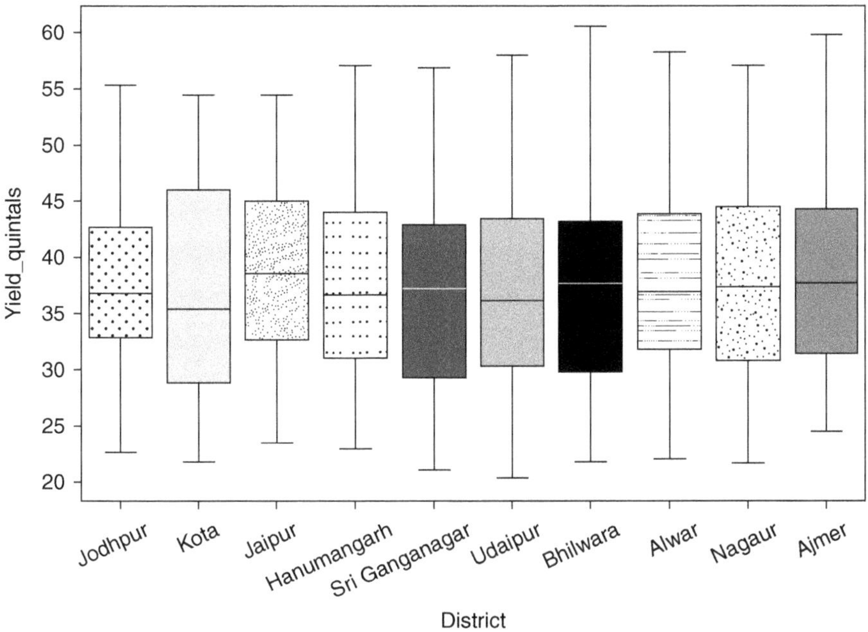

Figure 16.11 Boxplot of districts

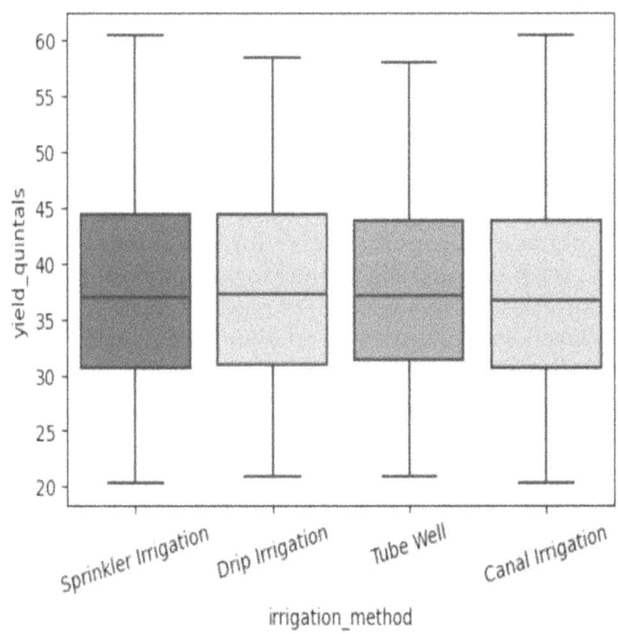

Figure 16.12 Boxplot of irrigation methods

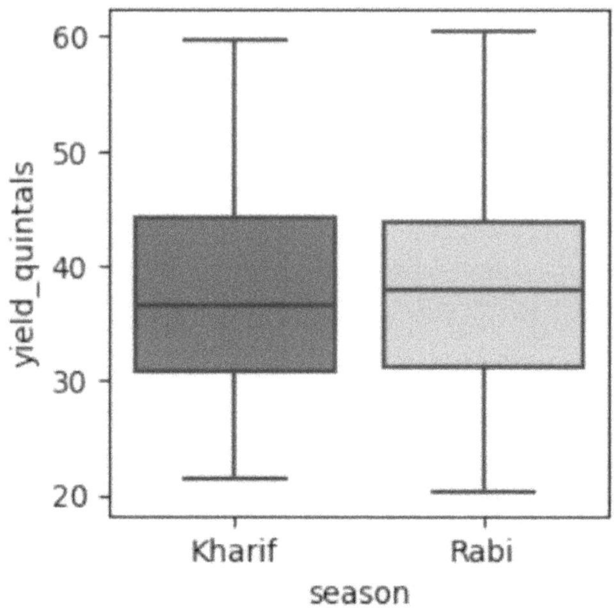

Figure 16.13 Box plot of seasons ("Kharif" and "Rabi")

instance, is an ideal technique for evaluating the connection between advertising expenditure and sales for a business.

Artificial Neural Networks

The ANN stands out as a widely used ML technique designed to mimic human brain functionality. Each layer includes a minimum of one node and is composed of a layer for input, a layer for output, and an assortment of hidden layers. Deep learning algorithms include several hidden layers and can be used for classification as well as regression. In ANNs, the number of nodes and layers is established by trial and error, with one hidden layer being sufficient for simple tasks. While there are broad guidelines for the number of nodes, it should be based on desired performance because there is no hard and fast rule.

Figure 16.14 illustrates the mathematical operations of neurons, involving multiplication of inputs by weights, summation, and application of a transfer function to the previous computation's results.

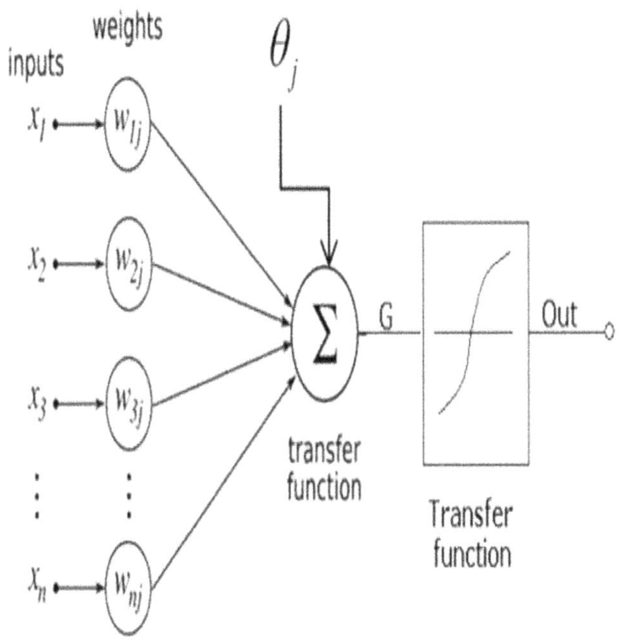

Figure 16.14 Diagram of an ANN

Source: Water quality prediction using ML methods (2018) taken from IWA Publishing

Support Vector Regressor

Regression analysis uses an ML technique called SVR. The main objective of SVR is to minimize prediction errors by finding an equation that matches the association among the input parameters and a continuous target variable.

Unlike SVMs, which are employed in classification tasks, SVR seeks to identify the hyperplane in a continuous space that most closely matches the data points. To do this, a feature space with high dimensions is mapped with the input variables, and a hyperplane that maximizes the margin, that is, the gap from the hyperplane and the closest data points is found. At the same time, the goal is to reduce prediction error to improve the regression analysis's accuracy.

SVR can manage nonlinear correlations between input variables and goal variables by mapping the data to a higher-dimensional space using a kernel function. This makes it an effective tool for regression tasks involving complex interactions between input variables and target variables.

Random Forest Regressor

Random forest is an ML paradigm used for regression or classification that employs ensemble learning techniques. In ensemble learning, predictions from multiple ML algorithms, specifically decision trees in the case of random forest, are combined to produce a more accurate and robust prediction compared to a single model. This approach is particularly effective in overcoming the overfitting issues associated with individual decision trees. Random forest is known for its robustness and precision, often outperforming single models, especially in scenarios with nonlinear relationships.

16.5.5 Feature Engineering

Feature engineering refers to the process of using domain knowledge to select and transform the most relevant variables from raw data when creating a predictive model using ML or statistical modeling. All three datasets are merged and features such as "district," "crop," "season," "area_ha," "yield_quintals," "soil_type," "pH_level," "organic_matter," "N_kg_ha," "P_kg_ha," "K_kg_ha," "irrigation_method," and "water_availability_ltr_ha" are chosen for engineering to predict yields(quintals) by enriching the dataset for modeling.

16.5.5.1 Machine Learning Algorithms used for Engineering

Linear Regression Model

The most popular modeling method, linear regression, is predicated on the idea that an independent variable (X) and a dependent variable (Y) have a linear relationship. Regression lines, also called best-fit lines, are used. $Y = c + m^*X + e$ is the definition of the linear connection, where c represents the intercept, m the line's slope, and e the error term.

There are two types of linear regression models: simple and complex. A model that is simple has one independent variable and one dependent variable.

Decision Tree Regression Model

Regression analysis or categorization can be done with decision trees. It works by breaking the data up into progressively smaller subgroups arranged in a tree-like fashion. By considering the category that the set of traits falls into, it estimates the final value of a collection of characteristics. Decision trees come in two varieties:

- trees of classification
- trees of regression.

A classification tree is employed to address issues with classification, whereas regression trees are used to address prediction problems. By using all the independent variables in a regression tree, following the fitting of a regression model to the target variable, the data are subsequently partitioned into distinct points corresponding to each independent variable. A sum of squared errors (SSE) is the result of squaring the difference between the expected and actual values in various situations.

The benefits of decision trees are as follows:

1. It's easy to understand them.
2. While data standardization is not necessary, users must prepare their data in some way before building a decision tree.

The disadvantages of decision trees are as follows:

1. Overfitting noisy data is a common problem with decision trees. Overfitting on noise is more likely to occur in a deeper tree.

XGBoost Regression Model

A quick and open-source version of the gradient booster technique is called extreme gradient boosting (XGBoost). Therefore, XGBoost is an algorithm, a project that is open source, and a library written in Python. Its objectives are to be significantly more successful and computationally effective (i.e., quick to execute) than existing open-source versions.

- It is engineered to be exceptionally potent and computationally efficient, potentially surpassing the performance of existing open-source versions.
- Models are fitted using the gradient descent optimization process and an arbitrary differentiable loss function. This is the reason for the technique's name, "gradient boosting," as fitting the model minimizes the loss gradient, much like a neural network does.
- Model performance and execution speed are XGBoost's two main justifications.
- Regarding issues with regression predictive modeling and classification, tabular or structured datasets are not as effective as XGBoost. Research suggests that it is the preferred strategy used by contest winners on the Kaggle competitive data science platform.

16.6 RESULTS AND DISCUSSIONS

A WQI is a singular numerical representation that encapsulates the comprehensive assessment of water quality at a specific location and time,

Table 16.1 Results

Sl. No	*Algorithm*	*MSE*	*R2 score*
1	Linear regression	68.2324	0.003
2	Decision tree regressor	2.1666	0.99999
3	XGBoost regressor	2.2521	0.9671

considering a range of factors influencing water quality. The datasets in this research had no missing values. The interquartile range was used to find outliers in data. However, in the datasets, there were very few outliers and they neither led to erroneous calculations nor affected the final result.

Because we had to predict the yield, we had to consider the other independent variables, such as soil characteristics, soil type, pH level, season, water consumption, and water availability. Hence, keeping this in mind, we merged the datasets into one single dataset and trained this dataset with different ML algorithms. Table 16.1 shows the R2 score and mean square error (MSE) values for each algorithm.

A control chart is a visual depiction illustrating the evolution of a process over time, also recognized as Shewhart charts or process-behavior charts. Essential components of a control chart include a center line representing the average, an upper line denoting the upper control limit, and a lower line indicating the lower control limit. When all data points fall within these boundaries without exhibiting discernible patterns, it suggests the presence of only common causes of variation, and the process is deemed to be "in control."

To have a good yield it becomes necessary to check the quality of the features on which the yield is dependent. In order to keep a check on the quality we can make use of control charts to ensure if the process is in control or not. To interpret and integrate the control charts with ML algorithms, data normalization is necessary. Each dataset in this project was considered separately and the data were normalized. To this normalized data, ANN, SVR, and RFR models were trained, and control charts for each dataset were interpreted.

The results obtained after training the models and integrating them with control charts for soil data are as shown in Figures 16.15–16.17.

The results obtained after training the models and integrating them with control charts for water usage data are as shown in Figures 16.18–16.20.

The results obtained after training the models and integrating them with control charts for crop-production data are as shown in Figure 16.21–16.23.

By the results obtained for integrating control charts with the merged dataset, we can conclude that most of the data points were in control and features chosen for prediction were apt. To obtain greater accuracy and achieve better model performance, hyperparameter tuning is necessary.

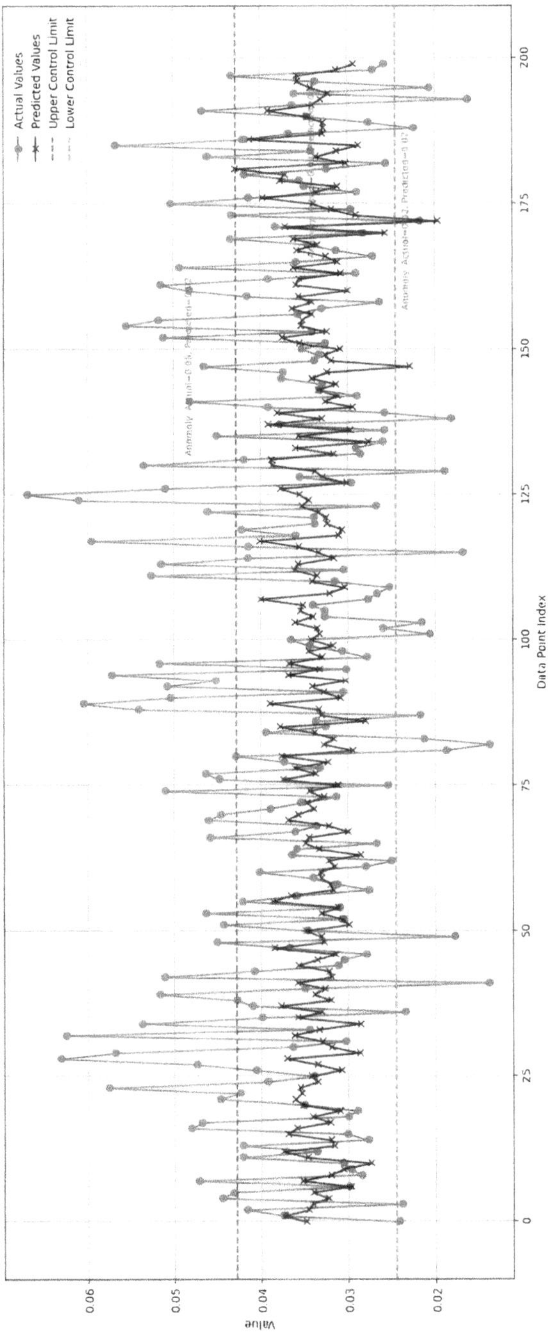

Figure 16.15 Interpretation of control charts for soil data (ANN)

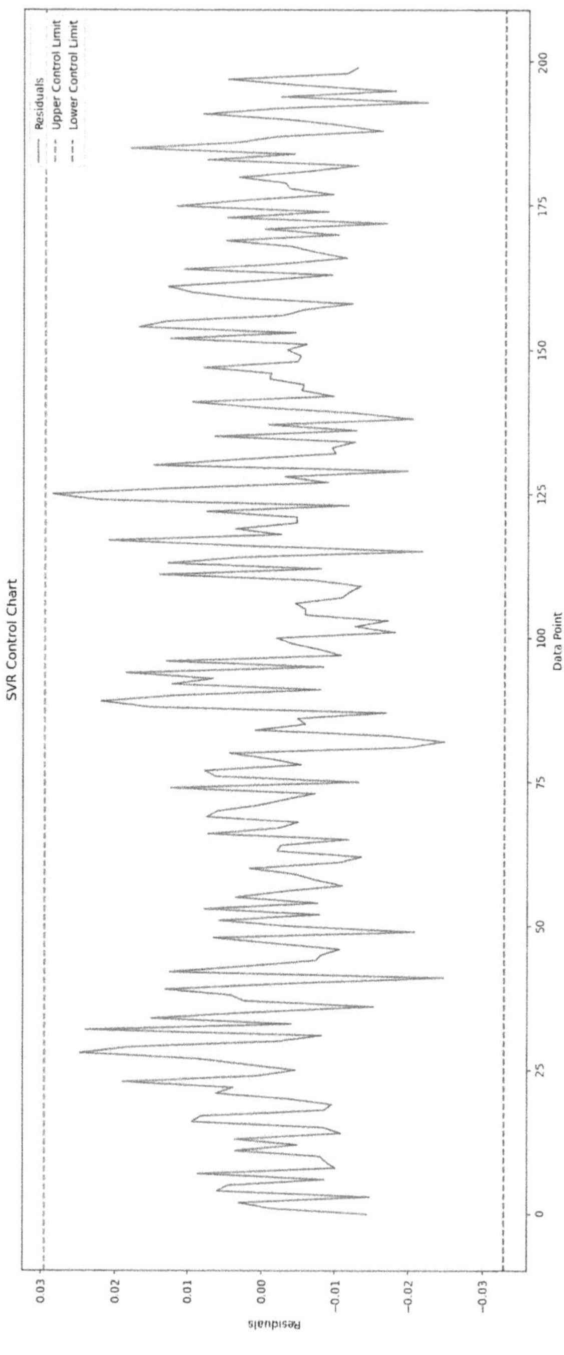

Figure 16.16 Interpretation of control charts for soil data (SVR)

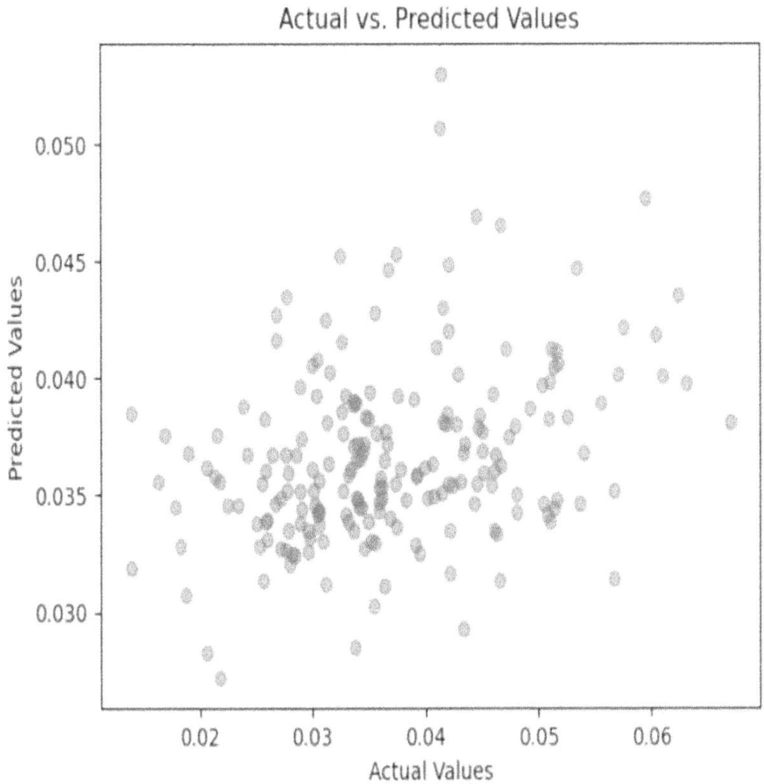

Figure 16.17 Residual plot (soil data)

The results obtained by ANN and SVR were more accurate for soil data, water usage data, and crop-production data. ANN was the best-fit model and most appropriate selection for each dataset compared to SVR and RFR models for predicting and checking if the process of the target variables was in control. This is because SVR did not help every time to bring the process into control.

Through the RFR model, we can conclude that there was a very weak correlation between the features and target variable for the soil data, while there was a highly positive correlation between the features and target variable for water usage data and crop production data.

The R2 scores and MSE values obtained by the RFR model for each feature corresponding to the dataset is given in Table 16.2.

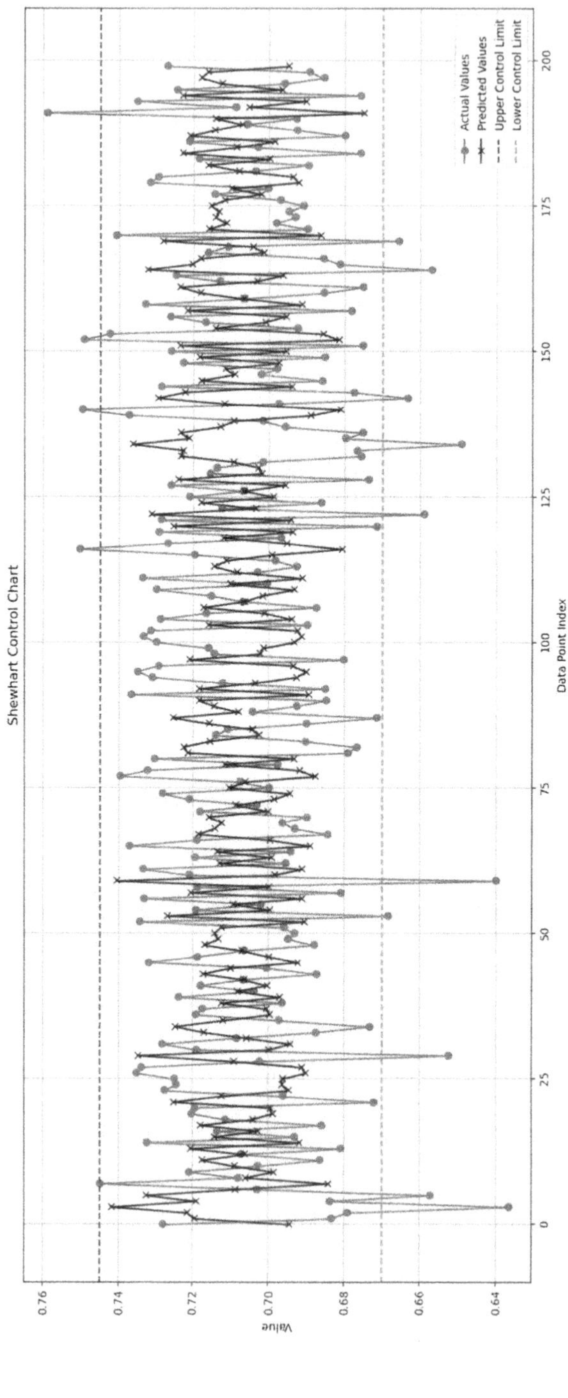

Figure 16.18 Interpretation of control charts for water usage data (ANN)

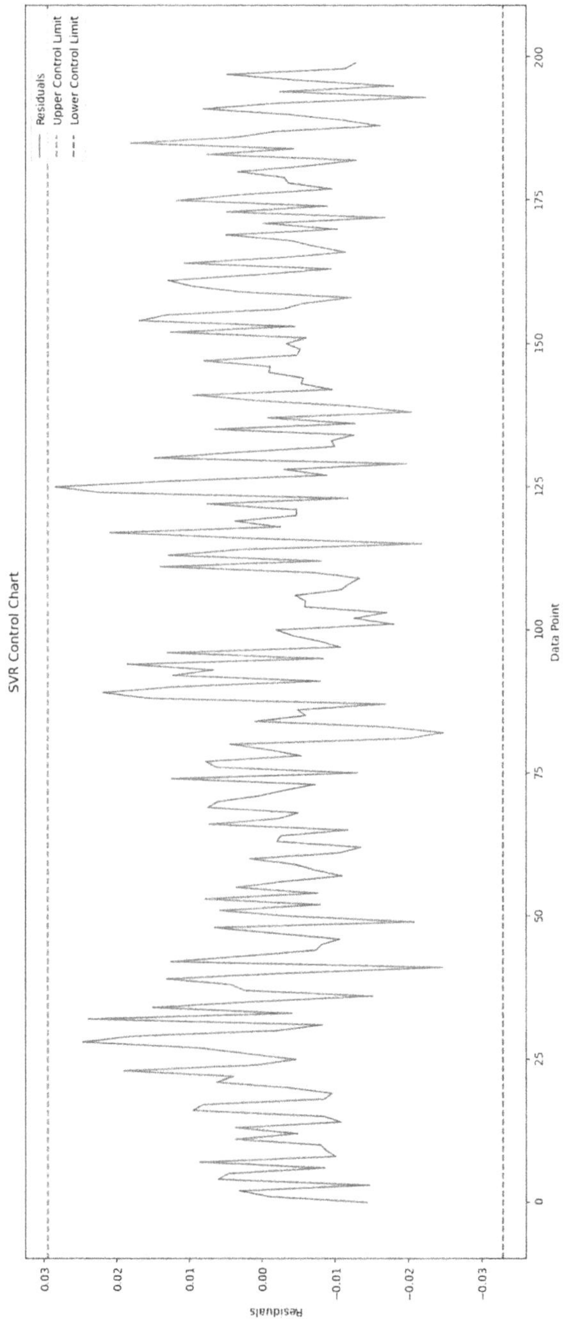

Figure 16.19 Interpretation of control charts for water usage data (SVR)

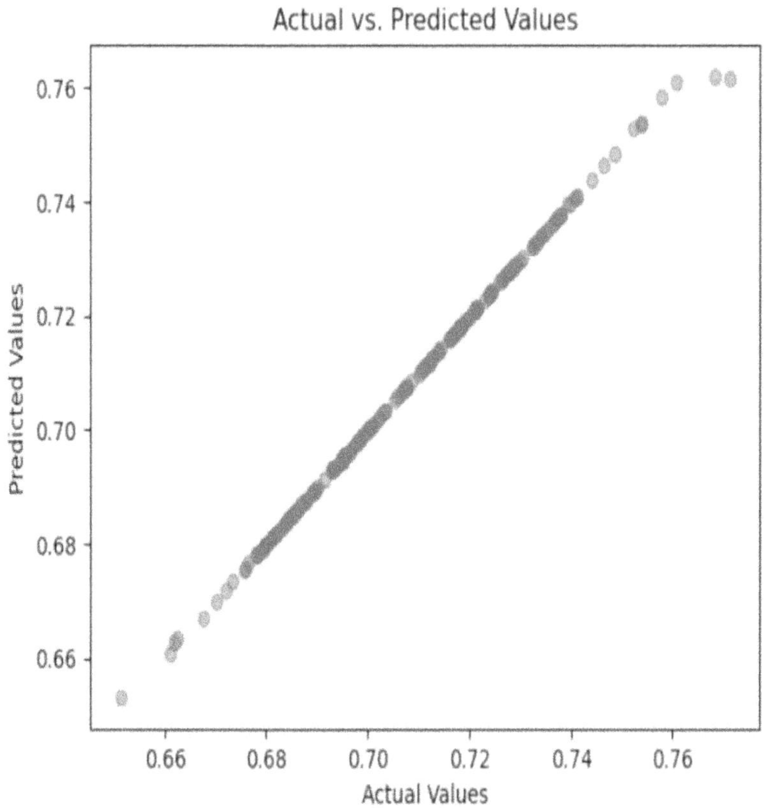

Figure 16.20 Residual plot (water usage data)

16.7 CONCLUSION

The primary goal is to achieve a more efficient and sustainable agricultural system by minimizing wastage and optimizing the allocation of resources. This involves employing predictive models or technologies to forecast crop yields, allowing farmers to adjust their irrigation and fertilization strategies accordingly. In this work, we examined the complex interactions that exist between crop yields and agricultural variables in several parts of Rajasthan. Using a comprehensive examination of data and the utilization of sophisticated ML algorithms and control charts, we arrived at noteworthy deductions that have the potential to improve the region's agricultural land. The predictive models demonstrated impressive accuracy in predicting agricultural yields based on data, especially the XGBoost regressor and decision tree regressor. A control chart is a graph that is used to study the evolution

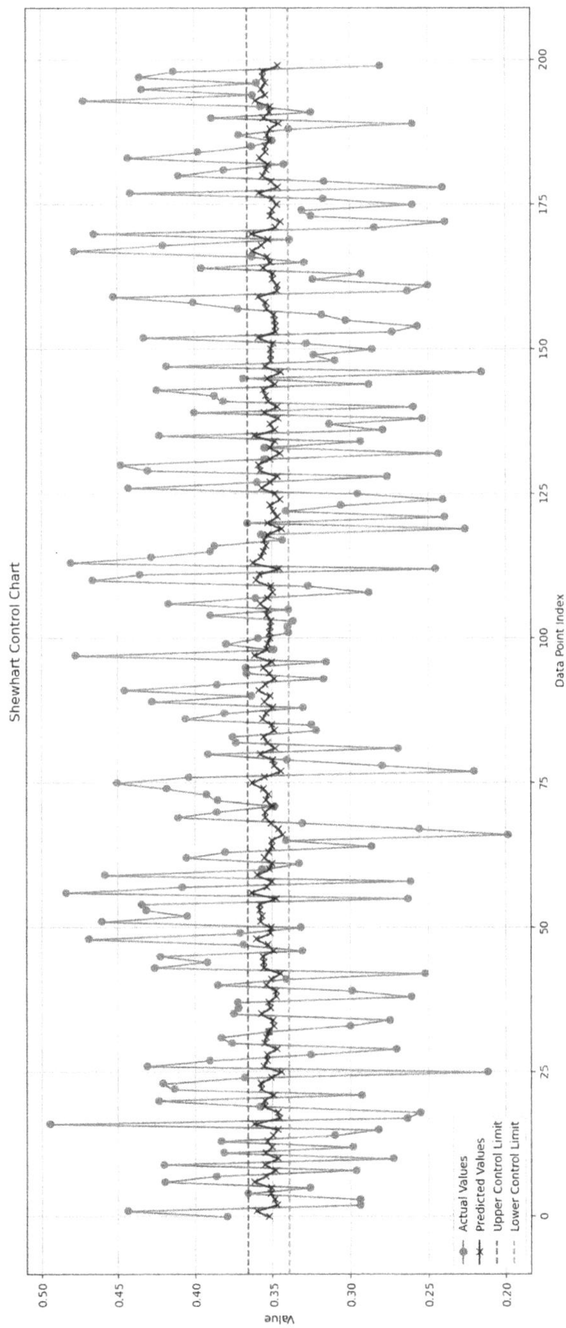

Figure 16.21 Interpretation of control charts for crop production (ANN)

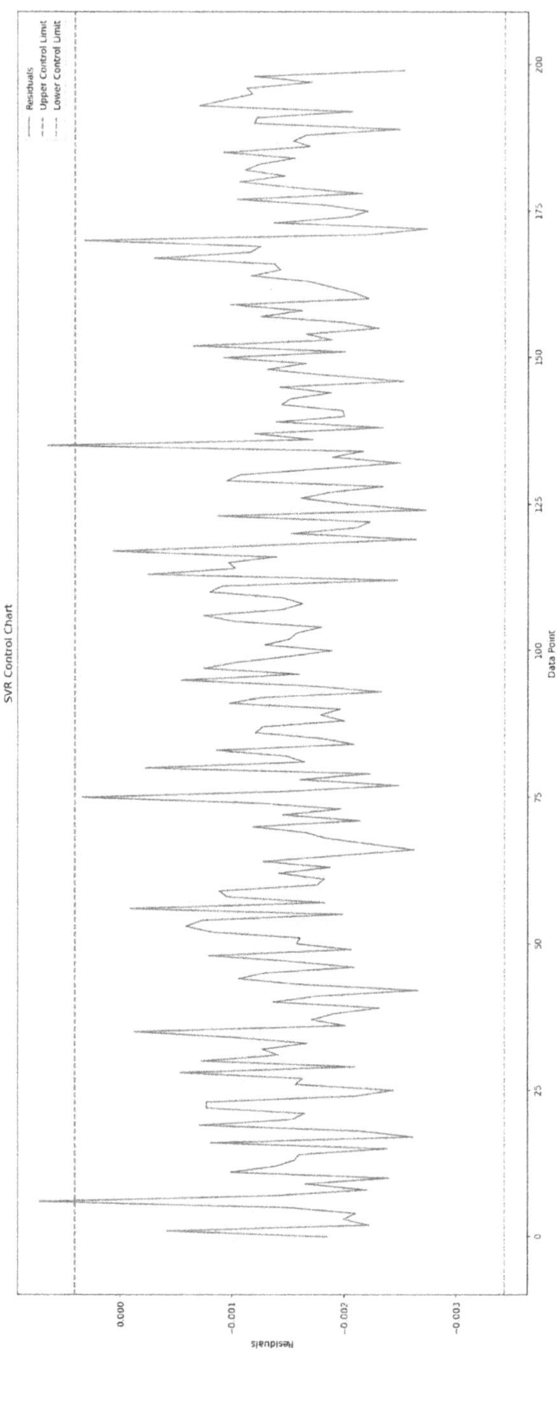

Figure 16.22 Interpretation of control charts for crop production (SVR)

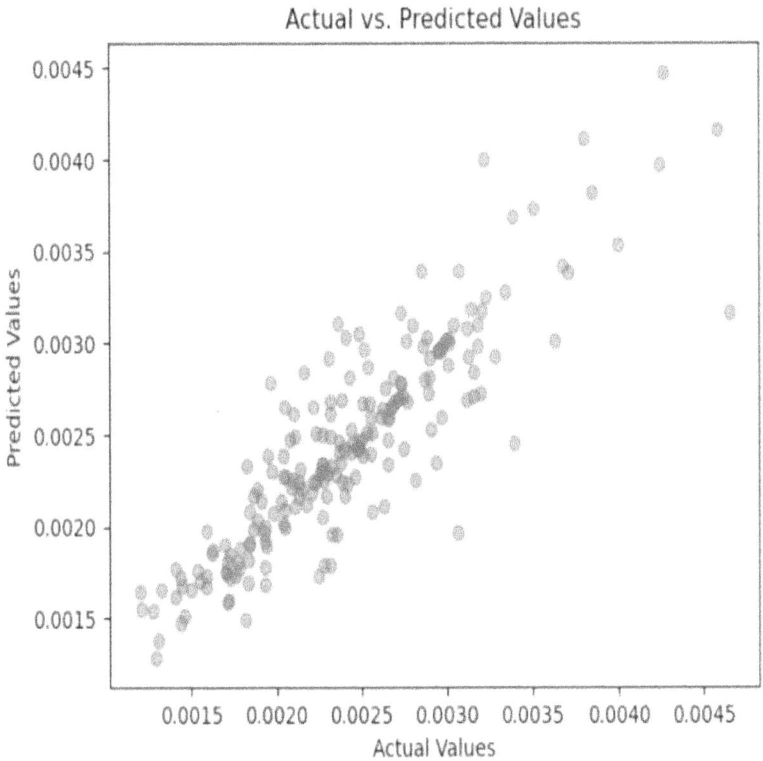

Figure 16.23 Residual plot (crop production)

Table 16.2 Dataset and results

Dataset	*Feature*	*R2 score*	*MSE*
Soil data	Soil type	-0.071807	0.001882
	Organic matter	0.0986465	0.065168
	pH level	0.0804098	0.000233
Water data	Water consumption	0.998611	0.000637
	Water availability	0.9977722	0.000284
Crop production data	Yield	0.76946	0.000313
	Production	0.999805	0.007618

of a process over time. Shewhart charts and process-behavior charts are other names for these charts. Additionally, the ANN was the best-fit model and most appropriate selection for each dataset compared to SVR and RFR models for predicting and checking if the process of the target variables was in control. Through the RFR model, we estimated the relationship between various features and the target variable. Using ML models, we were able to identify important variables that affect agricultural results. Further refinements, including feature engineering and hyperparameter tuning, may further enhance the performance of these models, ensuring their continued relevance as essential tools for agricultural analysis in the region.

16.8 LIMITATIONS AND FUTURE WORK

The consequences of our research are significant. Targeted initiatives based on insights into water quality can be beneficial for farmers in Rajasthan. By modifying their methods to correspond with ideal water conditions, they can raise crop yields and raise the standard of agricultural sustainability in general. Additionally, our research emphasizes how important it is to make well-informed decisions when drafting policies, which supports the incorporation of data-driven strategies into agricultural planning.

Although our study offers insightful information, there are certain drawbacks. Regional differences and data limitations provide issues that may be resolved by further study. We suggest more research on the long-term impacts of water quality on various types of crop, as well as the development of real-time monitoring technologies to enable dynamic farming technique adaptation.

As this study comes to an end, Rajasthan's agriculture sector is on the edge of a major change. Equipped with knowledge supported by data, academics, farmers, and policymakers may work together to bring in a sustainable agricultural era. Through the use of cutting-edge technology and the application of predictive models, we may improve food security, reduce obstacles, and clear the path for a more sustainable and prosperous future. In conclusion, this study not only clarifies the connection between crop yields and water quality but also acts as a stimulus for cutting-edge agricultural techniques. Let the knowledge gained from this study serve as a guide for our future actions as we work together to create a resilient and prosperous agricultural environment for Rajasthan.

BIBLIOGRAPHY

Abbasi, B. A neural network applied to estimate process capability of nonnormal processes. *Expert Syst. Appl.* 2009, 36, 3093–3100.

Abou Zakhem, B.; Hafez, R. Heavy metal pollution index for groundwater quality assessment in Damascus Oasis, Syria. *Environ. Earth Sci.* 2015, 73, 6591–6600. https://doi.org/10.1007/ s12665-014-3882-5.

Adak, M.D.G.; Purohit, K.M.; Datta, J. Assessment of drinking water quality of river Brahmani. *Indian J. Environ. Prot.* 2001, 8 (3).

Apsemidis, A.; Psarakis, S.; Moguerza, J.M. A review of machine learning kernel methods in statistical process monitoring. *Comput. Ind. Eng.* 2020, 142, 106376.

Arkat, J.; Niaki, S.T.A.; Abbasi, B. Artificial neural networks in applying MCUSUM residuals charts for AR(1) processes. *Appl. Math. Comput.* 2007, 189 (2), 1889–1901.

Arumugam, K.; Elangovan, K. Hydrochemical characteristics and groundwater quality assessment in Tirupur region, Coimbatore district, Tamil Nadu, India. *Environ. Geol.* 2009.

Balakrishnan, P. Groundwater quality mapping using geographic information system (GIS): A case study of Gulbarga City, Karnataka, India. *Afr. J. Environ. Sci. Technol.* 2012, 5, 1069–1084.

Bouwer, H. Integrated water management: Emerging issues and challenges. *Agric. Water Manag.* 2000.

Chang, S.I.; Ho, E.S. A two-stage network approach for process variance change detection and classification. *Int. J. Prod. Res.* 1999, 37, 1581–1599.

Cheng, C.-S.; Cheng, H.-P. Identifying the source of variance shifts in the multivariate process using neural networks and support vector machines. *Expert Syst. Appl.* 2008, 35, 198–206.

Corrêa, H.; Corrêa, C.A. *Administração da Produção e Operações: Manufatura e Serviços: Uma Abordagem Estratética*; Atlas, São Paulo, Brazil, 2008.

Costa, A.F.B.; Epprecht, E.K.; Carpinetti, L.C.R. *Controle Estatístico da Qualidade*, 2nd ed.; Atlas, São Paulo, Brazil, 2010.

Demircioglu Diren, D.; Boran, S.; Cil, I. Integration of machine learning techniques and control charts in multivariate processes. *Sci. Iran.* 2020, 27, 3233–3241.

Escobar, C.A.; Morales-Menendez, R. Machine learning techniques for quality control in high conformance manufacturing environment. *Adv. Mech. Eng.* 2018, 10, 1–16.

Gupta, S.; Gupta, S.K. A critical review on the water quality index tool: Genesis, evolution and future directions. *Ecol. Inform.* 2021, 63, 101299.

Hadian, H.; Rahimifard, A. Multivariate statistical control chart and process capability indices for simultaneous monitoring of project duration and cost. *Comput. Ind. Eng.* 2019, 130, 788–797.

Haghiabi, A.H.; Nasrolahi, A.H.; Parsaie, A. Water quality prediction using machine learning methods. *Water Qual. Res. J.* 2018, 53 (1), 3–13.

Hosseinifard, S.Z.; Abdollahian, M.; Zeephongsekul, P. Application of artificial neural networks in linear profile monitoring. *Expert Syst. Appl.* 2011, 38, 4920–4928.

Humphreys, W.F. Hydrogeology and groundwater ecology: Does each inform the other? *Hydrogeol. J.* 2009.

Jafar Ahamed, A.; Loganathan, K.; Ananthakrishnan, S. A comparative evaluation of groundwater suitability for drinking and irrigation purposes in Pugalur area, Karur district, Tamil Nadu, India. *Arch. Appl. Sci. Res.* 2013.

Juwana, I.; Muttil, N.; Perera, B.J.C. Uncertainty and sensitivity analysis of West Java Water Sustainability Index – A case study on Citarum catchment in Indonesia. *Ecol. Indic.* 2016, 61, 170–178.

Khan, R.A.; Juahir, H.; Yusoff, M.K.; Zain, S.M.; Hanida, T.I.T. *Using Principal Component Scores and Artificial Neural Networks in Predicting Water Quality Index*; INTECH Open Access Publisher: Rijeka, Croatia, 2012.

Kim, S.B.; Jitpitaklert, W.; Park, S.K.; Hwang, S.J. Data mining model-based control charts for multivariate and autocorrelated processes. *Expert Syst. Appl.* 2012, 39 (2), 2073–2081.

Mohammad, R. BOD, COD, and TSS predictions from DO measurement results for the Surabaya River, Indonesia. *J. Civ. Eng. Plan. Des.* 2022, 1, 1–7.

Mohammadzadeh, M.; Yeganeh, A.; Shadman, A. Monitoring logistic profiles using variable sample interval approach. *Comput. Ind. Eng.* 2021, 158, 107438.

Mohd Amiruddin, A.A.A.; Zabiri, H.; Taqvi, S.A.A.; Dendena Tufa, L. Neural network applications in fault diagnosis and detection: An overview of implementations in engineering-related systems. *Neural Comput. Appl.* 2020, 32, 447–472.

Niaki, S.T.A.; Abbasi, B. Detection and classification mean-shifts in multiattribute processes by artificial neural networks. *Int. J. Prod. Res.* 2008, 46, 2945–2963.

Niaki, S.T.A.; Abbasi, B. Fault diagnosis in multivariate control charts using artificial neural networks. *Qual. Reliab. Eng. Int.* 2005, 21, 825–840.

Perez, V.V.; Diacenco, A.A.; Paulista, P.H. Análise das Sete Ferramentas Estatísticas da Qualidade Utilizadas nos Sistemas Produtivos. *Rev. Univap*. 2016, 22, 807.

Ramos, E.M.L.S. *Aperfeiçoamento e Desenvolvimento de Ferramentas do Controle Estatístico da Qualidade—Utilizando Quartis Para Estimar o Desvio Padrão*. Ph.D. Thesis, Universidade Federal de Santa Catarina, Florianópolis, Brazil, 2003.

Sabahno, H.; Amiri, A. New statistical and machine learning based control charts with variable parameters for monitoring generalized linear model profiles. *Comput. Ind. Eng.* 2023.

Sabahno, H.; Amiri, A.; Castagliola, P. A new adaptive control chart for the simultaneous monitoring of the mean and variability of multivariate normal processes. *Comput. Ind. Eng.* 2021, 151, 106524.

Sabahno, H.; Niaki, S.T.A. New machine-learning control charts for simultaneous monitoring of multivariate normal process parameters with detection and identification. *Mathematics* 2023, 11 (16), 3566.

Saeedi, M.; Abessi, O.; Sharifi, F.; Meraji, M. Development of groundwater quality index. *Environ. Monit. Assess.* 2010.

Salehi, M.; Bahreininejad, A.; Nakhai, I. On-line analysis of out-of-control signals in multivariate manufacturing processes using a hybrid learning-based model. *Neurocomputing* 2011, 74, 2083–2095.

Samohyl, R.W. *Controle Estatístico de Qualidade*, Elsevier, São Paulo, Brazil, 2009.

Simonet, B. Quality control in qualitative analysis. *TrAC Trends Anal. Chem.* 2005, 24, 525–531.

Simsek, C.; Gunduz, O. IWQ index: A GIS-integrated technique to assess irrigation water quality. *Environ. Monit. Assess.* 2007.

Śliz, P.; Piotr, B. Assessment of the stability and reliability of the water treatment plant in Nowy Sącz using control cards. *J. Water Land Dev.* 2022, 251–256.

Subba Rao, N. Groundwater quality: Focus on fluoride concentration in rural parts of Guntur district, Andhra Pradesh, India. *Hydrol. Sci. J.* 2003.

Tal, A. Making conventional agriculture environmentally friendly: Moving beyond the glorification of organic agriculture and the demonization of conventional agriculture. *Sustainability* 2018, 10, 1078.

Tuerhong, G.; Kim, S. B. Gower distance-based multivariate control charts for a mixture of continuous and categorical variables. *Expert Syst. Appl.* 2014, 41 (4), 1701–1707.

Turuta, T.B. Aplicação de Cartas de Controle como Ferramenta de Melhoria Frente às Dificuldades Operacionais de Laboratórios Acreditados na ABNT NBR ISO/IEC 1705. Master's Thesis, *Instituto de Química de São Carlos da* Universidade de São Paulo, São Carlos, Brazil, 2015.

Uddin, M.G.; Nash, S.; Olbert, A.I. A review of water quality index models and their use for assessing surface water quality. *Ecol. Indic.* 2021, 122, 107218.

Weese, M.; Martinez, W.; Megahed, F.M.; Jones-Farmer, L.A. Statistical learning methods applied to process monitoring: An overview and perspective. *J. Qual. Technol.* 2016, 48, 4–24.

Yeganeh, A.; Chukhrova, N.; Johannssen, A.; Fotuhi, H. A network surveillance approach using machine learning based control charts. *Expert Syst. Appl.* 2023, 219, 119660.

Yeganeh, A.; Johannssen, A.; Chukhrova, N.; Abbasi, S.A.; Pourpanah, F. Employing machine learning techniques in monitoring autocorrelated profiles. *Neural Comput. Appl.* 2023, 35, 16321–16340.

Yeganeh, A.; Pourpanah, F.; Shadman, A. An ANN-based ensemble model for change point estimation in control charts. *Appl. Soft. Comput.* 2021, 110, 107604.

Yeganeh, A.; Shadman, A.; Shongwe, S.C.; Abbasi, S.A. Employing evolutionary artificial neural network in risk-adjusted monitoring of surgical performance. *Neural Comput. Appl.* 2023, 35, 10677–10693.

Zan, T.; Wang, M.; Fei, R.Y. Pattern recognition for control charts using AR spectrum and fuzzy ARTMAP neural network. *Adv. Mater. Res.* 2010, 97–101, 3696–3702.

Zhu, F.; Heinen, R.; van der Sluijs, M.; Raaijmakers, C.; Biere, A.; Bezemer, T.M. *Species-specific plant–soil feedbacks alter herbivore-induced gene expression and defense chemistry in Plantago lanceolata.* 2018.

Chapter 17

IoT-Based E-Bicycle Monitoring System

S. S. Harish Raaghav, M. M. Deepak, O.V. Gnana Swathika

17.1 INTRODUCTION

As the number of people living on the globe increases on a daily basis, there has also been an uptick in the number of products that are manufactured utilizing fuel. Along with this is the increase of fuel prices and environmental factors. It is important to understand that it is far wiser and better to use a bicycle rather than a motor vehicle for short distance traveling. This is because this vehicle requires fossil fuel to run, and fossil fuel is a limited source of energy that will run out after some period of time. To keep up with this need, the revolution for environmentally friendly cycles has made them the most depended upon mode of transportation. Imagine how much more helpful the bicycle would be if even a modest amount of effort was put in by man in order to ride it across hard terrain.

In this project, the normal geared bicycle will be combined with an electric power motor as the foundation. The system has been adjusted in such a manner that riders have the ability to pick which mode they like, meaning that they could either choose for the bicycle to be driven entirely by the electric motor or pedal the bicycle themselves. This decision is made possible by the system's modification. The mounting of the motor and support system onto a geared bicycle was conceived with the intention of minimizing the amount of extra labor that riders would have to put forth in order to transport the additional weight of the motor and support assembly. Because it has been constructed so that users of any age may rely on it, the device has been widely distributed. When thinking about how to implement the initiative, the primary focus is on finding ways to improve transit between different colleges.

An electric bicycle (e-bicycle) typically consists of a battery that is connected to an electric motor in the bicycle's gearbox system. This battery is capable of being charged using a standard wall outlet found in most homes. Using a throttle and controller that are positioned on the handlebar, the rider has the ability to adjust the amount of power that is delivered from the motor, which determines the speed. The phrase "e-bike" refers to a variety of different electrically powered two-wheelers, some of which are

DOI: 10.1201/9781003537878-17

operated by merely rotating a throttle. This is because the name “e-bike” is a generic term. This bicycle is created and manufactured at a cost that is far lower than the initial cost; hence, it is now within the financial reach of everybody. It is common knowledge that the use of motor vehicles that run on fossil fuels contributes to an ever-increasing level of pollution caused by the release of hazardous gases; as a result, the health of 4.3 million more individuals is at risk.

17.1.1 Motivation

The problem of pollution caused by an ever-increasing number of automobiles in metro towns and metropolitan zones serves as a driving force behind the identification of the need of locating and customizing an e-bicycle in order to find a solution to the problem. When every social class is taken into account, it is unreasonable to expect everyone to acquire motorized transportation of any kind (scooters, mopeds, or motorbikes). In addition to that, solar panels are provided so that additional amounts of electricity may be generated. E-bicycles have a number of benefits, the most notable one being that they are cost effective. This is because the majority of the costs associated with e-bicycles are associated with their construction, while the costs associated with their use are limited to the cost of charging the battery. Due to financial restraints, people with a middle class income cannot purchase such a locomotive in their location because electric scooters that are now on the market are far more expensive. Along with development of existing technologies, the respective theory must also be applied in order to design and create a product that has the potential to be sold at a larger frequency, has very cheap production costs, and is of high quality. In order to put all these ideas into action, it is necessary to create a design and product in such a way that they could compete with other “e-bikes” already on the market. This would allow us to incorporate all of the aforementioned concepts. The fundamental concept is to power the bicycle’s movement using an attached motor, that is, a motor that is driven by a battery and that can be turned on when riding on tough terrain and switched off and pedaled during motion on flat terrain to have the battery recharged during that period. Figure 17.1 shows an example of an e-bicycle that is currently on the market.

17.1.2 Objectives

The objectives are as follows:

- to develop and design an e-bicycle powered with a lead-acid battery and monitored using Internet of Things (IoT) technology
- ESP32-based microcontroller is used as an IoT device to process and transmit data regarding the user’s location, temperature, and battery power

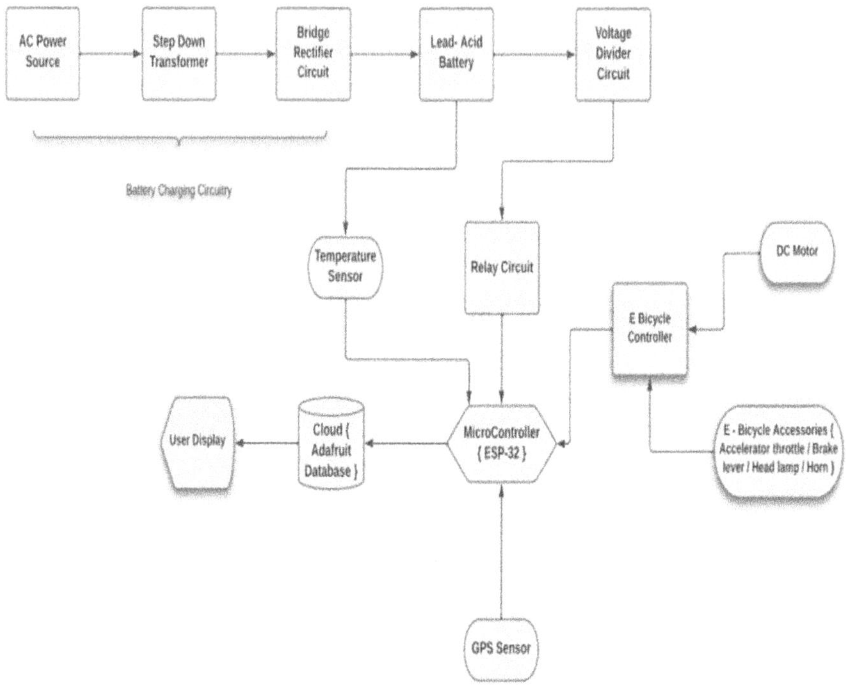

Figure 17.1 Block diagram of the proposed model

- to lessen the amount of work done by humans
- to improve the comfort of people and to make maintenance easier
- to overcome the difficulties posed by fossil fuels
- to lessen the impact of pollutants
- because it is possible to pedal the bicycle, increasing the positive effects on one's health.

17.1.3 Scope of the Work

In recent years, the government of India has made substantial efforts to encourage the use of electric vehicles (EVs) across the nation. The Faster Adoption and Manufacturing of (Hybrid) and Electric Vehicles (FAME) scheme in India was introduced by the government in 2018. It offers financial incentives for the purchase of electric two-wheelers, three-wheelers, and four-wheelers.

In India, not only is the government providing help but there is also a growing demand for EVs. The EV market in India was anticipated to reach 1.5 million units by 2025, as stated in research compiled by the Society of Manufacturers of Electric Vehicles. Concerns about air pollution and the need to lessen India's dependency on fossil fuels are the primary forces behind this desire.

Despite these encouraging developments, India is still facing a number of obstacles on the path to the mainstream adoption of EVs. The need for a charging infrastructure is a significant obstacle. Even if the number of charging stations is on the rise, there is still a significant lack of infrastructure to accommodate the expanding fleet of EVs. Another difficulty is the high initial cost of purchasing an EV, which may be a disincentive for many people who are considering making the purchase.

Despite these obstacles, it seems that EVs in India have a promising future. In future years, an increase in the number of electric cars on the road will be witnessed thanks to the continuous backing of the government and the growing demand from consumers. EVs have the potential to play a big part in India's transformation to a country with a more environmentally friendly and sustainable mode of transportation as the underlying technology continues to advance and prices come down.

This chapter talks about the introduction, motivation, objectives, and scope for the proposed methodology in a detailed manner. The overall project description is explained in detail for the various modules, such as electrical and mechanical. Additionally, the design approach of the proposed e-bicycle is explained. Finally, the outcomes are analyzed and conclusions are drawn accordingly.

17.2 PROJECT DESCRIPTION

17.2.1 Overview of the Project

The fundamental concept is to power the bicycle's movement using an attached motor, that is, an electrical motor that is driven using a battery unit and that can be turned on/off during tough terrain and switched off and pedaled during motion on flat terrain to have the battery recharged during that period. Additionally, an ESP-based microcontroller is installed, which serves as an IoT device to transmit user information such as location, battery percentage, and battery voltage to the Adafruit portal. Users logging on to the portal can witness the results in real time.

17.2.2 Modules of the Project

The proposed e-bicycle shown in Figure 17.1 has many functional parts, such as the motor, battery, charging circuitry, e-bicycle accessories (accelerator throttle, brake, headlamp, and horn), ESP32-based microcontroller, various sensors (temperature, Global Positioning System (GPS), and voltage), and Adafruit Cloud platform.

All these functional parts are categorized into three primary modules: mechanical unit, electrical unit, and software unit.

17.2.2.1 Mechanical Unit

Cycle Chain and Setup

Bicycles are propelled by the use of a roller chain that connects the pedals to the driving wheel. Most bicycle chains are forged from carbon or alloy steel, whereas some are nickel-plated for corrosion resistance or for aesthetic purposes. Figure 17.2(a) shows the bicycle chain and setup.

Lock Setup and Accelerator Throttle

A lock is a mechanical or electronic closure protected by a physical object (e.g., a key, security token, or coin), secret information (e.g., a password or combination of numbers and letters), or a combination thereof. Like a door chain, it can only be opened from one side. Figure 17.2(b) shows the lock setup and accelerator throttle. The object used to manipulate (lock or unlock) a lock is a key. A conventional key, which is a small piece of metal,

Figure 17.2 illustrates the e-bicycle parts

Figure 17.2 (Continued)

comprises a bit or blade that slides into the lock's keyway to distinguish between different keys and a shackle that intentionally protrudes so that the user can apply tension to it. The control that the driver uses to control the power of the vehicle is variously called the accelerator pedal or accelerator.

17.2.2.2 Electrical Unit

DC Motor (MY1016Z2)

This e-bicycle project uses a DC motor (MY1016Z2) suitable for high power and high torque. DC motors are commonly used in e-bicycles and produce a lot of torque at low speeds, even when the bicycle is going uphill. Below is a list of specifications for this engine (MY1016Z2):

1. 24 V 250 W
2. rated speed: 360 RPM
3. torque units: 2.5 Nm to 6.2 Nm
4. efficiency is less than 75%
5. dimensions (width × length): 101 × 69.

DC electrical motors are the first type to be widely used because they can be powered or supplied from the existing DC system. The speed of an DC electrical motor can be varied over a wide range by the adaptation of a variable supply voltage or by adjusting the magnitude of the field winding current. Household appliances, toys, and gadgets all use small DC motors. Universal motors are lightweight and can be driven by direct current. They are used in equipment and devices with practical strength. Large DC motors are used to drive steel rolling mills, drive hoists and hoists, and drive EVs.

Advances in power electronics have made it possible to use AC motors instead of DC motors in many applications.

Working Principle: The DC motor shown in Figure 17.3 is a device that converts electrical energy into its equivalent mechanical energy. DC motors are based on the theory that a conductor carrying current experiences a mechanical force when placed in a magnetic field.

Motor Driver/Controller (MY1016Z2)

Figures 17.4 and 17.5 show a motor driver used in an e-bicycle project to adjust the speed of the e-bicycle motor. It serves as a link between the driver and the engine. The driver responds to input from the microcontroller by slowing the motor down or vice versa. A list of motor drive specifications is presented below.

Figure 17.3 DC motor

Figure 17.4 Motor controller

Figure 17.4 (Continued)

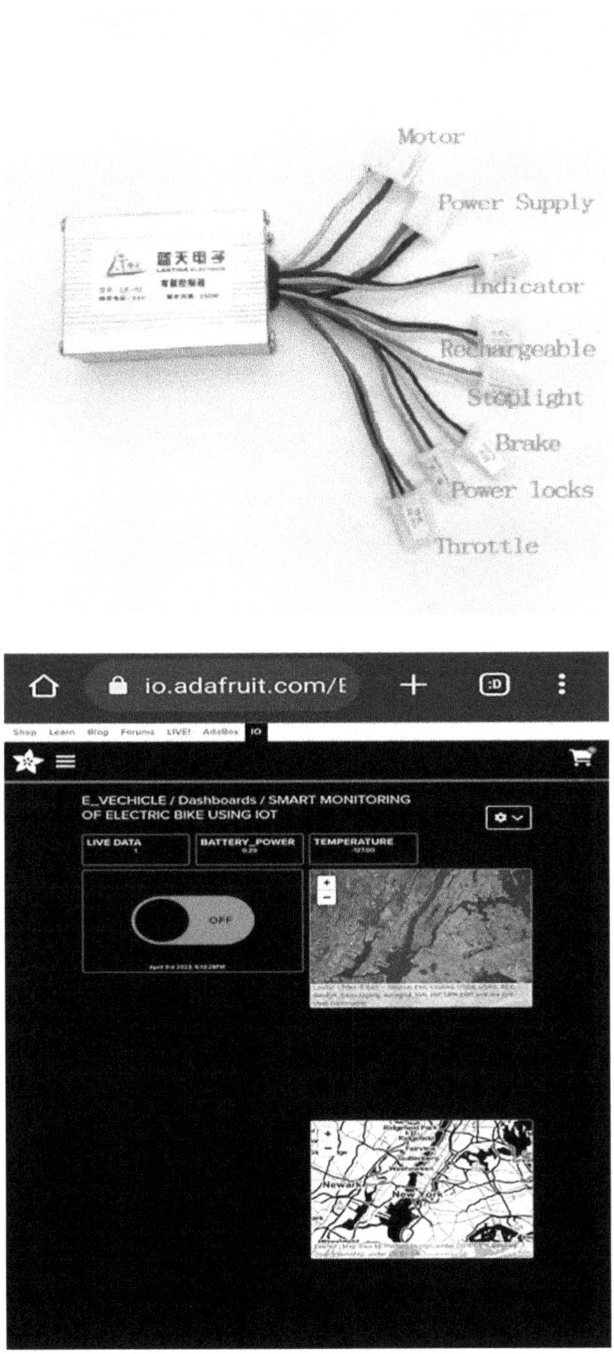

Figure 17.5 Detailed view of the controller connections

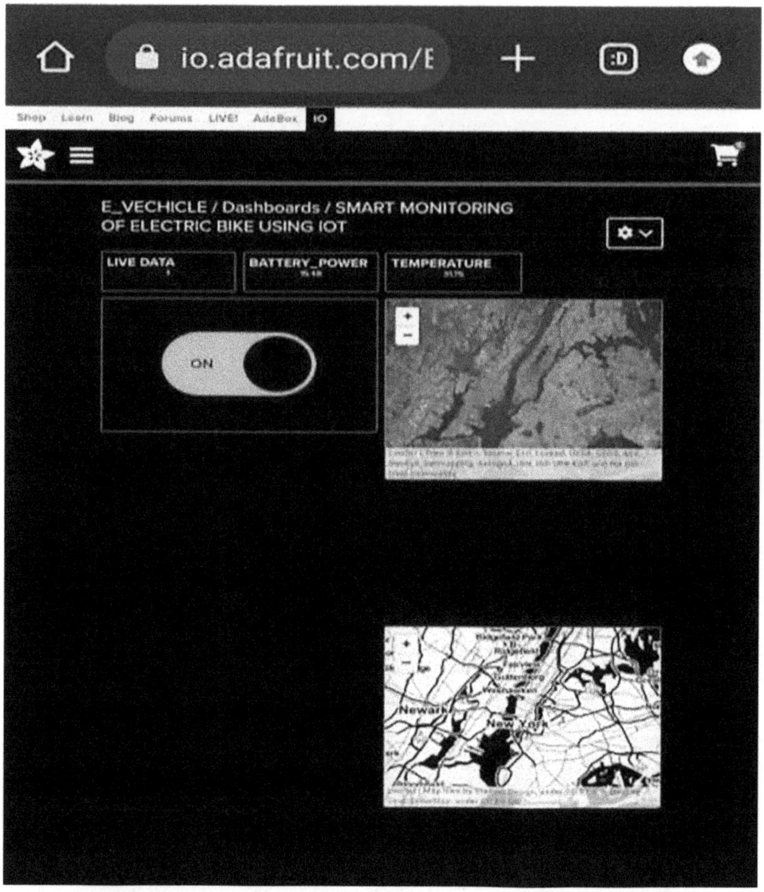

Figure 17.5 (Continued)

Technical Specification: Speed Control Basics: An e-bicycle speed controller is an electronic circuit that dynamically brakes the bicycle and controls the speed of the electric motor. This control unit supplies the motor with current from the battery box. Many types of controllers are used in the marketplace for both brushless and brushed motors. Adaptive e-bicycles use conversion kits, the central element of which is the controller.

Function: The e-bicycle motor receives signals of various voltages from the speed controller. These gauges show how the rotor is moving relative to the starting spool. Using different mechanisms determines how well the speed controller can perform its task. Hall-effect sensors help to locate rotors on e-bicycles designed specifically for the task.

If neither the speed controller nor the adaptive bicycle speed controller contains such a sensor, then the electromotive force (EMF) of the undriven coil is estimated to determine the rotor orientation. The electric speed control mechanism differs according to whether the e-bicycle is designed with accessibility in mind. A custom bicycle is a standard bicycle with an electric drive system attached. Custom bicycles offer better acceleration and more features at a higher price than non-custom bicycles.

Step-Down Transformer (0V-230V/SKU: 1156867)

The step-down transformer shown in Figure 17.6 transforms high voltage and low current by converting higher voltage (HV) and lower current from the primary winding of the transformer to lower-voltage (LV) and higher-current values in the secondary winding of the respective transformer. Connect the primary side to the LV and the LV to the secondary part.

The primary or first winding of the transformer converts electrical energy into magnetic energy, which is converted into electrical energy on the secondary part or winding side of the transformer. A transformer is a type of electrostatic device. Step-down transformers are used in many different types of electrical systems.

Figure 17.6 Step-down transformer

Figure 17.7 Lead-acid battery

Battery (OL 1.3-12)

The e-bicycle uses two rechargeable 12 V, 1.3 A lead-acid batteries, as shown in Figure 17.7, wired in series. Essentially, they store the electrical energy produced and use it to power the motor. A battery has a positive electrode, the cathode, and a negative electrode, the anode. When connected to an external circuit, the terminal marked negative is the source of electrons and the terminal marked positive has more potential energy. When bound together, electrons flow, providing energy to external devices. Rechargeable batteries can be charged multiple times.

With its low cost and well-established production base, the lead-acid battery has been around for over 100 years and will continue to be a market force. The internal chemistry of the two strains, valve-regulated lead-acid (VRLA) batteries and flooded lead-acid batteries, is identical. The two types differ most when it comes to system-level design problems. Unlike VRLA batteries, flooded lead-acid batteries require three things: upright position to prevent electrolyte leakage, ventilated atmosphere to disperse gas generated during cycling, and regular maintenance of electrolytes. Due to these fluctuations, the increased labor and follow-up costs must be compared with the lower price of submerged lead-acid. Both VRLA and submerged grades of lead-acid batteries are available in 'deep cycle' and 'shallow cycle' variants.

Flat-cycle VRLA batteries are often used for auto starting, lighting, ignition (SLI) batteries that need to produce high power pulses for short durations. Deep cycling is used because batteries in the stationary power market often discharge slowly over long periods of time.

Relay (SLA-05-VDC)

The electrical control switch is a relay. It consists of a row of make-contact terminals and a row of input terminals for one or more control signals. A switch can contain any number of connections with different contact configurations, such as B. NO contact, NC contact, or a combination of both.

Technical Specification:

1. trigger voltage (VDC): 5;
2. trigger current (mA): 20
3. switching voltage (VAC): 250@30A
4. length (mm): 50
5. switching voltage (VDC): 30@30A
6. width (mm): 33
7. height (mm): 24
8. weight (gm): 36.

Voltage Divider (SKU: 44161)

A voltage divider in electronics is a passive linear circuit that produces an output (V_{out}) that is a fraction of the input (V_{in}). It is also called a potential divider. The division of the input voltage into the voltage divider section creates a voltage division. An example of a voltage divider circuit is the connection of two resistors in series, to which the input voltage is applied and the output voltage is developed from that connection. The divider circuit is designed to step down the voltage to 5 V.

Sensors

Three sensors are used in the e-bicycle: GPS sensor, voltage sensor and temperature sensor, as shown in Figure 17.8. The sensors and how they are utilized are explained in the forthcoming explanations.

GPS Sensor (Ublox NEO-6M): GPS is a type of satellite navigation system that gives accurate time readings in addition to positional information. This system is accessible to everyone who has a GPS receiver and at least four GPS satellites that are not blocked in any way. The GPS receiver determines the location through the precise timing of the signal transmitted by the GPS satellites. GPS is now widely used and has become an integral part of smartphones.

Figure 17.8 GPS sensor

Voltage Sensor (SKU:29802): This sensor calculates, determines, and measures voltage. This sensor can detect AC or DC voltage levels. The input of this sensor is voltage and the output is, for example, switch, the analog voltage signal, current signal, and audio signal. Some sensors provide outputs such as sine waves and pulse waveforms.

Temperature Sensor (EC-1265): Figure 17.9 indicates a device that can detect and measure differences in temperature, as well as transform that data into an electrical signal, known as a temperature sensor.

17.2.2.3 Software Unit

ESP32-Based Microcontroller

The ESP32 family of system-on-a-chip microcontrollers is characterized by their cheap cost and low power consumption. They also have integrated Wi-Fi and dual-mode Bluetooth. The ESP32 series utilizes a Tensilica Xtensa LX6 microprocessor in both dual-core and single-core variations, an Xtensa LX7 microprocessor in dual-core variations, or a single-core RISC-V

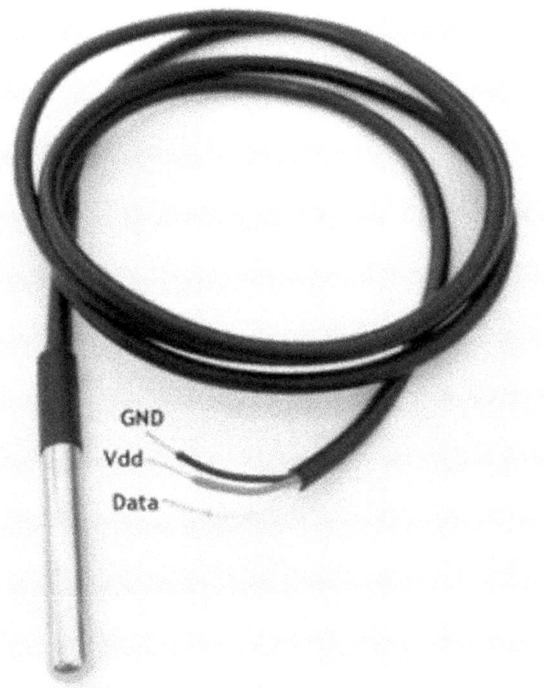

Figure 17.9 Temperature sensor

microprocessor. Additionally, the ESP32 series incorporates built-in antenna switches, RF baluns, power amplifiers, low-noise receive amplifiers, filters, and power-management modules. Espressif Systems, a Chinese business with headquarters in Shanghai, is responsible for the design and development of the ESP32. TSMC, using its 40 Nm technology, is the manufacturer of the ESP32. ESP32 is the microcontroller that will replace the ESP8266 in the future.

Figure 17.10 shows the ESP32 microcontroller which is often developed for use in mobile devices, wearable technology, and IoT applications, such as Nabto. Furthermore, with the introduction of an ESP32 IoT Starter Kit by Mongoose OS, the ESP32 has garnered a reputation for being the perfect chip or module for IoT developers, as well as enthusiasts. But the low-cost device may also be employed in a variety of other production systems, and its capabilities and resources have developed noticeably over the course of the previous four years.

Figure 17.10 ESP32-based microcontroller

Adafruit IO

A cloud provider that focuses primarily on IoT installations on the cloud is called Adafruit IO, as shown in Figure 17.11. Adafruit IO is compatible with a variety of devices, including Arduino, Raspberry Pi, ESP2866, and ESP microcontrollers.

IoT developers have chosen Adafruit IO over other IoT cloud services for a number of reasons, including the following:

- strong application programming interface (API): provides libraries for a variety of programming languages, which also provides support for built-in user interfaces
- dashboard: the ability to comprehend data via the use of graphs and charts helps users to make superior choices
- privacy: strong encryption techniques inside the cloud platform, which stores the user's data
- documentation and community support: the presence of a large number of blogs and an exceptionally helpful community makes it possible for the goods to undergo ongoing improvement.

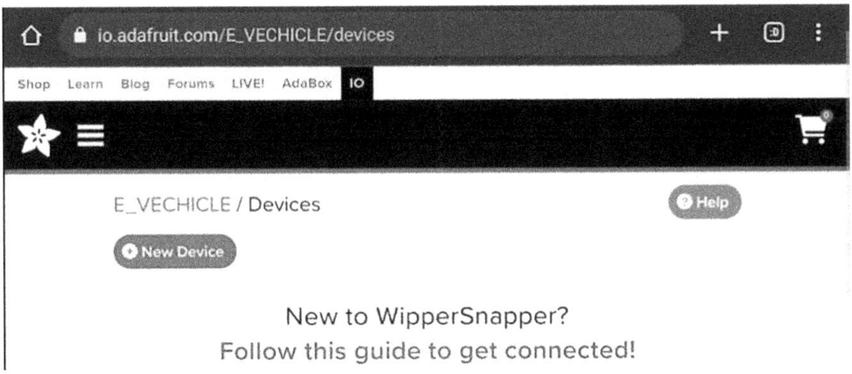

Figure 17.11 Adafruit login portal view

17.3 DESIGN OF AN IOT-BASED E-BICYCLE MONITORING SYSTEM

17.3.1 Design Approach

The first unit in the block diagram is the battery charging circuitry, which essentially comprises three blocks so the first one gets the input voltage from domestic 230 V AC power, which is fed as input into a step-down transformer in order to step down the input voltage from 230 V AC to 24 V AC power, which is followed by the rectifier circuit, which converts the 24 V AC power to 24 V DC current. This is provided as the input to the lead-acid battery unit. The lead-acid battery unit has two 12 V lead-acid batteries connected in series to attain the operational voltage of 24 V. This power is fed into the voltage divider, which reduces the voltage from 24 V to 12 V, which is provided as the input to the relay circuitry and other units.

The heart of the project is the microcontroller. It essentially controls all the major functional units of the project, such as e-bicycle peripherals and various sensors, and it sends the collected data from the e-bicycle to the cloud database.

The e-bicycle peripherals comprise the e-bicycle controller, DC motor, and e-bicycle accessories. This controller unit sends control signals to the motor functioning according to the user's toggle, and powers up and controls all the e-bicycle accessories, such as the accelerator throttle, brakes, headlamp, and horn. The functionality of the sensors is as follows: the GPS sensor is used to track the live location of the user; and the temperature and voltage sensor is deployed in the lead-acid battery in order to obtain the essential parameters while it is functioning. All the connected data from the e-bicycle are sent to the Adafruit cloud server and by logging onto the Adafruit portal, the user can visualize the results.

17.3.1.1 Simulation Analysis

Figure 17.12 shows the respective simulation circuit built using Proteus software for the proposed work.

17.3.1.2 Hardware Analysis

Figure 17.13 shows the core circuit connections of the project. The first board toward the right-hand side is the voltage divider block. Adjacent to it is the ESP32-based microcontroller and to its left is the relay circuit block. Beneath the ESP32-based microcontroller unit is the GPS board connected along with its antenna toward its left side. These are the core circuit connections utilized in the proposed work.

As shown in Figure 17.14, a 24 V DC motor is deployed at the back wheel of the e-bicycle, which essentially delivers rotational power to the e-bicycle and retains the default pedal mechanism for the bicycle. This helps it to drive in emergency conditions, such as battery run off.

17.3.1.3 Adafruit IO Interface

In order to visualize the results, the Adafruit cloud service is utilized, which is an open-source cloud service provider. An account is created in this service and deployed to the ESP module with its credentials. So this is how it looks when logging is done to the Adafruit portal, and using this, e-bicycle parameters such as live data, battery power, battery temperature, bicycle on/off functionality, and the inbuilt map, are accessed.

Regarding live data, this is nothing but the number of incremental e-bicycle input data sent by the ESP module to the cloud server, so this value exponentially increases over the period of operation. If this value is not incrementing and if it gets stuck over a period of time, this means that the ESP module has lost its connection with the cloud server, so it is necessary to check the network connections with the ESP modules. This helps to debug errors that might also occur.

Then the battery power indicator is used, which essentially lets the user know the amount of voltage that is left in the battery units. The battery temperature indicator lets the user know the temperature of the battery unit when it is in operation.

Then comes the bicycle ON/OFF toggle condition. It is one of the most important features of the project. Whenever the user wants to switch the bicycle ON, the user should log into the Adafruit portal, as shown in Figure 17.15, and toggle the respective switch to the ON state. Only then will

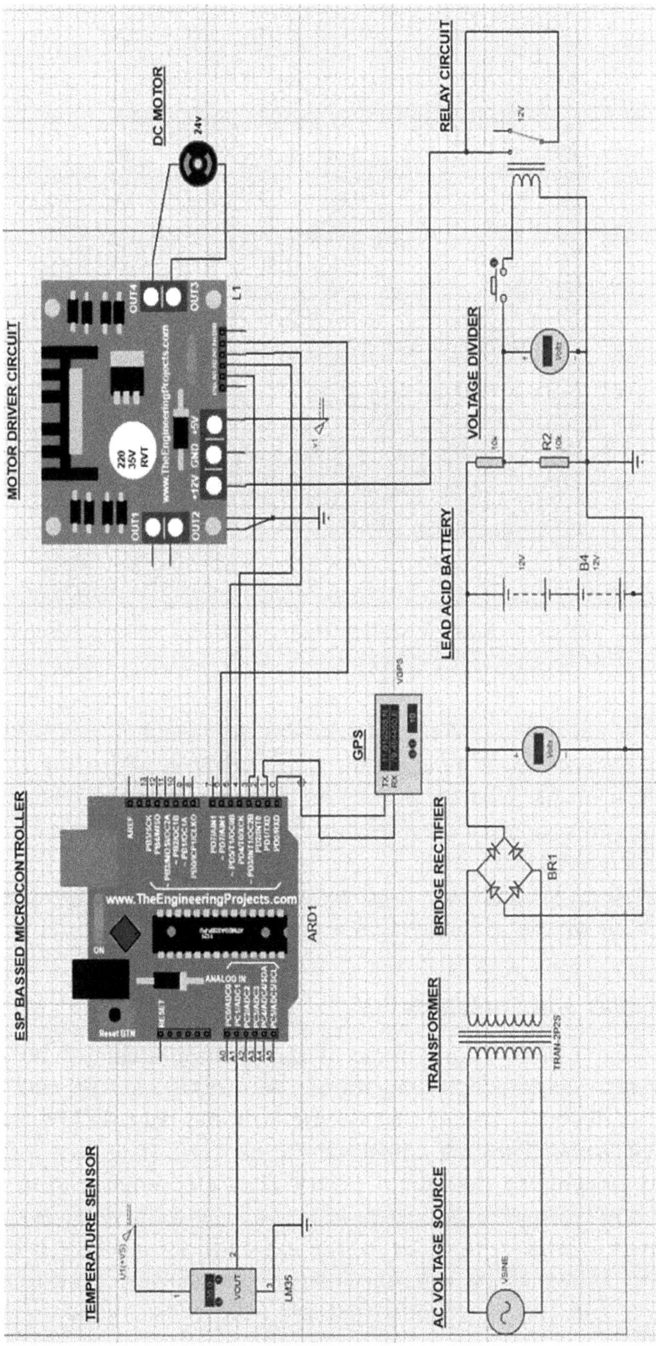

Figure 17.12 Proteus simulation file

Figure 17.13 Core circuit connections

the control signals be sent to the microcontroller, which in turn activates the e-bicycle peripherals and its ignition. Thus, without activating this switch that is present in the Adafruit portal, the bicycle cannot be turned on. This gives the user great control over the bicycle and could prevent various illegal activities, such as theft.

Inbuilt MAP functionality is used, by which the real-time location data of the user who is driving the bicycle is accessed.

17.3.1.4 Codes and Standards

Protocols are an essential component of digital communication because they specify the manner in which data should be formatted, sent, and received. They are a collection of rules that decide how the data will be transferred across the network. They are a protocol.

It is also possible to describe a protocol as a communication standard that the two most important parties (the sender and the receiver) in a computer network adhere to in order to interact with one another. It details the types of data that may be sent, the instructions that are used to send and receive data, and the procedures that are followed to validate successful data transfers.

Figure 17.14 A & B: motor deployment

To put it another way, a protocol may be thought of in the same way as a language. Each language has its own set of norms, as well as its own lexicon. The rules, definitions, and implementations for a protocol are all unique to that protocol. If two individuals speak the same language, it makes it much simpler and more efficient for them to communicate with one another. In a similar vein, if two hosts are using the same protocol, it will be simple for them to connect to and interact with one another. As a result, protocols

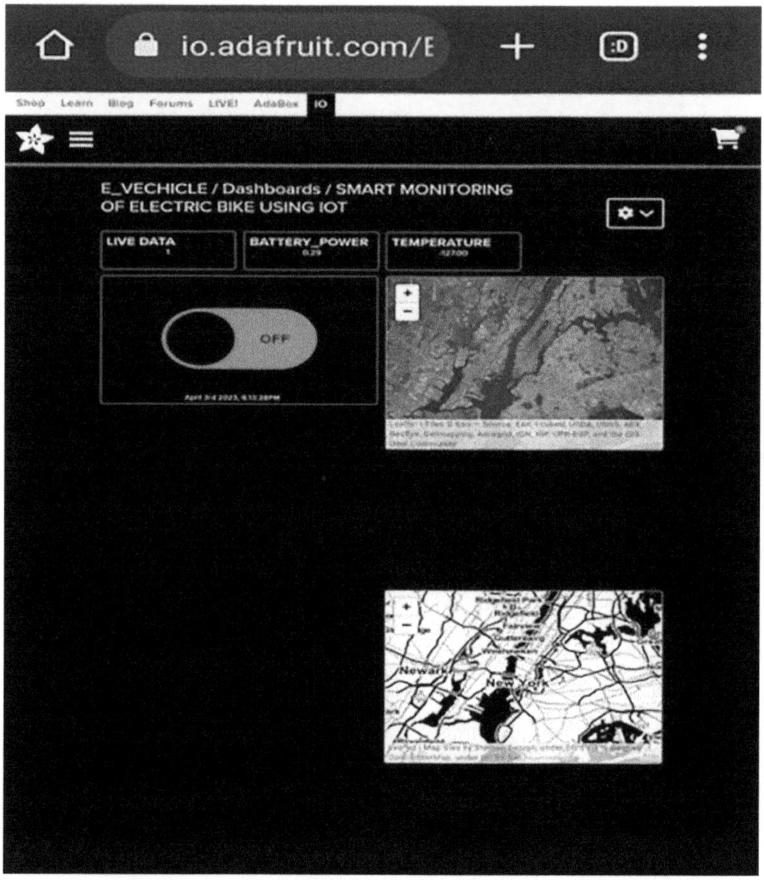

Figure 17.15 A & B: Adafruit control page for the e-bicycle

serve as a common language across the many network devices that take part in data exchange.

List of Standards and Protocols

The list of standards and protocols that are used in this work are as follows:

Hypertext Transfer Protocol (HTTP): HTTP is the protocol that underpins the World Wide Web and makes it possible for web browsers and web servers to communicate with one another. It is a client-server protocol, which means that a client, which is often a web browser, makes requests for resources to a server, which then answers with the desired material. Examples of resources

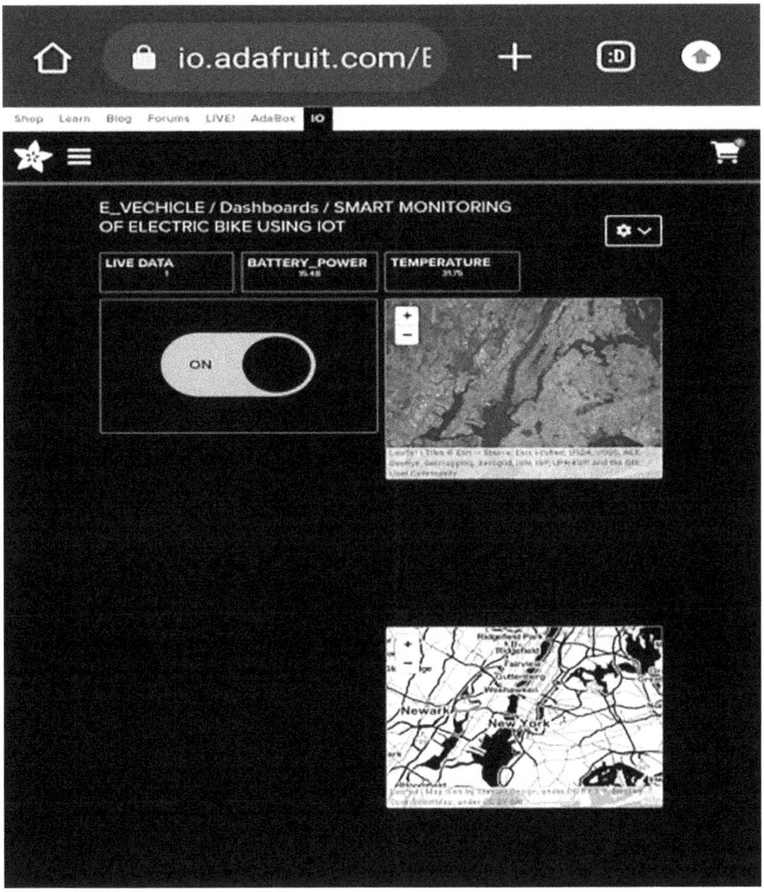

Figure 17.15 (Continued)

include web pages, photos, and videos. The request-response paradigm is the one that is used by HTTP. In this model, clients submit requests in the form of Uniform Resource Locators (URLs), and servers answer with HTML information that can be displayed by browsers. To carry out the many operations required on online resources, a wide range of techniques are used, often known as HTTP verbs, such as GET, POST, PUT, and DELETE. Because HTTP uses a stateless and connectionless architecture, each request may be considered independent of each answer and HTTP does not retain any information on prior requests or replies. Web surfing, web services, and other applications that need communication between clients and servers via the internet are just some of the many extensive uses of this protocol.

Extensible Markup Language (XML): XML is a markup language that may be used to describe and arrange data in a manner that is both understandable by humans and machines. It is a flexible language that gives users the ability to construct their own unique markup elements and properties, which makes it very adaptable for a broad variety of uses. Tags are used in XML to define elements. An element may include data or additional nested elements inside of it. Because XML offers a standardized approach to the data format and representation that is not reliant on the underlying platform, it is often used for the process of data exchange across various types of systems. The usage of XML is widespread across many sectors, including online services, data storage and retrieval, configuration file management, and document exchange, to name a few of these applications.

Constrained Application Protocol (COAP): COAP is a communication protocol that is both lightweight and efficient. It was developed specifically for restricted networks and devices, such as those used in applications for the IoT. COAP is a good option for low-power, low-bandwidth, and unstable network conditions because it is based on the RESTful architecture and runs over the User Datagram Protocol (UDP). It offers straightforward request-response exchanges, comparable to those offered by HTTP, and it allows for the management of resources via methods such as GET, POST, PUT, and DELETE. COAP also offers group communication and asynchronous alerts, which makes it an effective choice for IoT applications that have restricted resources. COAP is commonly used in IoT installations, which need resource-limited devices to communicate and interact with one another in a way that is both lightweight and effective.

Bluetooth Low Energy (BLE)

Bluetooth is a standard or benchmark for wireless technology that enables data to be sent between mobile and stationary devices over relatively short distances. It is optimized for the continuous streaming of data, in contrast to BLE, which is optimized for the transmission of brief bursts of data. Bluetooth has around 100 times higher power consumption than BLE, which has the following characteristics:

1. short distance gearbox
2. designed to support the uninterrupted flow of data.

The remaining protocols, such as Wi-Fi and Low-Rank Adaptation (LoRa) can be found in the references.

17.3.1.5 Realistic Constraints

Realistic constraints for the proposed work are as follows:

- environmental constraints, such as seepage of water into the system
- battery takes a long time to hit full charge
- thermal runaway of battery units
- overcurrent and undercurrent conditions that might affect the functional units
- frequent battery monitoring is important
- e-bicycle laws and regulations are yet to be laid down concretely.

17.3.1.6 Alternatives and Tradeoffs

Using Lithium-Ion Batteries Instead of Lead-Acid Batteries

In the 1970s, an idea for what is now known as a lithium-ion battery was originally envisioned, and by the 1990s, it had begun to receive broad usage. Every lithium-ion battery has what's known as "deep cycle" capability, which means that it may be completely depleted as well as charged. If the depth of each discharge is kept to no more than 80 percent of the battery's rated capacity, the battery's lifespan will be greatly extended.

When it comes to applications involving deep discharge, lithium-ion has a much longer cycle life than lead acid. As the average temperature of the environment rises, the discrepancy becomes even more pronounced. It is possible to lengthen the cycle life of each chemical by imposing restrictions on the depth of discharge (DoD), discharge rate, and temperature, however lead acid is often quite sensitive to all these parameters.

In terms of their impact on the surrounding ecosystem, lead-acid batteries pale in comparison to their lithium-ion counterparts. Because lead-acid batteries need several times more raw material than lithium-ion batteries to attain the same level of energy storage, the mining process for lead-acid batteries has a far greater negative effect on the surrounding ecosystem. Additionally, the lead processing sector has a very high energy consumption rate, which results in significant levels of environmental damage. Even though lead is very dangerous to human health, the risk to humans is almost nonexistent due to the techniques of production and the packing of batteries.

The use of lithium comes with its own set of environmental concerns. The extraction of lithium carbonate, copper, aluminum, and iron ore are necessary for the construction of the primary components of a lithium-ion battery. The mining of lithium in particular uses a lot of resources, but considering that lithium only makes up a small percentage of the total mass of a battery cell, the environmental repercussions caused by aluminum and copper are far more severe.

Use of Regenerative Braking Technology to Increase Energy Efficiency

Regenerative braking systems (RBSs) are a form of kinetic energy recovery system. These systems convert the kinetic energy of an item in motion into potential or stored energy in order to slow down the vehicle, which in turn results in an improvement in the vehicle's fuel economy. These systems are also known as kinetic energy recovery systems in certain contexts. In RBSs, the conversion of energy may be accomplished in a variety of ways, including through a spring, flywheel, or electromagnetic or hydraulic system. In more recent times, there has also been the development of an RBS that is a mix of electromagnetic and flywheel technology. Because each type of RBS employs a unique strategy for the conversion or storage of energy, its respective efficiencies and ranges of applicability are very variable.

Adding a Secondary Battery Unit as a Standby

The best course of action is to constantly be ready for whatever life throws at you. Once you have a second battery, you will be prepared for anything that may come your way. Your e-bicycle's battery abruptly died on you? It's no big deal; simply switch it out for the other available option. Your first battery isn't going to start up? Even if this is unfortunate, you will still be able to ride thanks to the second battery you have. You couldn't find your batteries? It's time to move on to the second one! And when your battery eventually approaches the end of its life, which typically occurs somewhere around 60% of its capacity, you will already be prepared with your backup battery that you purchased earlier.

The technology that goes into making batteries has steadily improved over the years, which has led to a reduction in the overall cost of producing batteries.

Any rider of an e-bicycle would benefit from the addition of a second battery. It helps you extend your riding range, enables you to charge your battery less often, ensures that you are ready for anything that may come your way, and enables you to save some money.

17.3.2 Design Specifications

Table 17.1 indicates the design aspects of the proposed system.

17.4 PROJECT DEMONSTRATION

17.4.1 Introduction

The overarching idea is to propel the bicycle forward with the assistance of a motor that is mounted on it, that is, an electrical motor that is powered by

Table 17.1 Design aspects

Component	*Specification*	*Explanation*
Motor (MY1016Z2)	Refer to Section 17.2	This e-bicycle project uses a DC motor (MY1016Z2) suitable for high power and high torque. DC motors are commonly used in e-bicycles and produce a lot of torque at low speeds, even when the bicycle is going uphill. Below is a list of specifications for this engine (MY1016Z2).
Motor driver (MY1016Z2)	Refer to Section 17.2	A motor driver is used in the e-bicycle project to adjust the speed of the e-bicycle motor. It serves as a link between the driver and the engine. The driver responds to input from the microcontroller by slowing the motor down or vice versa. A list of motor drive specifications is shown below.
Relay (SLA-05-VDC)	1. Trigger voltage (VDC): 5 2. Trigger current (mA): 20 3. Switching voltage (VAC): 250@30A 4. Length (mm): 50 5. Switching voltage (VDC): 30@30A 6. Width (mm): 33 7. Height (mm): 24 8. Weight (gm): 36	It is an electromechanical relay used as an electrical switch. The contact rating for the SLA-05-VDC relay is 10 A at 250 V AC or 30 V DC. It has applications in areas such as industrial automation, control systems, and automotive systems. It is cost effective and the perfect choice for the switching application.
Temperature sensor	EC-1265	EC-1265 is a type of sensor used to measure the voltage output that is directly proportional to the temperature measured. It can measure from -55°C to 150°C. It is commonly used in temperature controllers, environmental monitoring systems, heating, ventilation, and air conditioning (HVAC) systems, and its linear output voltage.
GPS module	Ublox Neo-6M GPS	It is an electronic device used to track data on the live location. It can be powered with communicating devices such as a universal serial port and universal asynchronous receiver transmitter, and is also powered with the help of a microcontroller with a 5 V input.
ESP Wi-Fi module	ESP32-based microcontroller	The ESP module is also known as the ESP32-based microcontroller. The Wi-Fi enabled module is used to transfer data to the cloud with the development of the IoT. Low power consumption is required to power up the module. It also deals with the application of smart home devices, remote monitoring systems, and home automation.

(continued)

Table 17.1 (Cont.)

Component	*Specification*	*Explanation*
Voltage sensor	SKU-1265	This sensor calculates, determines, and measures the voltage supply.
Lead-acid battery	Refer to Section 17.2	It uses two rechargeable 12 V, 1.3 A lead-acid batteries wired in series. Essentially, it stores the respective electrical energy produced and uses it to power the motor. A battery has a positive electrode, the cathode, and a negative electrode, the anode. When connected to an external current-carrying circuit, the terminal that is marked as negative acts as the respective source of the electrons and the terminal marked positive has more potential energy.
Voltage divider	SKU: 44161	It is also called a potential divider. The division of the input voltage into the voltage divider section creates a voltage division. In this work the divider circuit steps down the voltage to 5 V.
Bicycle	Hercules Cycle and Motor Company	A Hercules bicycle is used as the core framework. A 24 V DC motor is used on the back wheel of the e-bicycle, which essentially delivers rotational power.

an respective battery unit and that can be turned on while traveling over difficult terrain and turned off, with the bicycle being pedaled, when traveling over flat terrain in order to replenish the battery. Additionally, an ESP-based microcontroller is added that acts as an IoT device to broadcast user information to the Adafruit portal. This information includes the user's location, as well as the percentage and voltage of the battery. Logging on to the portal enables the user to see the results in real time.

17.4.2 Simulation Results

Figure 17.16 shows the simulation results obtained from the project. It also shows the MAP functionality, which opens when the circuit is executed. The remaining results are as shown in Figure 17.17.

17.4.3 Hardware Results

Figure 17.18 shows the final working model of the e-bicycle.

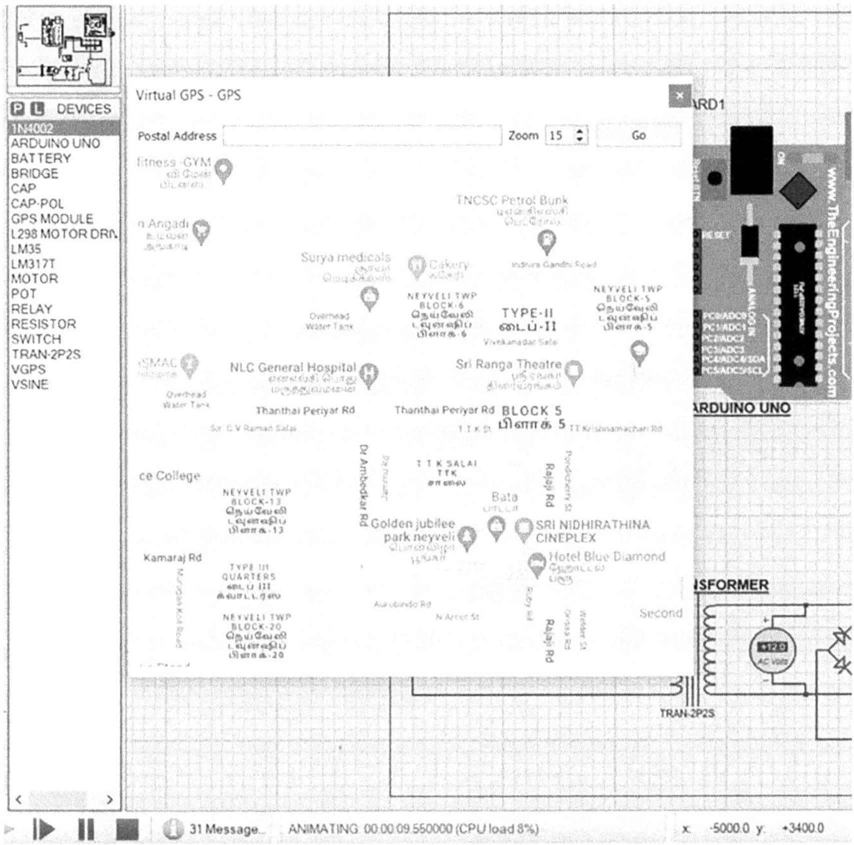

Figure 17.16 Simulation result

17.5 CONCLUSION

In conclusion, the project of creating and manufacturing an e-bicycle is a complicated and demanding endeavor that takes a significant amount of study, design, development, and marketing. Despite this, all went according to plan. This initiative is a fantastic opportunity to provide a product that is not only beneficial to the environment but also lucrative and easy to use. Thus, the goal of creating an e-bicycle powered with a lead-acid battery and monitored using IoT technology by the use of an ESP32-based microcontroller as the IoT device to process and transmit data regarding the user's location, temperature, and battery power, and visualizing those results in Adafruit portal, has been successfully realized.

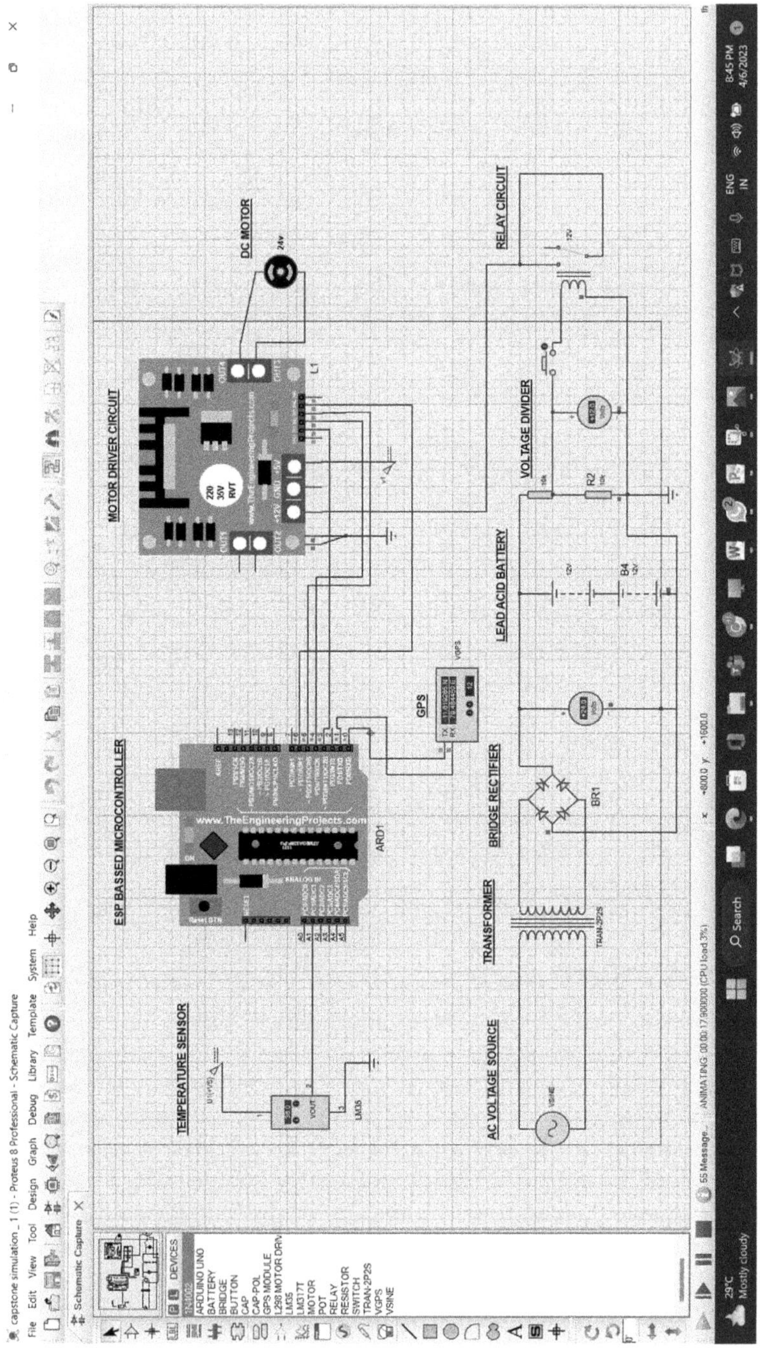

Figure 17.17 Simulation output

Figure 17.18 Final working model of the e-bicycle

17.5.1 Scope of the Work

It will be necessary to conduct further research and development in the future in order to continue increasing the specific energy ratio and charge energy density of the battery used in automobiles, while at the same phase bringing the cost of technology down. It will be necessary to continue developing and implementing infrastructure in order to facilitate the mass adoption of electric cars.

Efforts such as those mentioned above may help to alleviate the range anxiety for customers and perhaps modify general attitudes toward electric cars. As a consequence, EVs will have higher market and sales penetration, which will lead to a significant transformation in the automotive industry.

BIBLIOGRAPHY

A. Arno, K. Toyoda and I. Sasase, "Accelerometer Assisted Authentication Scheme for Smart Bicycle Lock," *2015 IEEE 2nd World Forum on Internet of Things (WF-IoT)*, 2015, pp. 520–523, doi: 10.1109/WF-IoT.2015.7389108.

A. C. Dias and O. Postolache, "Cyclist Performance Assessment based on WSN and Cloud Technologies," *2018 International Conference and Exposition on Electrical And Power Engineering (EPE)*, 2018, pp. 1041–1046, doi: 10.1109/ICEPE.2018.8559950.

A. Duz and M. Corno, "Flexible Pricing Strategies in Electric Free-Floating Bicycle Sharing," in *IEEE Access*, 9, pp. 152972–152983, 2021, doi: 10.1109/ACCESS.2021.3127568.

A. Gandhi, S. R. Mistri and A. K. Gangrade, "Design and Performance Testing of an E-Bicycle," *2021 3rd International Conference on Sustainable Technologies for Industry 4.0 (STI)*, 2021, pp. 1–6, doi: 10.1109/STI53101.2021.9732615.

A. K. Das, A. M. Joshi and S. Dhal, "A Machine Learning Based Bike Recommendation System Catering to User's Travel Needs," *2020 IEEE 17th India Council International Conference (INDICON)*, 2020, pp. 1–6, doi: 10.1109/INDICON49873.2020.9342509.

A. Z. Aliyadin, A. Purwadi and S. Hidayat, "Performance Analysis and Design of 250 Watt Outer Rotor BLDC Motor for Urban Electric Bicycles," *2022 7th International Conference on Electric Vehicular Technology (ICEVT)*, 2022, pp. 195–199, doi: 10.1109/ICEVT55516.2022.9925011.

C. Ambika Bhuvaneswari and M. Muthumari, "Design and Realization of Radio Communication Using LoRa & XBee Module for an e-Bike," *2018 IEEE International Conference on Computational Intelligence and Computing Research (ICCIC)*, 2018, pp. 1–4, doi: 10.1109/ICCIC.2018.8782420.

C. M. I. Juare, L. L. Rocillo, D. L. E. Suterio and E. Q. Raguindin, "Tilting Advisory and Prevention (TAP) Device for Bicycles," *TENCON 2018 – 2018 IEEE Region 10 Conference*, 2018, pp. 2561–2565, doi: 10.1109/TENCON.2018.8650223.

C. Roger and H. E. Omari, "Solar-E-Cycles, Empowering People Project 2014–2019," *2019 7th International Renewable and Sustainable Energy Conference (IRSEC)*, 2019, pp. 1–8, doi: 10.1109/IRSEC48032.2019.9078297

C. Xiong et al., "Static Balancing of Robotic Bicycle through Nonlinear Modeling and Control," *2018 3rd International Conference on Robotics and Automation Engineering (ICRAE)*, 2018, pp. 24–28, doi: 10.1109/ICRAE.2018.8586765.

D. K. P. Gudavalli, B. S. Rani and C. V. Sagar, "Helmet Operated Smart E-Bike," *2017 IEEE International Conference on Intelligent Techniques in Control, Optimization and Signal Processing (INCOS)*, 2017, pp. 1–5, doi: 10.1109/ITCOSP.2017.8303138.

D. Lee, S. Park and M. Hahn, "A Study on Bicycle Crash Risk Based on GPS Trajectories Collected from Bidirectional Traffic," *2014 IEEE Asia Pacific Conference on Circuits and Systems (APCCAS)*, 2014, pp. 383–386, doi: 10.1109/APCCAS.2014.7032800.

D. Zanic, R. Gotal and S. Hrabar, "Multi-band Antenna System for Fully Connected Smart E-bicycle," *2021 International Symposium ELMAR*, 2021, pp. 43–48, doi: 10.1109/ELMAR52657.2021.9550900.

E. H. -C. Lu and Z. -Q. Lin, "Rental Prediction in Bicycle-Sharing System Using Recurrent Neural Network," in *IEEE Access*, vol. 8, pp. 92262–92274, 2020, doi: 10.1109/ACCESS.2020.2994588.

F. H. Hidayatullah, M. Abdurohman and A. G. Putrada, "Accident Detection System for Bicycle Athletes Using GPS/IMU Integration and Kalman Filtered AHRS Method," *2021 International Conference Advancement in Data Science, E-learning and Information Systems (ICADEIS)*, 2021, pp. 1–6, doi: 10.1109/ICADEIS52521.2021.9702085.

F. P. Batet, R. V. Martín, L. Igualada and C. Corchero, "An Optimal Solution for a Smart Charging Station of Light Electric Vehicles," *2022 International*

Conference on Smart Energy Systems and Technologies (SEST), 2022, pp. 1–6, doi: 10.1109/SEST53650.2022.9898421.

F. Pellitteri, V. Boscaino, R. Miceli and U. K. Madawala, "Power Tracking with Maximum Efficiency for Wireless Charging of E-Bikes," *2014 IEEE International Electric Vehicle Conference (IEVC)*, 2014, pp. 1–7, doi: 10.1109/IEVC.2014.7056208.

F. Tabei, B. Askarian and J. W. Chong, "Accident Detection System for Bicycle Riders," in *IEEE Sensors Journal*, 2021, 21(2), pp. 878–885, doi: 10.1109/JSEN.2020.3021652.

I. -H. Peng, P. -C. Lee, C. -K. Tien and J. -S. Tong, "Development of a Cycling Safety Services System and its Deep Learning Bicycle Crash Model," *Journal of Communications and Networks*, 2022, 24(2), pp. 246–263, doi: 10.23919/JCN.2022.000007.

J. D'Souza, K. J. Burnham and J. E. Pickering, "Modelling and Simulation of an Autonomous Vehicle Ethical Steering Control System (ESCS)," *2022 26th International Conference on Methods and Models in Automation and Robotics (MMAR)*, 2022, pp. 76–80, doi: 10.1109/MMAR55195.2022.9874320.

K. Draženović, A. Perić, Ž. Jakopović and V. Šunde, "Modeling and Predictive Control of LLC Resonant Converter for Solar Powered E-Bicycle Charging Station," *2021 International Conference on Electrical Drives & Power Electronics (EDPE)*, 2021, pp. 136–143, doi: 10.1109/EDPE53134.2021.9604053.

K. F. I. Faruque, N. Nawshin, M. F. Bhuiyan, M. R. Uddin, M. Hasan and K. M. Salim, "Design and Development of BLDC controller and its Implementation on E-Bike," *2018 International Conference on Recent Innovations in Electrical, Electronics & Communication Engineering (ICRIEECE)*, 2018, pp. 1461–1465, doi: 10.1109/ICRIEECE44171.2018.9008597.

K. -W. Chun, Z. -H. Chen and C. -C. Sun, "IOT Solution for Public Bicycle System," *2016 IEEE International Conference on Consumer Electronics-Taiwan (ICCE-TW)*, 2016, pp. 1–2, doi: 10.1109/ICCE-TW.2016.7520911.

L. Stilo, H. Lugo, D. S. Velandia, P. P. Conway and A. A. West, "Personalised Controller Strategies for Next Generation Intelligent Adaptive Electric Bicycles," *IEEE Transactions on Intelligent Transportation Systems*, 2021, 22(12), pp. 7814–7825, doi: 10.1109/TITS.2020.3009400.

M. Gogola, "Are the e-Bikes more Dangerous than Traditional Bicycles?," *2018 XI International Science-Technical Conference Automotive Safety*, 2018, pp. 1–4, doi: 10.1109/AUTOSAFE.2018.8373344.

M. Oberle, M. Stöhr, A. Brandstetter and S. Gärtner, "Design and Development of a Cyber-Physical System E-Bike Retrofit Prototype," *2022 IEEE 27th International Conference on Emerging Technologies and Factory Automation (ETFA)*, 2022, pp. 1–8, doi: 10.1109/ETFA52439.2022.9921733.

N. Alfiany, N. Hamid, G. Jati, M. A. Ma'sum and W. Jatmiko, "Kinematics and Dynamics Analysis of an Autonomous Three-Wheeled Bicycle Modeling," *2019 4th Asia-Pacific Conference on Intelligent Robot Systems (ACIRS)*, 2019, pp. 17–21, doi: 10.1109/ACIRS.2019.8935966.

N. C. D. Silva et al., "An Experimental Study on Smart Electric Bicycle with Weight Minimized Power Pack," *2021 3rd International Conference on Electrical Engineering (EECon)*, 2021, pp. 80–86, doi: 10.1109/EECon52960.2021.9580942.

O. B. Piramuthu, "Connected Bicycles—State-of-the-Art and Adoption Decision," *IEEE Internet of Things Journal*, 2017, 4(4), pp. 987–995,, doi: 10.1109/JIOT.2017.2697203.

O. Yang and W. M. Griggs, "Connected e-Bicycle Platoons at Unsignalised Intersections," *2021 IEEE International Intelligent Transportation Systems Conference (ITSC)*, 2021, pp. 1875–1882, doi: 10.1109/ITSC48978.2021.9564869.

P. M. Santos, M. Rosa, L. R. Pinto and A. Aguiar, "Cooperative Bicycle Localization System via Ad Hoc Bluetooth Networks," *2020 IEEE Vehicular Networking Conference (VNC)*, 2020, pp. 1–4, doi: 10.1109/VNC51378.2020.9318328.

P. R. Reddy, K. S. S. Gowda, S. Charitha and R. Mahalakshmi, "Review and Redesign of Pedal Energy – Solar Power Augmented Hybrid Bicycle," *2020 Third International Conference on Smart Systems and Inventive Technology (ICSSIT)*, 2020, pp. 376–380, doi: 10.1109/ICSSIT48917.2020.9214286.

P. R. Reddy, K. S. S. Gowda, S. Charitha and R. Mahalakshmi, "Review and Redesign of Pedal Energy – Solar Power Augmented Hybrid Bicycle," *2020 Third International Conference on Smart Systems and Inventive Technology (ICSSIT)*, 2020, pp. 376–380, doi: 10.1109/ICSSIT48917.2020.9214286.

R. de Souza e Almeida, R. Cherullo, D. Radetic, G. Francy, M. Silva-Calpa and A. Barbosa Raposo, "Exploring Motor and Sensory Devices in a Bicycle Simulator," *2019 21st Symposium on Virtual and Augmented Reality (SVR)*, 2019, pp. 232–241, doi: 10.1109/SVR.2019.00048.

R. Gonzalez, L. Fajardo and D. Florez, "A Prototype to Manage the Share of Assisted Bicycles on Bogota as a Creative form of Public Transportation," *2018 ICAI Workshops (ICAIW)*, 2018, pp. 1–6, doi: 10.1109/ICAIW.2018.8555003.

R. Pinto, M. Moreira, J. Bonifácio and A. E. Santo, "An IEEE 1451 Vehicular Sensor Network Development," *2020 IEEE International Instrumentation and Measurement Technology Conference (I2MTC)*, 2020, pp. 1–6, doi: 10.1109/I2MTC43012.2020.9129224.

S. Guidon, H. Becker and K. Axhausen, "Avoiding Stranded Bicycles in Free-Floating Bicycle-Sharing Systems: Using Survival Analysis to Derive Operational Rules for Rebalancing," *2019 IEEE Intelligent Transportation Systems Conference (ITSC)*, 2019, pp. 1703–1708, doi: 10.1109/ITSC.2019.8916869.

S. Huber, S. Lißner, P. Lindemann, K. Muthmann, A. Schnabel and J. Friedl, "Modelling Bicycle Route Choice in German Cities Using Open Data, MNL and the BikeSim Web-App," *2021 7th International Conference on Models and Technologies for Intelligent Transportation Systems (MT-ITS)*, 2021, pp. 1–6, doi: 10.1109/MT-ITS49943.2021.9529273.

S. Islam, R. I. Hossain, S. S. Amin, A. K. M. Ferdous and A. K. M. A. Malek Azad, "Implementation of Torque Sensor Technology in Electric Bicycle with Solar Charging Kit," *2020 IEEE Region 10 Symposium (TENSYMP)*, 2020, pp. 718–721, doi: 10.1109/TENSYMP50017.2020.9230825.

S. K. Gharghan, R. Nordin, M. Ismail and J. A. Ali, "Accurate Wireless Sensor Localization Technique Based on Hybrid PSO-ANN Algorithm for Indoor and Outdoor Track Cycling," *IEEE Sensors Journal*, 2016, 16(2), pp. 529–541, doi: 10.1109/JSEN.2015.2483745.

S. Nath, S. Sinha, L. M. Gladence, Y. BevishJinila and V. Rajalakshmi, "Health Analysis of Bicycle Rider and Security of Bicycle Using IoT," *2017 International*

Conference on Communication and Signal Processing (ICCSP), 2017, pp. 0802–0806, doi: 10.1109/ICCSP.2017.8286475.

S. Radrizzani, G. Panzani, M. Corno and S. M. Savaresi, "A Concurrent Controllers Integration Strategy to Enhance the Riding Experience in Bicycle Driveline Emulators," *2021 IEEE Conference on Control Technology and Applications (CCTA)*, 2021, pp. 1043–1048, doi: 10.1109/CCTA48906.2021.9659131.

T. -S. Lan, H. -W. Zhang, J. -S. Gao and X. -J. Dai, "Design and Optimizated Simulation of Bicycle Frame Structure," *2020 IEEE Eurasia Conference on IOT, Communication and Engineering (ECICE)*, 2020, pp. 301–304, doi: 10.1109/ECICE50847.2020.9301921.

T. Uchim and S. Chiba, "Proposal of a Smart Bicycle Tourism Improving Tourability and Safety for Tourists by Mesh LPWA Communication System," *2020 Eighth International Symposium on Computing and Networking Workshops (CANDARW)*, 2020, pp. 298–301, doi: 10.1109/CANDARW51189.2020.00065.

V. K. K, A. Singh, D. M. M. S, M. T. Ali and Appaji, "IoT Based Parameters Calculation of Electric Bicycle using OpenModelica Simulation Tool with Data Analytics Technology," *2022 IEEE International Conference on Distributed Computing and Electrical Circuits and Electronics (ICDCECE)*, 2022, pp. 1–4, doi: 10.1109/ICDCECE53908.2022.9792637.

V. Kumar, A. Singh, D. Machamma, M. T. Ali and Appaji, "IoT Based Parameters Calculation of Electric Bicycle using OpenModelica Simulation Tool with Data Analytics Technology," *2022 IEEE International Conference on Distributed Computing and Electrical Circuits and Electronics (ICDCECE)*, 2022, pp. 1–4, doi: 10.1109/ICDCECE53908.2022.9792637.

X. -H. Huang, Z. -H. Chen, A. Ahamad and C. -C. Sun, "ADAS E-Bike: Auxiliary ADAS Module For Electric Power-assisted Bicycle," *2022 IET International Conference on Engineering Technologies and Applications (IET-ICETA)*, 2022, pp. 1–2, doi: 10.1109/IET-ICETA56553.2022.9971513.

X. Huang, F. Han, Y. Han, S. Wang, T. Liu and J. Yi, "Motion Control of an Autonomous Wheel-Leg Bikebot," *2022 IEEE 18th International Conference on Automation Science and Engineering (CASE)*, 2022, pp. 2341–2346, doi: 10.1109/CASE49997.2022.9926543.

Y. Gu, M. Liu, M. Souza and R. N. Shorten, "On the Design of an Intelligent Speed Advisory System for Cyclists," *2018 21st International Conference on Intelligent Transportation Systems (ITSC)*, 2018, pp. 3892–3897, doi: 10.1109/ITSC.2018.8569725.

Y. -X. Zhao, Y. -S. Su and Y. -C. Chang, "A Real-Time Bicycle Record System of Ground Conditions Based on Internet of Things," *IEEE Access*, 2017, 5, pp. 17525–17533, doi: 10.1109/ACCESS.2017.2740419.

Chapter 18

Critical Review: Net Zero-Energy Building

Hari Krishnan M. B., Teenu Raj M., Aayush Karthikeyan, K. T. M. U. Hemapala, O.V. Gnana Swathika

18.1 INTRODUCTION

This chapter mainly concerns the net zero-energy building (NZEB) and its definition. The difference between the energy source and energy carriers is discussed. The definition of the primary energy factor required for carrying energy has gained special consideration. It has led to the discussion of net zero, zero, and plus energy buildings. The idea of a zero-energy building (ZEB) has gained popularity recently because it entails annually offsetting a building's primary energy consumption with energy produced from renewable sources. A ZEB has a lot of promise, with the efficiency of photovoltaics (PV) placed on walls used in construction. In this study, we have merged practical and theoretical analyses, and compared PV systems that are often put on roofs to those that are wall-mounted. Depending on the season, the PV system mounted on walls displayed an odd performance characteristic. In the summer, the vertical wall received less irradiation, and it experienced three peaks that the standard model found difficult to account for. Using vertical spectrum variation, we created a thorough model of the PVs installed on the wall's output power.

A detailed examination of NZEBs and their recent evolution in hot and muggy climates is provided in this study. This study details the key design principles, technological options, and energy performance of 34 NZEB instances throughout the world. We discovered that in hot and muggy areas, passive design and technology, like daylighting and organic ventilation, are frequently used for NZEBs, with further renewable and energy efficient technologies. Additionally, the data show that not every NZEB is an energy-efficient building and that structures that incorporate enough renewable energy sources can still qualify as NZEBs while having high energy demand intensities. In-depth case study analysis is used in this study to assess the energy execution of NZEB and compile best practices for extremely effective NZEBs. This assessment provides crucial practical information and policy suggestions for the construction of NZEBs in hot and muggy areas.

DOI: 10.1201/9781003537878-18

Globally, the shift from conventional structures to sustainable NZEBs has been accomplished. The key challenge still lies in providing thermal comfort for inhabitants while still being energy efficient. Combining a desiccant dehumidifier with a conventional vapor compression system, sometimes known as a "hybrid air conditioner," has become a potential room cooling technique with the introduction of new kinds of materials that are porous and microporous. The NZEB is a remedy to environmental issues like CO_2 emissions. An NZEB might be achievable if there is sufficient energy that is renewable that can be used. For usage in an NZEB, building-service systems that are powered by energy sources that are renewable have received careful attention. Additionally, this chapter offers a review of NZEB evaluation through a literature review. For the methodology section, which discusses the usage of life cycle assessment (LCA), a review of commonly used research techniques, instruments, and performance metrics is presented.

A common answer to the growing energy and environmental issues is the NZEB. The findings of this study indicate that enabling NZEB cooperation has the potential to significantly increase performance. The study's findings further examine the advantages of NZEB collaborations in each of three areas. Traditional controls need to be replaced with new collaborative ones in order to increase NZEB performance at the building-group level. ZEBs have to be planned and built affordably, ideally without incurring additional costs. To demonstrate that ZEBs may be made with no increase in cost, a case study of this technology's use is presented. A future course of action is suggested for higher market penetration of ZEBs after the existing condition of ZEB economics is assessed. The design and construction of ZEBs can be brought into the mainstream of commercial building by showcasing how to combine ZEB technology and design methods, creating a comprehensive efficiency bundle that can be used for very little (or nil, in some circumstances) added capital expenditure.

18.2 IMPLEMENTATION OF A ZEB

The Research Council of Oman announced a competition in 2011. For different climatic conditions in Oman, academic institutions created five green buildings. This study describes how one of those buildings was designed using sustainable energy systems. Oman has a climate that the American Society of Heating, Refrigerating and Air-Conditioning Engineers (ASHRAE) Standard classifies as being extremely hot. In this regard, this study intends to consider the energy system's design to be employed in the article from a variety of view points, including grid interaction, potential for global warming, and other design choices.

When there are no programmable renewable energy sources present, the utilization of cells at the building level significantly reduces the need for the power grid. An excellent load match and high energy efficiency are provided by fuel cells.

Schools use electricity for lights, thus being one of the largest consumers of it. There are many school buildings in Greece that have rather ancient lighting systems that use luminaires with deteriorated translucent diffusers and T8, and occasionally T12, fluorescent lights. The public sector refused to invest in light-emitting diodes (LEDs) due to the economic crisis, and while there is enough sunshine throughout the year, control technology adoption has been delayed. This study investigates a variety of lighting solutions (with AC and DC power supplies), as well as two daylight harvesting systems in a typical Greek public school classroom. It is obvious that the mandated use of daylight controls is the route to zero energy in school buildings.

A practical solution to meet the requirements of the European Energy Performance Directive for the construction industry is solar district heating systems (SDHSs), giving rise to the nearly zero-energy building programs in Europe. To ensure service quality, however, present techniques must overcome a number of technological and financial obstacles Hence, it is important to satisfy the NZEB and positive energy building (PEB) goals. The study provides a technical, environmental, economic, and social sustainability analysis for the integration of SDHS in residential areas. In order to help decision-making stakeholders and policymakers create a sustainability index, the article suggests using a machine learning system that combines multiple criterion decision-making with multiple objectives optimization. The optimization findings demonstrate SDHS's capacity to serve the neighborhood of 500 homes with a solar fraction of up to 95%. Overall, the research provides an assessment of communities with renewable energy sources and a recommendation to incorporate the SDHS to achieve the aims of global sustainability.

A crucial evaluation of the important initiatives taken by the European Union to advance the NZEB and PEB design models has been done in light of the most current developments. There have also been some additions regarding cutting-edge materials and technology, like smart glass, integrated PV systems, phase change materials (PCMs), and vacuum insulation panels (VIPs). Finally, a study on single-family residential NZEB is presented to demonstrate how the building exterior and systems are designed in such a thoughtful way that renewable energy sources almost entirely cover the structure's energy needs, and for the generation of excess energy that could be used by urban networks (PEB potential).

The purpose of the study is to offer retrofit regulations that, given Brasilia's climatic circumstances, achieve the aim of making four office buildings NZEBs. A total of 240 office buildings in the Brazilian capital's downtown area were morphologically defined as a result of this study. The overall energy use for lighting, air conditioning, and office equipment was then predicted using simulations in DesignBuilder and Diva, along with the production of the total amount of energy used by office equipment, lighting, and air conditioning. The modified model's overall final energy usage was found to have decreased by 46%. As a result, the paper offers the information required to create a set of principles and tactics to apply in a vigorous building complex and building refurbishment in order to obtain the NZEB target in the climate zone of Brasilia.

The article also outlines a monitoring campaign's findings conducted in spring and fall to assess the performance of the operational energy of a current building that uses very low energy (nZEB). It is situated in Benevento, a medium-sized South Italian city with a characteristic Mediterranean climate. For the purpose of determining the best way to regulate internal and external shade, as well as the ventilation system, the compilation and evaluation of suitable quantitative indicators based on data with hourly time steps is suggested.

An NZEB home generates and consumes an equal amount of energy, but because the patterns of generation and load are inefficient, it exchanges a sizable quantity of energy with the grid. A local battery energy storage system is installed in order to lower the annual power expenditure. By solving an optimization issue with the help of statistics for a South Australian's hourly load and PV output residence, retail pricing (RP), yearly battery payment rate, and feed-in tariff information, the ideal battery capacity is discovered. The study looks into the difficult issue of lowering the expense required to transform a normal building into a building that uses almost little energy (nZEB). This can be achieved by properly scaling the battery storage system and domestic renewable energy sources (DRES), as well as by enhancing the functionality of the building electric microgrid. To accurately identify the optimum magnitude of the DRES and battery storage system (BSS), a novel approach that depends on the genetic algorithm (GA) is suggested.

The depiction of nearly zero-energy buildings, which presents the biggest potential to enhance conserving energy in Europe, is the main topic of the article. Additionally, it evaluates the execution of NZEBs in Europe and contrasts the definitions of NZEBs in the United States and the EU. Discussions involving NZEBs are assessed. These include the differences between primary energy and energy, as well as between energy carriers and sources of energy. The data show that even though NZEBs have received more attention in recent years, there is still debate surrounding them and inconsistent implementation. The life cycle energy analysis (LCEA) of a residential structure is the topic of the study. The normalization process followed the standards of Product Category Rule (PCR) 2014:02 for buildings. Although the border was clearly stated by the PCR, uncertainty arising from life cycle inventory (LCI) methods and gaps in the system boundaries show that more standardization is required. The rise in embodied energy in buildings suggests that the current analytical framework for energy efficiency requirements may require a full life cycle energy study.

A NZEB has certain structural problems that need to be resolved. In this study, the performances of seven well-known multi-objective evolutionary optimization methods are compared, including multi-objective particle swarm optimization, a regulated non-dominated sorting genetic algorithm (NSGA) with a passive archive, and others (MOPSO). The performance of these algorithms was compared using a number of factors. As the number of generations increased, so did the standard of the answers that were produced. The study also discovered that 1400–1800 assessments were the bare minimum needed to stabilize the building energy model's optimization outcomes.

The next article explores potential office building fenestration design options that balance economic effectiveness and energy efficiency. Alternative strategies for achieving the nZEB level are also examined in the investigation. It was necessary to use PV panels to generate electricity in order to reach nZEB. Although current nZEB solutions are not cost effective, this should change soon. The paper offers design recommendations for high-performance office building facades in its conclusion.

In an article, a brief review is provided of the energy consumption patterns of buildings, useful techniques for saving energy, and the impact of a change in climate. Initial comparisons of building energy depletion between the US, China, and the EU are made using data from international energy reports. ZEB technologies are categorized into technology for passive energy reduction, energy-efficient building services, and renewable energy production. These technologies' viability is examined. Additionally, the impact of future changes in building energy demand due to global climate change is also explored. The paper's major goal is to demonstrate how the ZEB principle might aid in the development of smart cities in Europe. The Strategic Energy Technology Plan (SET-Plan European)'s "Smart Cities & Communities Initiative" aids cities and regions in their efforts to advance toward a 40% decrease in emissions of greenhouse gases due to the adoption of alternative energy sources by the year 2020. Additionally, previous research on the function of ZEBs in countries with smart energy has been evaluated.

Furthermore, the major goal of the next paper is to examine existing structures, analyze them, and provide a general overview of how to transform current structures into ideal NZEBs. According to the survey, buildings have a significant influence on the consumption of energy on a global level, consuming around 40% of total energy and emerging as a significant primary energy consuming component of the global structure. The demand and fuel supply strategies, as well as conversion accounting, are suitable to achieve a ZEB aim and can be largely described by the ZEB definition. In light of energy conservation and cost effectiveness, it is determined that the Zero Energy Project's use of solar energy is the optimal energy source. The article examines current development in the creation of materials used for thermal energy storage for applications related to ZEBs in the construction industry. With a specific emphasis on their technological attributes and growth stage, the thermochemical and latent energy storage materials that are designed for a low-to-medium application of temperature are examined and evaluated with respect to their volumetric heat capacities. The pros, cons, and difficulties of the various possibilities are explored. To optimize operating conditions, efficiency, prices, and system designs, more study is required.

The new issues brought about by the revision of the Energy Performance of Buildings Directive 2010/31/EU (EPBD) and its supplemental rule served as the motivation for the development of this study. The approach combines the usage of the generic optimization application GenOpt with TRNSYS, that is, simulation software for dynamic energy. The goal is to determine the French single-family dwelling typology's cost-optimal level. Following market research performed on the French market, cost functions were developed for every parameter, with the worldwide cost function (EN15459 Standard) serving as the optimization's intent. The results of a cross-comparison study on the sociological and technological obstacles to nZEB deployment in seven Southern European nations are summarized in the paper. The report examines the current state and provides a general overview of nZEB's possibilities in Southern Europe. The study concludes that the majority of Southern European nations are ill-equipped to implement nZEB, particularly when it comes to the risk and opportunity of retrofitting buildings that were built years ago. It also makes recommendations for steps that can be taken to turn the gaps found into chances for the development of future climate-adaptive, high-performance buildings.

A simulation is based on an enhanced framework for budget-effective decisions and energy-saving practices in new construction. Starting from a base configuration, it uses a search technique that is sequential and combines cost and energy simulations to determine the most efficient mix of energy measures that are renewable and energy efficient. The calculations take into account how climate change can affect the projected cooling loads.

The report also demonstrates the viability of European standards for new NZEBs situated in various cities. It explains how to accomplish the NZEB concept in 14 places around Europe most effectively and affordably. The article focuses on a case study of a stadium that utilizes zero energy. The recently repaired and remodeled Dacia Arena in northern Italy is the subject of the chosen case study. In order to determine the energy demand for the new plant, the consumption of energy was first evaluated for Scenario 0, which is the dynamic state (current condition), and Scenario 1 (refurbished state). Second, two fictitious system designs were proposed and assessed. To summarize, the energy assessments performed on the two layout options for the stadium plant showed that the suggested strategies were workable and could be applied to other facilities.

In order to shorten the total charging time and increase system performance, this study suggests an enhanced battery management system (BMS) that offers dynamic equalization in the battery cells. Lithium-ion (Li-ion) energy storage systems used in virtually ZEBs (nZEBs) and electric vehicle (EV) microgrids can be controlled using the recommended control strategy. For both nZEB and EV microgrids, the functionality and efficiency of the proposed BMS are confirmed by simulation models that use MATLAB Simulink software. To demonstrate the operational gains that the new BMS approach can offer, certain simulation results are shown. The paper offers a tour of NZEB evaluation using literature analysis. To clarify the study purpose and technological boundary for NZEB evaluation, an overview of NZEB definitions and energy-efficient metrics is provided. The methodology section then includes a summary of frequently used research methods, tools, and performance indicators. This section also covers the use of LCA in NZEB evaluation and the function LCA plays in fostering a clear-cut defined NZEB. Last but not least, the paper indicates that future developments in NZEB evaluation could highlight potential trends in energy storage, the smart grid, and load balancing.

In order to support the decision-making process for ZEBs, the study introduces an energy-focused software tool that takes into account the Egyptian environment, while also offering informational help. To develop a decision assistance tool that would allow an architectural plan to quickly and easily assess the energy efficiency and thermal comfort of early design choices, energy simulation software is coupled with sensitivity analysis modeling to establish a residential benchmark. A case study and usability testing are used to state the tool's ability to validate the findings it generates and to enhance decision-making. The current work presents an effective, open-source, and quick simulation-based optimization strategy for investigating energy-saving devices (energy-saving materials (ESMs)). The technique is used to determine the nZEB energy performance levels and cost-optimal grades for a case study of a single-family home in Finland. Through a three-stage optimization process, numerous alternatives for covering the building characteristics, heat compensating units, low and high temperature systems, and different-sized thermal and PV solar panel systems are investigated as design options. This necessitates comparing a vast array of potential combinations of energy-supply systems, including renewable energy sources, and energy-saving measures.

In this study three levels of energy demand and three alternative energy supply systems are examined and explained: (1) heat pump of ground source and thermal solar collectors; (2) ambient air and solar source heat pumps; and (3) district heating grid as an addition to PV installation. The findings show that a net ZEB must be constructed with a minimum level of energy usage, with the remaining energy demand being entirely covered by renewable energy sources. Additionally, for the nZEB in Denmark, the district heating grid is a more expensive source of heat than a heat pump. In order to evaluate the lifespan costs of various renewable energy source choices and to determine the most cost-effective mix of energy productivity and renewable energy generation, the article applies lifespan cost analysis and adopts a private economy point of view. Two on-site solutions are among the five technologies included in the analysis, that is, PV energy and micro-combined heat and power (CHP), and three off-premises choices, that is, an off-premises windmill, a share in a wind farm, and redeeming green energy from a grid powered entirely by renewable resources. The results show that in order to construct a net ZEB that is as affordable as possible, energy resourcefulness must be given high importance in the case of locally renewable supply choices.

18.3 IMPACTS OF ZEB

The next study concerns construction with net zero energy in Morocco's housing portfolio. In each of Morocco's six climatic zones, the results of retrofitting a residential structure to attain zero energy balance are examined. Among the design factors taken into account are building architecture, window model, ratio of window-to-wall, wall and roof coating, and infiltration speed. In order to find the ideal option that maintains balance between the whole building lifespan cost, energy consumption savings, and temperature comfort, some optimization is performed with the design mentioned before, like carrying out passive energy efficiency operations. The study describes and evaluates an innovative energy handling system for microgrid-connected buildings that includes electric automobiles as active elements. By taking as the basis of both constructing a building and movability consumption, the energy balance of the analyzed energy design, building to vehicle and vehicle to building, is examined. Through a case study analysis, three alternative management system scenarios are examined to evaluate the contribution of EVs as an electricity resource among structures combining PV panels and electrical storage. For that purpose, a lively simulation structure in MATLAB is created to look at the energy requirements and loads of the structure, and as a result of the planned EV's energy use level.

The next study describes the current state-of-the-art of ZEB's overall performance and similar technical solutions, analyzing them to extend the potential in energy usage. A collection of case studies involving ZEBs from the literature is presented, together with an analysis of the performance of often-used solutions. The technologies involved are explained, as well as their applications for achieving the ZEB standards. A debate is recommended, highlighting the main aspects that advantage further examination and identifying the main factors in making the zero-energy aim the new building standard. For temperate and tropical regions, a study is conducted on hybrid microgrid NZEBs. A shopping mall building is studied. Climate and weather data were collected for a year. PV modules and converters were used in the proposed hybrid microgrid NZEB. The thermal load, on the other hand, is a feature of a grid-connected hybrid system. Utilizing the net present cost's economic characteristics, the project's cost-effectiveness is confirmed, along with the paying period, and operating costs. The outcomes show that the study is economical and has a paying period of nearly 1 year in Thailand and over 2 years in Pakistan. Additionally, the decrease in each unit cost of electricity is over 30% and over 25% in Thailand and Pakistan, respectively.

The next article analyzes the most recent definition of ZEBs, compiles existing definitions from around the world, and pinpoints the geographic limits of almost zero energy structures located in China. It provides a broad outline of the recent situation, as well as comprehensive legislation and building regulations for buildings in China that use almost no energy. This paper suggests how to produce and realize China's ZEBs. The government will be guided by the findings of this report in order to adopt more efficient and effective building standards and laws to lower the building carbon emissions in the industry.

The energy effectiveness practices (EEPs) and NZEBs' renewable energy technologies, as well as an analysis of their use and applicability, are covered in this study. More research is being performed in this area, including life cycle energy analysis (LCEA), changes in climatic conditions, smart building operation guidance systems, energy storehouse systems, and public policy concerns. In order to support the ongoing development of NZEBs in China's varied climate zones, the document provides relevant references.

The next paper discusses numerous retrofitting possibilities for buildings, including lean manufacturing, eco-friendly technologies, and renewable energy. The academy of Civil Engineering at University Technology Malaysia (UTMM50)'s building was chosen as a research study for an in situ energy audit of Building Energy Intensity (BEI). This involved identifying the different types of machinery and equipment utilized in the building, as well as the time of day and amount of power they consume when operating normally. The study's findings recommended retrofit solutions that may be used to create buildings that use zero energy. It is a key component of lowering energy consumption through sustainability initiatives and economic stimulus programs.

How climate change may affect the extent of virtually ZEBs and the assessment of durability in terms of energy balance are the important topics of the article. The case study focuses on a genuine case called BNZEB, a one-story house constructed in Benevento. Both a short-term and medium-term study are conducted using five years of observed meteorological data: the first one involves comparing energy behavior to the initial operating year. Then, energy efficiency is assessed in light of the CCWorldWeatherGen tool's medium-term climatic projections.

Creating a cutting-edge knowledge base and presenting essential design and performance characteristics that identify these technologies, an analytic framework of the technological choices and techniques to achieve ZEBs was conducted with the ultimate goal of supporting building-related climate change reduction options. All pertinent literature published between January 2013 and August 2019 was evaluated. This literature demonstrated that because released information is unorganized and heterogeneous, it cannot be used to predict climate and architectural typologies. The cases demonstrate that buildings can help to mitigate climate change by using suitable design to reduce the energy presented in the materials that were used in the building.

The performance of practical ZEBs and the findings of regional peculiarities are critical for similar upcoming projects. The late evaluation and innovators of ZEBs from the US, China, and the EU are compared in this study. According to the findings, the energy performance of cases is determined by the ZEB meaning, energy drivers, standards, local policies, adoption of technology, and their ratio of adoption.

NZEBs are becoming more and more well-liked as a means of lowering operating energy utilization and limiting the emission of greenhouse gas in the construction sector. Despite the advances in NZE research, there are still considerable obstacles standing in the way of its widespread adoption and implementation, especially when it comes to high-rise residential structures, which are increasingly making up a significant component of Australia's housing supply. The uncertainty associated with choosing acceptable solution strategies that might produce quality performance is one of the problems related to the attainment of NZEB performance. Following an assessment of the effect of various innovations across cities in Australia's five climate condition zones, a framework of solutions that can reach the target for each climate condition zone is presented.

The next article examines how Swedish law has been adapted to incorporate the Energy Performance of Buildings Directive (EPBD) of the EU, which concludes by making ZEBs almost a requirement. The power inefficiency of resourceful factors, the balancing of interests, and the various perceptions and assessments of what actually is crucial for the housing division structure calculation processes can be seen from studying the negotiations and controversies surrounding the adaptation of the Swedish building code. The goal of the study is to examine the relationship between buildings and the electrical grid, and the secular correlation between building load and PV generation. The nursing of energy use and interior microclimate states is also addressed, along with measurements of a number of performance characteristics using four distinct heating, ventilation, and air conditioning (HVAC) configurations. In regions with similar climatic conditions, the discussion of the energy balance can be used as a guide to improve management and design techniques.

The increased energy use and pollutant emissions from buildings have been addressed by the concept of NZEBs. The goal in the study is to analyze, differentiate, and optimize six solutions of renewable energy sets for NZEB design in three climatic conditions. The advice is to use a solution established in Indore (which has a hot environment) that consists of a cooling system used as a heat pump air source and the generation of domestic hot water (DHW) as flat plate solar collectors. The utilization of a biodiesel generator in Tromso (a cold climate) has the promise of producing both electricity and steam for heating and DHW consumption.

The adoption of required ZEBs has been outlined in a roadmap by the Korean government because the construction industry is a significant source of carbon emissions. We looked at the possibility for residential structures to save energy in order to assess whether enacting mandatory ZEBs would make it possible to achieve carbon neutrality. A holistic approach was necessary because governing individual element technologies alone has its limitations. Consequently, it's also vital to create an agreement to motivate consumers to adopt energy-saving practices. The study can be helpful as a guide when creating building energy policies when ZEBs become legally required.

The minimized sizing of a hybrid PV-wind system for NZEBs is determined in order to reach its lowest loss caused by the power supply probability and cost while meeting NZE balancing requirements. The oversizing/undersizing of renewable energy generators results in increased energy out and in, which is transported between the construction and the grid, ultimately increasing the stress on the electrical system. NSGA-II is utilized to optimize three objective functions: the annualized cost of the system (ACS), the loss of power supply probability (LPSP), and the total energy transferred (TET). By considering two major issues, that is, (1) renewable energy available less in cities and (2) the difficulties of location on transport energy consumption, the paper first explores the potential to expand the term ZEB to base-level marks. The primary variables that affect the energy balance are noted. At the local level, the possibility of "energy mutualization" is underlined. Thus, the article illustrates the potential of an integrated strategy connecting building and transportation energy consumptions.

The primary goal of the study is to achieve clarity about the impact of NZEBs on upcoming energy systems through studying its interactions among on-site generation and building loads, commonly referred to as load match, and the ensuing in/out transport contact with the local electricity grid, also known as grid contact. The cover factors depict the production pattern of various renewable energy systems, the daily and seasonal effects, and the employed operation/control procedures.

In order to achieve a net zero-energy balance, this study suggests why buildings should use different energy forms, especially solar energy, as a key source of energy PV installation, and solar thermal systems will finally fill up the space on roofs and facades in projects with aggressive energy reaches or restricted usable places for domestic energy generation. The findings demonstrate that constructions using just high-efficiency PV modules are very near to achieving a zero-energy balance, but this conclusion mainly depends on the definition of nZEB, the boundary conditions, and the energy system layout of this structure.

User patterns that reflect large differences in behavior without making models unnecessarily complex are required to obtain more realistic energy demand simulations. In order to achieve this, it is important to consider three states: (1) at home and will be awake, (2) will be sleeping, or (3) will be absent. The approach used to create this outgrown model, which was totally based on Belgian's use survey from 2005, is described in the publication. Seven typical occupancy patterns were found using hierarchical clustering.

The next study uses computer simulations to assess the constructiveness of converting Brazilian office buildings into ZEB. The model's creation process is broken down into three stages. The standard model (referring to the building model), which shows the present building structure of Brazilian office constructions, is evaluated as the first phase. The creation of an ideal example (low-energy building) in accordance with the Brazilian system of energy efficiency labeling is the second phase. The findings demonstrate that Fortaleza requires more effective PV modules due to the higher energy use, particularly for cooldown. A reduction in final energy usage of over 50% is also made possible by use of the Brazilian building energy efficiency standard method.

The article focuses on issues with energy storage brought on by microgeneration in individual homes. The research study features a minor wind and solar power generation system installed in a net zero-energy structure. The microgeneration units and the NZEBs are both connected to a utility grid. The paper proposes that when there is a mix of minor solar and wind energy in the ratio of 70/30, the condition in favor of wind, self-utilization reaches its peak.

The construction of integrated PV systems is a strong and adaptable instrument for meeting the future's rising demand for ZEBs. The progress of solar PV technology for building integration and design is thoroughly reviewed in the paper. It emphasizes how solar PV cells and building-integrated PV (BIPV) products are categorized for use in building design. BIPV offers the chance to contribute significantly to a new age of distributed power generation. In recent years, the idea of ZEB has garnered global-scale attention and is currently considered the ideal solution for building design in the future. The type of renewable energy source used and the energy infrastructure utilized are the most crucial factors that should be mainly focused on before developing the latest ZEB definition. The majority of ZEB definitions currently in use are reviewed in the work, along with several approaches to potential ZEB computation algorithms.

18.4 CONCLUSION

Many countries are seeking ZEBs as their future building energy target because they have already established ZEBs. ZEBs have the potential to remarkably reduce energy, as well play a big part in the renewable energy field. However, in order not to fall short of the presumption, there is need to understand the definition and framework of ZEBs. This chapter presents a critical review of ZEB, thus showing the importance of saving energy by a shift to ZEBs. The conclusion is that ZEB is still in progress and needs more time to develop and be accessible to a common person. ZEB is very much likely to be the future in all developing countries.

BIBLIOGRAPHY

Abdou, N., Mghouchi, Y.E., Hamdaoui, S., Asri, N.E. and Mouqallid, M., 2021. Multi-objective optimization of passive energy efficiency measures for net-zero energy building in Morocco. *Building and Environment*, 204, p.108141.

Abidin, N.I.A., Zakaria, R., Pauzi, N.N.M., Mustaffar, M. and Bandi, M., 2019, September. Building energy intensity measurement for potential retrofitting of zero energy building in higher learning institution. In *IOP Conference Series: Materials Science and Engineering* (Vol. 620, No. 1, p. 012070). IOP Publishing.

Aerts, D., Minnen, J., Glorieux, I., Wouters, I. and Descamps, F., 2014. A method for the identification and modelling of realistic domestic occupancy sequences for building energy demand simulations and peer comparison. *Building and Environment*, 75, pp.67–78.

Alawode, A. and Rajagopalan, P., 2022. Feasibility of net zero energy high rise apartment buildings in Australia. *Solar Energy*, 231, pp.158–174.+B23.

Al-Saadi, S.N. and Shaaban, A.K., 2019. Zero energy building (ZEB) in a cooling dominated climate of Oman: Design and energy performance analysis. *Renewable and Sustainable Energy Reviews*, 112, pp.299–316.

Arif, S., Taweekun, J., Ali, H.M., Yanjun, D.A.I. and Ahmed, A., 2021. Feasibility study and economic analysis of grid connected solar powered net zero energy building (NZEB) of shopping mall for two different climates of Pakistan and Thailand. *Case Studies in Thermal Engineering*, 26, p.101049.

Ascione, F., Borrelli, M., De Masi, R.F. and Vanol, G.P., 2020. Nearly zero energy target and indoor comfort in Mediterranean climate: Discussion based on monitoring data for a real case study. *Sustainable Cities and Society*, 61, p.102349.

Ascione, F., Borrelli, M., De Masi, R.F. and Vanoli, G.P., 2020. Hourly operational assessment of HVAC systems in Mediterranean Nearly Zero-Energy Buildings: Experimental evaluation of the potential of ground cooling of ventilation air. *Renewable Energy*, 155, pp.950–968.

Ascione, F., De Masi, R.F., Gigante, A. and Vanoli, G.P., 2022. Resilience to the climate change of nearly zero energy-building designed according to the EPBD recast: Monitoring, calibrated energy models and perspective simulations of a Mediterranean nZEB living lab. *Energy and Buildings*, 262, p.112004.

Attia, S., Eleftheriou, P., Xeni, F., Morlot, R., Ménézo, C., Kostopoulos, V., Betsi, M., Kalaitzoglou, I., Pagliano, L., Cellura, M. and Almeida, M., 2017. Overview and future challenges of nearly zero energy buildings (nZEB) design in Southern Europe. *Energy and Buildings*, 155, pp.439–458.

Attia, S., Gratia, E., De Herde, A. and Hensen, J.L., 2012. Simulation-based decision support tool for early stages of zero-energy building design. *Energy and Buildings*, 49, pp.2–15.

Barone, G., Buonomano, A., Calise, F., Forzano, C. and Palombo, A., 2019. Building to vehicle to building concept toward a novel zero energy paradigm: Modelling and case studies. *Renewable and Sustainable Energy Reviews*, 101, pp.625–648.

Barone, G., Buonomano, A., Calise, F., Forzano, C. and Palombo, A., 2019. Building to vehicle to building concept toward a novel zero energy paradigm: Modelling and case studies. *Renewable and Sustainable Energy Reviews*, 101, pp.625–648.

Barone, G., Buonomano, A., Calise, F., Forzano, C. and Palombo, A., 2019. Building to vehicle to building concept toward a novel zero energy paradigm: Modelling and case studies. *Renewable and Sustainable Energy Reviews*, 101, pp.625–648.

Belussi, L., Barozzi, B., Bellazzi, A., Danza, L., Devitofrancesco, A., Fanciulli, C., Ghellere, M., Guazzi, G., Meroni, I., Salamone, F. and Scamoni, F., 2019. A review of performance of zero energy buildings and energy efficiency solutions. *Journal of Building Engineering*, 25, p.100772.

Belussi, L., Barozzi, B., Bellazzi, A., Danza, L., Devitofrancesco, A., Fanciulli, C., Ghellere, M., Guazzi, G., Meroni, I., Salamone, F. and Scamoni, F., 2019. A review of performance of zero energy buildings and energy efficiency solutions. *Journal of Building Engineering*, 25, p.100772.

Cabeza, L.F. and Chàfer, M., 2020. Technological options and strategies towards zero energy buildings contributing to climate change mitigation: A systematic review. *Energy and Buildings*, 219, p.110009.

Cao, X., Dai, X. and Liu, J., 2016. Building energy-consumption status worldwide and the state-of-the-art technologies for zero-energy buildings during the past decade. *Energy and Buildings*, 128, pp.198–213.

Costa, J.F.W., Amorim, C.N.D. and Silva, J.C.R., 2020. Retrofit guidelines towards the achievement of net zero energy buildings for office buildings in Brasilia. *Journal of Building Engineering*, 32, p.101680.

D'Agostino, D. and Mazzarella, L., 2019. What is a nearly zero energy building? Overview, implementation and comparison of definitions. *Journal of Building Engineering*, 21, pp.200–212.

D'Agostino, D. and Mazzarella, L., 2019. What is a Nearly zero energy building? Overview, implementation and comparison of definitions. *Journal of Building Engineering*, 21, pp.200–212.

D'Agostino, D. and Parker, D., 2018. A framework for the cost-optimal design of nearly zero energy buildings (NZEBs) in representative climates across Europe. *Energy*, 149, pp.814–829.

Deng, S., Wang, R.A. and Dai, Y.J., 2014. How to evaluate performance of net zero energy building–A literature research. *Energy*, 71, pp.1–16.

Deng, S., Wang, R.A. and Dai, Y.J., 2014. How to evaluate performance of net zero energy building–A literature research. *Energy*, 71, pp.1–16.

Didoné, E.L., Wagner, A. and Pereira, F.O.R., 2014. Strategies towards net zero energy office buildings in Brazil with emphasis on BIPV. *Ambiente Construído*, 14, pp.27–42.

Feng, W., Zhang, Q., Ji, H., Wang, R., Zhou, N., Ye, Q., Hao, B., Li, Y., Luo, D. and Lau, S.S.Y., 2019. A review of net zero energy buildings in hot and humid climates : Experience learned from 34 case study buildings. ;*Renewable and Sustainable Energy Reviews*, ;*114*, p.109303.

Ferrara, M., Fabrizio, E., Virgone, J. and Filippi, M., 2014. A simulation-based optimization method for cost-optimal analysis of nearly Zero Energy Buildings. *Energy and Buildings*, 84, pp.442–457.

Good, C., Andresen, I. and Hestnes, A.G., 2015. Solar energy for net zero energy buildings–A comparison between solar thermal, PV and photovoltaic–thermal (PV/T) systems. *Solar Energy*, 122, pp.986–996.

Hamdy, M., Hasan, A. and Siren, K., 2013. A multi-stage optimization method for cost-optimal and nearly-zero-energy building solutions in line with the EPBD-recast 2010. *Energy and Buildings*, 56, pp.189–203.

Hamdy, M., Nguyen, A.T. and Hensen, J.L., 2016. A performance comparison of multi-objective optimization algorithms for solving nearly-zero-energy-building design problems. *Energy and Buildings*, 121, pp.57–71.

Harkouss, F., Fardoun, F. and Biwole, P.H., 2019. Optimal design of renewable energy solution sets for net zero energy buildings. *Energy*, 179, pp.1155–1175.

Kalder, J., Allik, A., Hõimoja, H., Jõgi, E., Hovi, M., Märss, M., Kurnitski, J., Fadejev, J., Lill, H., Jasinskas, A. and Annuk, A., 2017. Optimal wind/solar energy mix for residential net zero-energy buildings. In *International scientific conference Rural Development 2017* (pp. 621–626).

Koseoglou, M., Tsioumas, E., Jabbour, N. and Mademlis, C., 2018, December. An improved active Li-ion battery equalization scheme for enhancing the performance of a nearly zero energy building and electric vehicle microgrid. In *2018 IEEE 4th Southern Power Electronics Conference (SPEC)* (pp. 1–7). IEEE.

Kylili, A. and Fokaides, P.A., 2015. European smart cities: The role of zero energy buildings. *Sustainable Cities and Society*, 15, pp.86–95.

Liu, Z., Liu, Y., He, B.J., Xu, W., Jin, G. and Zhang, X., 2019. Application and suitability analysis of the key technologies in nearly zero energy buildings in China. *Renewable and Sustainable Energy Reviews*, 101, pp.329–345.

Liu, Z., Zhou, Q., Tian, Z., He, B.J. and Jin, G., 2019. A comprehensive analysis on definitions, development, and policies of nearly zero energy buildings in China. *Renewable and Sustainable Energy Reviews*, 114, p.109314.

Liu, Z., Zhou, Q., Tian, Z., He, B.J. and Jin, G., 2019. A comprehensive analysis on definitions, development, and policies of nearly zero energy buildings in China. *Renewable and Sustainable Energy Reviews*, 114, p.109314.

Lizana, J., Chacartegui, R., Barrios-Padura, A. and Valverde, J.M., 2017. Advances in thermal energy storage materials and their applications towards zero energy buildings: A critical review. *Applied Energy*, 203, pp.219–239.

Mademlis, C., Jabbour, N., Tsioumas, E., Kosseoglou, M. and Papagiannis, D., 2022, September. Reduction of the cost needed for converting a conventional building to a nearly zero energy building. In *2022 IEEE International Smart Cities Conference (ISC2)* (pp. 1–7). IEEE.

Magrini, A., Lentini, G., Cuman, S., Bodrato, A. and Marenco, L., 2020. From nearly zero energy buildings (NZEB) to positive energy buildings (PEB): The next challenge-The most recent European trends with some notes on the energy analysis of a forerunner PEB example. *Developments in the Built Environment*, 3, p.100019.

Maheshwari, S., Chauhan, P., Tandon, S. and Sagar, S., 2017. A review study on net zero energy building. *International Research Journal of Engineering and Technology (IRJET)*, 4, pp.1567–1570.

Manni, M., Coccia, V., Nicolini, A., Marseglia, G. and Petrozzi, A., 2018. Towards zero energy stadiums: The case study of the Dacia arena in Udine, Italy. *Energies*, 11(9), p.2396.

Marique, A.F. and Reiter, S., 2014. A simplified framework to assess the feasibility of zero-energy at the neighbourhood/community scale. *Energy and Buildings*, 82, pp.114–122.

Marszal, A.J. and Heiselberg, P., 2011. Life cycle cost analysis of a multi-storey residential Net Zero Energy Building in Denmark. *Energy*, 36(9), pp.5600–5609.

Marszal, A.J., Heiselberg, P., Bourrelle, J.S., Musall, E., Voss, K., Sartori, I. and Napolitano, A., 2011. Zero Energy Building–A review of definitions and calculation methodologies. *Energy and Buildings*, 43(4), pp.971–979.

Marszal, A.J., Heiselberg, P., Jensen, R.L. and Nørgaard, J., 2012. On-site or off-site renewable energy supply options? Life cycle cost analysis of a Net Zero Energy Building in Denmark. *Renewable Energy*, 44, pp.154–165.

Nagaoka, A., Ota, Y., Sakai, K., Araki, K., Matsuo, H., Kadota, N., Maeda, K., Nakajima, A. and Nishioka, K., 2021. Performance evaluation and spectrum-based analysis of a wall-mounted photovoltaic system for zero-energy building. *Renewable Energy*, 174, pp.147–156.

Niskanen, J. and Rohracher, H., 2022. A politics of calculation: Negotiating pathways to zero-energy buildings in Sweden. *Technological Forecasting and Social Change*, 179, p.121630.

Niveditha, N. and Singaravel, M.R., 2022. Optimal sizing of hybrid PV–Wind–Battery storage system for Net Zero Energy Buildings to reduce grid burden. *Applied Energy*, *324*, p.119713.

Park, B.R. and Chung, M.H., 2023. Analysis of the additional energy-saving potential of residential buildings after mandatory zero-energy buildings to achieve carbon neutrality in South Korea. *Building and Environment*, *228*, p.109908.

Perlova, E., Platonova, M., Gorshkov, A. and Rakova, X., 2015. Concept project of zero energy building. *Procedia Engineering*, 100, pp.1505–1514.

Pikas, E., Thalfeldt, M. and Kurnitski, J., 2014. Cost optimal and nearly zero energy building solutions for office buildings. *Energy and Buildings*, 74, pp.30–42.

Salom, J., Marszal, A.J., Widén, J., Candanedo, J. and Lindberg, K.B., 2014. Analysis of load match and grid interaction indicators in net zero energy buildings with simulated and monitored data. *Applied Energy*, 136, pp.119–131.

Shamim, J.A., Hsu, W.L., Paul, S., Yu, L. and Daiguji, H., 2021. A review of solid desiccant dehumidifiers: Current status and near-term development goals in the context of net zero energy buildings. *Renewable and Sustainable Energy Reviews*, 137, p.110456.

Sharma, V., Haque, M.H. and Aziz, S.M., 2019. Energy cost minimization for net zero energy homes through optimal sizing of battery storage system. *Renewable Energy*, 141, pp.278–286.

Shukla, A.K., Sudhakar, K. and Baredar, P., 2016. A comprehensive review on design of building integrated photovoltaic system. *Energy and Buildings*, 128, pp.99–110.

Sun, Y., Huang, G., Xu, X. and Lai, A.C.K., 2018. Building-group-level performance evaluations of net zero energy buildings with non-collaborative controls. *Applied Energy*, 212, pp.565–576.

Torcellini, P., Pless, S. and Leach, M., 2015. A pathway for net-zero energy buildings: creating a case for zero cost increase. *Building Research & Information*, 43(1), pp.25–33.

Tumminia, G., Guarino, F., Longo, S., Aloisio, D., Cellura, S., Sergi, F., Brunaccini, G., Antonucci, V. and Ferraro, M., 2020. Grid interaction and environmental impact of a net zero energy building. *Energy Conversion and Management*, 203, p.112228.

Wang, R., Feng, W., Wang, L. and Lu, S., 2021. A comprehensive evaluation of zero energy buildings in cold regions: Actual performance and key technologies of cases from China, the US, and the European Union. *Energy*, 215, p.118992.

Chapter 19

Voltage Stability in Microgrids: A Critical Review

Asha Anu Kurian, O.V. Gnana Swathika

19.1 INTRODUCTION

A microgrid (MG) is a small-scale power system that may operate either independently or in conjunction with the main grid. Renewable energy sources (RESs, such as solar photovoltaic and wind turbines), energy storage devices (batteries and fuel cells), and loads ranging from domestic loads to industrial loads fall under the category of distributed energy resources (DERs). MGs can be configured in a variety of ways depending on the individual needs and resources available in the region. The major factors that trigger instability in MGs are the DER power limit, inadequate reactive power sharing, load voltage sensitivity, DC link capacitor responses, low steady state voltage, excessive power swing, and high DC link voltage.

The voltage range for an MG lies between 400 V and 69 kV. MGs can be tiny or large (tens of megawatts in capacity). The load current varies as per the impedance of the load connected and its dynamics. The ratio of resistance (R) to reactance (X) (R/X ratio) defines whether active or reactive power is dependent on frequency or voltage in the single machine infinite bus system (SMIB) power equation. MGs differ from bulk power systems in several ways, including smaller system size, increased use of RESs, more uncertainty, reduced inertia of the system, increased feeder R/X ratio, restricted short-circuit capacity and uneven three-phase loading. The stability criteria and classifications need to be evaluated again in light of the inherent differences between MGs and bulk power systems. Figure 19.1, shows the classification of stability in a smart grid. Voltage stability is defined by the joint task force of Institute of Electrical and Electronics Engineers (IEEE) and the International Council on Large Electric Systems (CIGRE) as "the ability of a power system to maintain steady voltages at all buses in the system after being subjected to a disturbance from a given initial operating condition". The main factor contributing to voltage instability is inadequate reactive power in the system. On detailed analysis, this can be traced to the fact that a reactive compensation limit exists for a critically loaded bus that is to be determined using the voltage stability indices method.

DOI: 10.1201/9781003537878-19

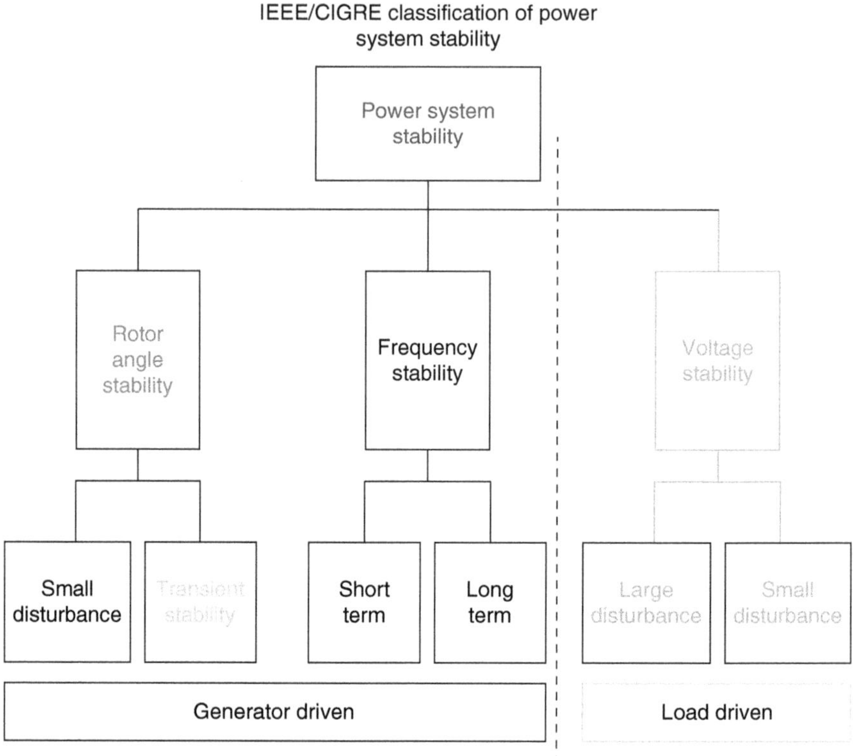

Figure 19.1 Classification of power system stability in a smart grid

The major components in the modeling of MGs are a combination of following components:

1. **Energy sources:** They comprise both conventional generators (like diesel generators) and RESs (like solar photo and wind turbines)
2. **Electrical machine and control system:** Doubly fed induction motors with optimized controllers can aid in dynamic voltage stability and improve the bus voltages under planned and unplanned islanding operations. Furthermore, load forecasting based on machine learning deployed systems enables the adaptive control of MG operation.
3. **Energy storage systems (ESS):** The solutions include battery energy storage lithium-ion (Li-ion) and vanadium redox batteries, hybrid battery energy storage, super capacitors, vehicle-to-grid topologies, and flywheel energy storage devices.
4. **Load modeling:** MG loads are categorized as residential, commercial, military base, community, and industrial that affect short-term voltage stability.

5. **Power electronics converter interfaces and controllers:** Nonlinear controllers, particularly sliding mode controllers (SMCs), may provide better dynamic and regulatory performance throughout a wider variety of operating conditions and this sort of nonlinear control technique is suitable for power converter management. The inverter configuration for photovoltaic (PV) systems is as follows:
   (i) central inverter (for a power rating up to 30 kW)
   (ii) string inverter (simple and high-energy profit configuration and high efficiency)
   (iii) multi-string inverter (combination of central and string inverter configuration): master-slave configuration exists; the multilevel inverter configuration enables the reduction of total harmonic distortion in large power plant applications.
   (iv) micro-inverter (up to 200 W): voltage variations are higher for a central inverter and string inverter; summarize the control strategy types, advantages and disadvantages. By utilizing a traditional synchronous reference frame (SRF) phase-locked loop, a current controller that is adaptive in nature may eliminate measurement errors in the synchronization unit, which often impact the performance of the current controller. When assessing the stability of the dynamic response of grid-feeding converters, the traditional PV curve should be used, taking into account both the power and current limitations
6. **Transformers and cables for distribution lines:** If the MG is in a grid-connected (GC) configuration, the reliable and safe transmission and distribution of electrical power is made possible by transformers and cables, which are essential infrastructure elements in MGs that are connected to the grid
7. **Devices for protection including circuit breakers and fuses:** In terms of voltage and frequency trip durations, breakers have to adhere to IEEE 1547 and IEC 61727 requirements. A key challenge nowadays is the frequent and necessary isolation/reconnection of loads without power deviations/fluctuations due to the rising frequency of blackouts in distribution systems caused by various natural and man-made tragedies. The feasible choice for a very reliable and effective power grid operation is a network of MGs. The utility grid is an important component that today's MG control must have. The expanding usage of distributed RESs in MGs makes this transition capability increasingly crucial. The MG capacity to do this should allow essential loads and facilities to have a steady supply of energy. Deploying MG control mechanisms is therefore necessary to provide a seamless transition between GC and islanded operational modes. During the switchover, the control method should be able to reduce the changes in voltage and current brought on by frequency

> and phase misalignment. The literature for mitigating voltage and frequency (V-f) fluctuations at the MG's point of common coupling (PCC) and the utility grid during the transition period provides an overview of the various MG control strategies that enable smooth transitions between GC and islanded modes of operation. It aims to spur the creation of better control methods in this area of MG-related research.

In GC mode, DERs typically use a grid-following control approach (synchronization is performed using an external trigger signal from the grid), but in islanding mode, at least one DER must employ a grid-forming strategy. In order to facilitate a seamless shift to the islanding mode and lessen the impact of disturbances, a control approach is required that consists of two parts: (a) a linear voltage controller that receives input from the output current feed forward and capacitor current feedback; and (b) a modified droop control that mimics the synchronous generator's inertia response.

Improving voltage stability in real-world MGs is a difficult task that calls for a blend of sophisticated control techniques, DER integration, energy storage options and voltage regulation hardware. For industrial MGs to identify any voltage stability problems, it is imperative to conduct the routine monitoring of voltage levels, load circumstances and equipment performance. Voltage regulation system dependability may be increased and equipment breakdowns can be avoided by putting into practice predictive maintenance procedures based on real-time data analysis. The wise use of data learning techniques will result in more advanced voltage regulation in an MG.

This chapter is organized as follows: Section 19.1 provides an introduction to the aspects of MG stability basics and voltage stability, Section 19.2 provides an outline to the stability modeling of MGs, Section 19.3 deals with methods of enhancing voltage stability in MGs, and Section 19.4 presents the conclusion.

19.2 AN OUTLINE OF THE STABILITY MODELING OF A MICROGRID

Figure 19.2 presents a flowchart for stability studies on MGs.

1. Identification of MG components
   List all components of the MG, including generators, loads, RESs (e.g., solar panels and wind turbines), energy storage systems (ESSs, e.g., batteries), inverters, and control devices. For a redundant power supply with less emissions, cost has to be compromised.
2. Comprehending the state-space averaged model's static and dynamic behavior

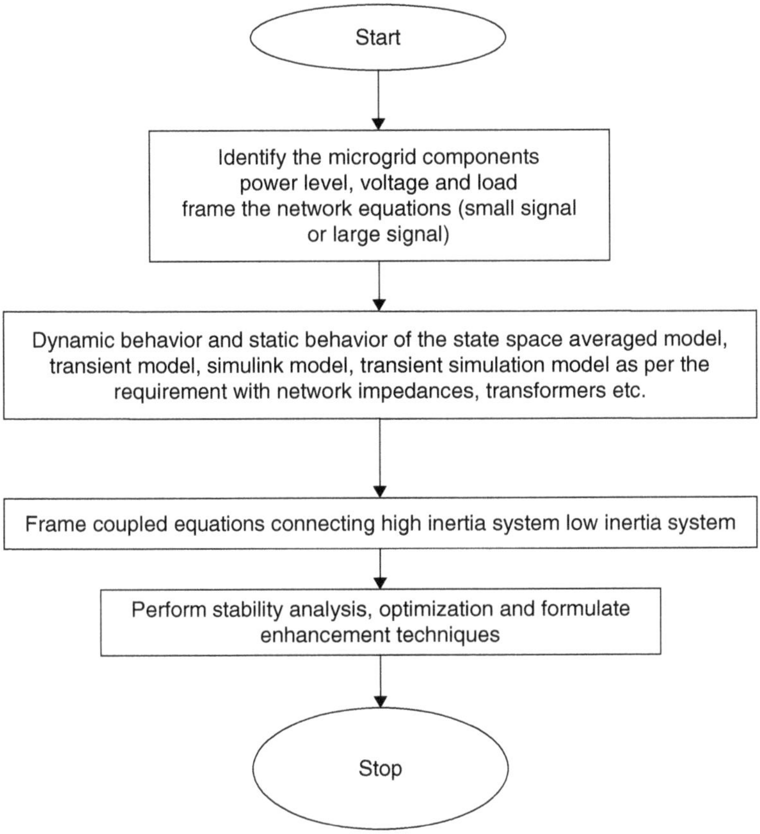

Figure 19.2 Stability studies on an MG

Understand the dynamic behavior of each component. For example, synchronous generators have electromechanical dynamics, whereas inverters have power electronics dynamics. An extraneous variable like the pollution factor has not been considered in the majority of models of the transmission line benchmark.

3. Formulate differential equations
   Write differential equations that describe the dynamic behavior of each component. This may involve mechanical equations for generators, electrical equations for inverters, and state-of-charge equations for batteries. Table 19.1 shows energy sources, significant maximum power point tracking (MPPT) methods and significant curves.

Table 19.1 Energy sources, significant MPPT methods, significant curves Energy sources, like wind, PV and fuel cells are considered, significant MPPT methods, significant curves like Λ-Cp characteristics for various β values (wind speed distributions Weibull and Rayleigh distribution), P-V curve, Voc, and Ioc plot. Polarization curve, power curve, efficiency curve, and VI curve

Distributed energy sources	*Significant MPPT methods*	*Significant curves*	*References*
Wind energy system	P&O, neural network, metaheuristic algorithms, fuzzy logic etc.	Λ-Cp characteristics for various β values. Wind speed probability distributions: Weibull and Rayleigh distribution	[25]
Photovoltaic systems	P&O, incremental conductance, advanced machine learning algorithms, deep learning techniques etc.	P-V curve, Voc, Ioc plot	[26]
Fuel cell	P&O (based on the operating point of the fuel cell, incremental conductance (compares instantaneous conductance of the fuel cell), neural network, fuzzy logic, etc.	Polarisation curve, power curve, efficiency curve, VI curve	[2]

4. Control systems
Incorporating the state-space modeling of the control loop into the dynamic equations is required to predict the actual plant response. For example, voltage control loops for inverters or excitation control systems for synchronous generators affect the dynamic response of the MG components and are essential for maintaining stability. Figure 19.3 presents various types of MG controller classifications.
5. Account for network effects
Consider the network effects, including the impedance of transmission lines and transformers. These affect the power flow and voltage dynamics within the MG. Include network equations that describe the electrical behavior of the MG, such as Kirchhoff's laws and network admittance matrices.
6. Framing the dynamic state equations
Combine the differential equations for each component with the network equations to form a set of coupled differential-algebraic equations (DAEs).These DAEs describe the dynamic behavior of the entire MG system, including interactions between components and network effects.

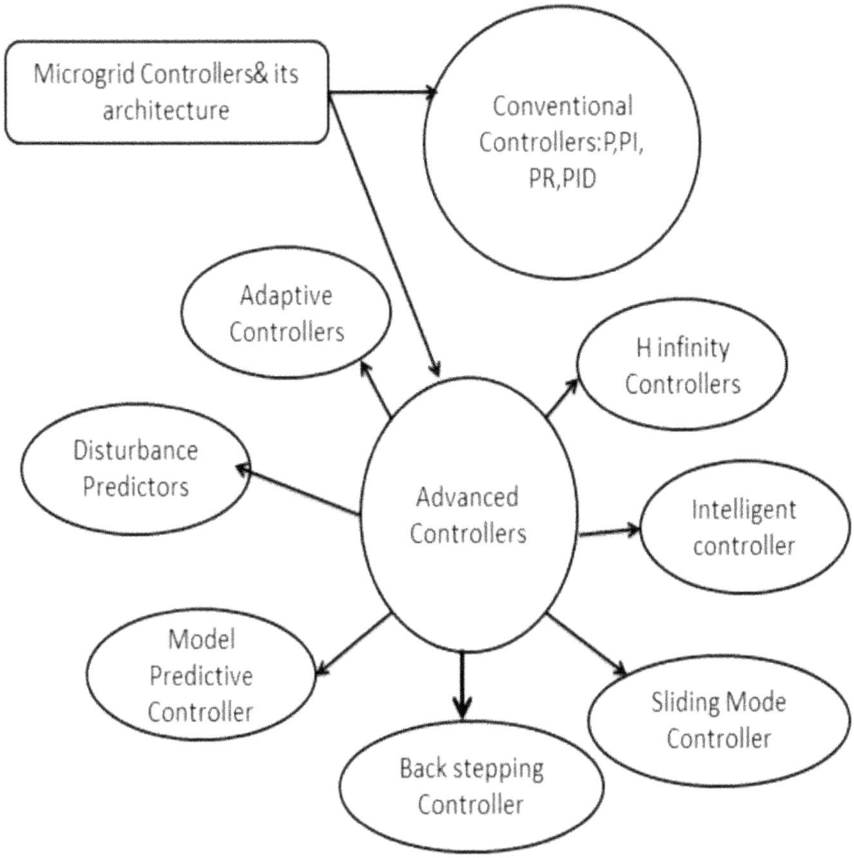

Figure 19.3 MG controller classifications

7. Numerical integration
   Use numerical integration techniques to solve the DAEs and simulate the transient response of the MG system over time. Common numerical methods include Runge–Kutta methods and backward differentiation formulas.
8. Validation and testing
   Validate the stability model by comparing simulation results with real-world data or results from other simulation tools. Test the model under various disturbance scenarios, including faults, load changes, and renewable energy fluctuations.

9. Sensitivity analysis
   Perform sensitivity analysis to assess the impact of parameter variations and uncertainties on the stability of the MG. Identify critical parameters that affect system stability and performance.
10. Optimization and control design
    Use the insights from the voltage stability model to optimize the control strategies and operating parameters. Design control algorithms to ensure the stable and reliable operation of the MG under different operating conditions.

Research on the distribution stability of MGs is especially important because of the increasing penetration of DER, which helps in the bidirectional flow of power in MGs. The grid adaptive power management strategy (GA-PMS) is designed to create current references for RESs, ESSs, and MG-connected converters. The algorithm also features smooth MG functioning in adverse situations, priority-based load shedding, and power quality norms at the local bus. Simulation and experimental experiments are used to evaluate and validate the performance of the proposed GA-PMS. The lack of complete relevant standardization from different regulation bodies for migrating to DC systems is yet another challenge in terms of using off-the-shelf devices, harming the stability and reliability of the system. A micro turbine with optimized combustion ensures uninterrupted operation during the season in a yacht. Single- and multi-objective optimization approaches are performed by addressing the practical and technological limits, uncertainty, and intermittency of RESs in energy management systems.

Monocrystalline PVs are mostly used. Efficiency is higher. They are cost effective and they have a longer lifespan. The stability offered by perovskite is higher, but its lifespan is 30 months; monocrystalline and polycrystalline panels have a longer lifespan.

19.3 METHODS OF ENHANCING VOLTAGE STABILITY

1. DERs: As dependable and sustainable sources of power, micro-hydro generators, wind turbines and solar panels are important DERs that support MG voltage stability. By integrating these DERs with cutting-edge control systems and energy storage options, the MGs overall resilience and dependability are increased by reducing power generation variations and maintaining stable voltage levels.
2. An ESS can stabilize voltage levels by supplying extra power during times of low generation or absorbing surplus energy during times of high generation. Charge controllers and intelligent adaptive control strategies can improve voltage stability issues.
3. Voltage regulation equipment: Voltage regulation equipment, such as voltage regulators, tap-changing transformers and static volt-ampere

reactive (VAR) compensators (SVCs) are used to adjust voltage levels within the MG, as needed. These devices help to maintain voltage stability under varying operating conditions.

4. Smart inverters: Voltage level regulation is possible with inverter-based DERs, such as solar and battery inverters, if they include intelligent control features. To help the MGs' voltage stability, these inverters may actively regulate reactive power production. Autonomous volt/VAR control and aggregated reactive power control are two forms of smart inverter functions that are taken into account while evaluating the cost value reconciliation (CVR) advantage through the use of voltage optimization methods.
5. Advanced MG controllers: MG controllers play a pivotal role in coordinating the operation of DERs and voltage regulation equipment. These controllers use advanced algorithms to optimize the use of available resources and maintain voltage stability across the MG. An additional control technique called the stochastic consensus-based control approach is proposed for an islanded MG to account for system and communication noise. This approach leads to reduced complexity, low computational load and good estimation accuracy for frequency and voltage restoration. The suggested methodology ensures robust mean-square synchronization for distributed generator (DG) voltage and frequency restoration, thereby demonstrating superior performance compared with existing methods. Controllers (both inner and outer loops) are crucial for stability because their design is extremely sensitive to gain levels.
6. Grid-forming inverters: In some MG configurations, grid-forming inverters are used to establish voltage and frequency reference signals, allowing the MG to operate in islanded mode without relying on the main grid. These inverters help to maintain voltage stability within the MG. Significantly increasing system stability is required for the tight control of voltage source converters. However, tight control results in incremental negative admittances reflected to the common DC link. These effects are to be mitigated by damping control, coordinated control, dynamic voltage support, and advanced control strategies, as stated in.
7. Islanding capability: MGs are often designed with islanding capability, allowing them to operate independently from the main grid during grid disturbances or outages. This enhances voltage stability by isolating the MG from external disturbances and maintaining stable voltage levels within the islanded system. MGs susceptible to voltage collapse are those with a high proportion of consistent power needs, especially in the event of unforeseen events, like weather-related power outages.

8. Load management: Load curtailment and demand response strategies can be implemented to manage voltage stability in the MG. By adjusting the operation of non-critical loads during periods of voltage instability, operators can balance supply and demand, and prevent voltage fluctuations. Islanding and load reduction choices in a GC MG are based on the frequency/voltage values of post-grid scenarios. In data-driven approaches, the normalized values of the active and reactive power of historical data are compared with predicted values, and load curtailment is performed based on the demand response model.
9. Reactive compensation: Because reactive power controls voltage stability, raising the reactive power to meet demand will also improve the voltage profile of the system. Using electronic reactor and capacitor switching (SVC) or voltage-source converter (static synchronous compensator (STATCOM)) waveform processing, SVC and STATCOM produce or absorb all the necessary reactive power.

19.4 CONCLUSION

This chapter emphasizes the importance of GC power converters in MGs for improving loadability and voltage stability, resulting in a dependable power supply. It also underlines the significance of cutting-edge technology and control systems for effective MG operation and the transition between GC and island modes. Hence, it can be summarized as follows: a combination of advanced control strategies, DER integration, energy storage, optimal placement of voltage regulation equipment, fuel cells, and flexible AC transmission system (FACTS) devices can be used to enhance voltage stability in MGs, ensuring reliable and resilient power supply to connected loads. STATCOM-based optimization control can mitigate power oscillations, leading to the stability of the multi-connected system.

BIBLIOGRAPHY

Aazami, R., Shoaei, M., Moradkhani, A., Shirkhani, M., ELrashidi, A. and AboRas, K.M., 2024. Deep neural networks based method to islanding detection for multi-sources microgrid. *Energy Reports*, 11, pp.2971–2982.

Ahmed, N.A., Alajmi, B.N., Abdelsalam, I. and Marei, M.I., 2022. Soft switching multiphase interleaved boost converter with high voltage gain for EV applications. *IEEE Access*, 10, pp.27698–27716.

Arango-Manrique, A., Lopez, L., Ramirez-Ortiz, J. and Oliveros, I., 2021. Data-driven load modeling to analyze the frequency of system including demand response: A Colombian study case. *IEEE Access*, 9, pp.50332–50343.

Bolognani, S. and Zampieri, S., 2013. A distributed control strategy for reactive power compensation in smart microgrids. *IEEE Transactions on Automatic Control*, 58(11), pp.2818–2833.

Boscaino, V., Ditta, V., Marsala, G., Panzavecchia, N., Tinè, G., Cosentino, V., Cataliotti, A. and Di Cara, D., 2024. Grid-connected photovoltaic inverters: Grid codes, topologies and control techniques. *Renewable and Sustainable Energy Reviews*, 189, p.113903.

Chen, J., Milano, F. and O'Donnell, T., 2019. Assessment of grid-feeding converter voltage stability. *IEEE Transactions on Power Systems*, 34(5), pp.3980–3982.

Chen, J., Milano, F. and O'Donnell, T., 2019. Assessment of grid-feeding converter voltage stability. *IEEE Transactions on Power Systems*, 34(5), pp.3980–3982.

Choobineh, M., Silva-Ortiz, D. and Mohagheghi, S., 2018. An automation scheme for emergency operation of a multi-microgrid industrial park. *IEEE Transactions on Industry Applications*, 54(6), pp.6450–6459.

Choobineh, M., Silva-Ortiz, D. and Mohagheghi, S., 2018. An automation scheme for emergency operation of a multi-microgrid industrial park. *IEEE Transactions on Industry Applications*, 54(6), pp.6450–6459.

D'silva, S., Shadmand, M., Bayhan, S. and Abu-Rub, H., 2020. Towards grid of microgrids: Seamless transition between grid-connected and islanded modes of operation. *IEEE Open Journal of the Industrial Electronics Society*, 1, pp.66–81.

Ding, F. and Baggu, M.M., 2018. Coordinated use of smart inverters with legacy voltage regulating devices in distribution systems with high distributed PV penetration—Increase CVR energy savings. *IEEE Transactions on Smart Grid*, 14(3), pp.1804–1813.

Eajal, A.A., Yazdavar, A.H., El-Saadany, E.F. and Ponnambalam, K., 2018, October. On the existence of voltage collapse in islanded microgrid. In *2018 IEEE Electrical Power and Energy Conference (EPEC)* (pp. 1–6). IEEE.

Farrokhabadi, M., Cañizares, C.A., Simpson-Porco, J.W., Nasr, E., Fan, L., Mendoza-Araya, P.A., Tonkoski, R., Tamrakar, U., Hatziargyriou, N., Lagos, D. and Wies, R.W., 2019. Microgrid stability definitions, analysis, and examples. *IEEE Transactions on Power Systems*, 35(1), pp.13–29.

Ganjian-Aboukheili, M., Shahabi, M., Shafiee, Q. and Guerrero, J.M., 2019. Seamless transition of microgrids operation from grid-connected to islanded mode. *IEEE Transactions on Smart Grid*, 11(3), pp.2106–2114.

Golestan, S., Guerrero, J.M. and Vasquez, J.C., 2017. A PLL-based controller for three-phase grid-connected power converters. *IEEE Transactions on Power Electronics*, 33(2), pp.911–916.

Golestan, S., Guerrero, J.M. and Vasquez, J.C., 2017. A PLL-based controller for three-phase grid-connected power converters. *IEEE Transactions on Power Electronics*, 33(2), pp.911–916.

Gui, Y., Blaabjerg, F., Wang, X., Bendtsen, J.D., Yang, D. and Stoustrup, J., 2020. Improved DC-link voltage regulation strategy for grid-connected converters. *IEEE Transactions on Industrial Electronics*, 68(6), pp.4977–4987.

Gui, Y., Blaabjerg, F., Wang, X., Bendtsen, J.D., Yang, D. and Stoustrup, J., 2020. Improved DC-link voltage regulation strategy for grid-connected converters. *IEEE Transactions on Industrial Electronics*, 68(6), pp.4977–4987.

Harrag, A. and Messalti, S., 2018. How fuzzy logic can improve PEM fuel cell MPPT performances?. *International Journal of Hydrogen Energy*, 43(1), pp.537–550.

Iqbal, A., Ayoub, A., Waqar, A., Ul-Haq, A., Zahid, M. and Haider, S., 2021. Voltage stability enhancement in grid-connected microgrid using enhanced dynamic voltage restorer (EDVR). *AIMS Energy*, 9(1), pp.150–177.

Kolantla, D., Mikkili, S., Pendem, S.R. and Desai, A.A., 2020. Critical review on various inverter topologies for PV system architectures. *IET Renewable Power Generation*, 14(17), pp.3418–3438.

Korada, N. and Mishra, M.K., 2016. Grid adaptive power management strategy for an integrated microgrid with hybrid energy storage. *IEEE Transactions on Industrial Electronics*, 64(4), pp.2884–2892.

Kundur, P., Paserba, J., Ajjarapu, V., Andersson, G., Bose, A., Canizares, C., Hatziargyriou, N., Hill, D., Stankovic, A., Taylor, C. and Van Cutsem, T., 2004. Definition and classification of power system stability IEEE/CIGRE joint task force on stability terms and definitions. *IEEE Transactions on Power Systems*, 19(3), pp.1387–1401.

Latorre, A., Soeiro, T.B., Geertsma, R., Coraddu, A. and Polinder, H. (2023) 'Shipboard DC systems—A critical overview: challenges in primary distribution, power-electronics-based protection, and power scalability', *IEEE Open Journal of the Industrial Electronics Society*, 4, pp. 259–286. https://doi.org/10.1109/OJIES.2023.3294999.

Li, Y., Vilathgamuwa, D.M. and Loh, P.C., 2004. Design, analysis, and real-time testing of a controller for multibus microgrid system. *IEEE Transactions on Power Electronics*, 19(5), pp.1195–1204.

Liu, J., Luo, W., Gao, Y., Yin, Y. and Sun, G., 2020. Sliding mode control of grid-connected power converters for microgrid applications. In *Distributed Control Methods and Cyber Security Issues in Microgrids* (pp. 3–27). Academic Press.

Liu, J., Luo, W., Gao, Y., Yin, Y. and Sun, G., 2020. Sliding mode control of grid-connected power converters for microgrid applications. In *Distributed Control Methods and Cyber Security Issues in Microgrids* (pp. 3–27). Academic Press.

Micallef, A., Apap, M., Spiteri-Staines, C. and Guerrero, J.M., 2015. Single-phase microgrid with seamless transition capabilities between modes of operation. *IEEE Transactions on Smart Grid*, 6(6), pp.2736–2745.

Mohammadi, F., Mohammadi-Ivatloo, B., Gharehpetian, G.B., Ali, M.H., Wei, W., Erdinç, O. and Shirkhani, M., 2021. Robust control strategies for microgrids: A review. *IEEE Systems Journal*, 16(2), pp.2401–2412.

Mohebbi, P., Aazami, R., Moradkhani, A. and Danyali, S., 2024. A novel intelligent hybrid algorithm for maximum power point tracking in PV system. *International Journal of Electronics*, 111(3), pp.537–560.

Monchusi, B.B., 2023, November. A comprehensive review of microgrid technologies and applications. In *2023 International Conference on Electrical, Computer and Energy Technologies (ICECET)* (pp. 1–7). IEEE.

Montoya, O.D., Gil-González, W., Grisales-Noreña, L., Orozco-Henao, C. and Serra, F., 2019. Economic dispatch of BESS and renewable generators in DC microgrids using voltage-dependent load models. *Energies*, 12(23), p.4494.

Ou, T.C., Lu, K.H. and Huang, C.J., 2017. Improvement of transient stability in a hybrid power multi-system using a designed NIDC (Novel Intelligent Damping Controller). *Energies*, 10(4), p.488.

Popavath, L.N. and Kaliannan, P., 2018. Photovoltaic-STATCOM with low voltage ride through strategy and power quality enhancement in a grid integrated wind-PV system. *Electronics*, 7(4), p.51.

Raju, D.K., Kumar, R.S., Raghav, L.P. and Singh, A.R., 2022. Enhancement of loadability and voltage stability in grid-connected microgrid network. *Journal of Cleaner Production*, 374, p.133881.

Ranamuka, D., Agalgaonkar, A.P. and Muttaqi, K.M., 2013. Online voltage control in distribution systems with multiple voltage regulating devices. *IEEE Transactions on Sustainable Energy*, 5(2), pp.617–628.

Ranjan, S., Das, D.C., Latif, A., Sinha, N., Hussain, S.S. and Ustun, T.S., 2021. Maiden voltage control analysis of hybrid power system with dynamic voltage restorer. *IEEE Access*, 9, pp.60531–60542.

Roy, P., He, J., Zhao, T. and Singh, Y.V., 2022. Recent advances of wind-solar hybrid renewable energy systems for power generation: A review. *IEEE Open Journal of the Industrial Electronics Society*, 3, pp.81–104.

Salehi, N., Martínez-García, H., Velasco-Quesada, G. and Guerrero, J.M., 2022. A comprehensive review of control strategies and optimization methods for individual and community microgrids. *IEEE Access*, 10, pp.15935–15955.

Shahabi, M., Haghifam, M.R., Mohamadian, M. and Nabavi-Niaki, S.A., 2009. Microgrid dynamic performance improvement using a doubly fed induction wind generator. *IEEE Transactions on Energy Conversion*, 24(1), pp.137–145.

Sharma, S., Jain, K.K. and Sharma, A., 2015. Solar cells: in research and applications—a review. *Materials Sciences and Applications*, 6(12), pp.1145–1155.

Van der Sande, R., Shekhar, A. and Bauer, P. (2025) 'Reliable DC shipboard power systems—Design, assessment, and improvement', *IEEE Open Journal of the Industrial Electronics Society*, 6, pp. 235–264. https://doi.org/10.1109/OJIES.2025.3532095

Worku, M.Y., Hassan, M.A. and Abido, M.A., 2019. Real time energy management and control of renewable energy based microgrid in grid connected and island modes. *Energies*, 12(2), p.276.

Xie, J., Alvarez-Fernandez, I. and Sun, W., 2020, August. A review of machine learning applications in power system resilience. In *2020 IEEE Power & Energy Society General Meeting (PESGM)* (pp. 1–5). IEEE.

Xu, Y., Zhang, R., Zhao, J., Dong, Z.Y., Wang, D., Yang, H. and Wong, K.P., 2015. Assessing short-term voltage stability of electric power systems by a hierarchical intelligent system. *IEEE Transactions on Neural Networks and Learning Systems*, 27(8), pp.1686–1696.

Zhou, Q., Shahidehpour, M., Paaso, A., Bahramirad, S., Alabdulwahab, A. and Abusorrah, A., 2020. Distributed control and communication strategies in networked microgrids. *IEEE Communications Surveys & Tutorials*, 22(4), pp.2586–2633.

Index

For Product Safety Concerns and Information please contact our EU representative GPSR@taylorandfrancis.com
Taylor & Francis Verlag GmbH, Kaufingerstraße 24, 80331 München, Germany

www.ingramcontent.com/pod-product-compliance
Ingram Content Group UK Ltd.
Pitfield, Milton Keynes, MK11 3LW, UK
UKHW022213190926
475702UK00003B/338

* 9 7 8 1 0 3 2 7 7 4 9 4 7 *